SOFTWARE ENGINEERING FOR EMBEDDED SYSTEMS

SOFTWARE ENGINEERING FOR EMBEDDED SYSTEMS

Methods, Practical Techniques, and Applications

SECOND EDITION

Edited by

Robert Oshana

Mark Kraeling

ELSEVIER

Newnes
An imprint of Elsevier

Newnes is an imprint of Elsevier
The Boulevard, Langford Lane, Kidlington, Oxford OX5 1GB, United Kingdom
50 Hampshire Street, 5th Floor, Cambridge, MA 02139, United States

Notices

Knowledge and best practice in this field are constantly changing. As new research and experience broaden our understanding, changes in research methods, professional practices, or medical treatment may become necessary.

Practitioners and researchers must always rely on their own experience and knowledge in evaluating and using any information, methods, compounds, or experiments described herein. In using such information or methods they should be mindful of their own safety and the safety of others, including parties for whom they have a professional responsibility.

To the fullest extent of the law, neither the Publisher nor the authors, contributors, or editors, assume any liability for any injury and/or damage to persons or property as a matter of products liability, negligence or otherwise, or from any use or operation of any methods, products, instructions, or ideas contained in the material herein.

Library of Congress Cataloging-in-Publication Data
A catalog record for this book is available from the Library of Congress

British Library Cataloguing-in-Publication Data
A catalogue record for this book is available from the British Library

ISBN: 978-0-12-809448-8

For information on all Newnes publications
visit our website at https://www.elsevier.com/books-and-journals

Publisher: Mara Conner
Acquisition Editor: Tim Pitts
Editorial Project Manager: Leticia M. Lima
Production Project Manager: Kamesh Ramajogi
Cover Designer: Miles Hitchen

Typeset by SPi Global, India

CONTENTS

Additional material is available online and you can refer it in
https://www.elsevier.com/books-and-journals/book-companion/9780128094488/

CONTRIBUTORS

Michael C. Brogioli Polymathic Consulting, Austin, TX, United States

Jagdish Gediya NXP Semiconductors, Automotive Division, Noida, India

Ruchika Gupta Software Architect, AMP & Digital Networking, NXP Semiconductor Pvt. Ltd., Noida, India

Pankaj Gupta Senior Software Staff, AMP & Digital Networking, NXP Semiconductor Pvt. Ltd., Noida, India

Joe Hamman Director, Platform Software Solutions at Integrated Computer Solutions, Waltham, MA, United States

Shreyansh Jain Digital Networking, NXP, Delhi, India

Mark Kraeling CTO Office, GE Transportation, Melbourne, FL, United States

Prabhakar Kushwaha NXP Semiconductors, Automotive Division, Noida, India

Jean J. Labrosse Founder and Chief Architect, Micrium LLC, Weston, FL, United States

Markus Levy NXP Semiconductors, Eindhoven, The Netherlands

Sandeep Malik Digital Networking, NXP, Delhi, India

Rob Oshana Vice President Software Engineering R&D, NXP Semiconductors, Austin, TX, United States

Mark Pitchford LDRA, Monks Ferry, United Kingdom

Jaswinder Singh NXP Semiconductors, Automotive Division, Noida, India; Digital Networking, NXP, Delhi, India; Software Director, AMP & Digital Networking, NXP Semiconductor Pvt. Ltd., Noida, India

Rajan Srivastava NXP Semiconductors, Automotive Division, Noida, India

Lindsley Tania Release Train Engineer LOCOTROL® Technologies GE Transportation, a Wabtec company

Zening Wang NXP Semiconductors, Microcontroller Division, Shanghai, China

ACKNOWLEDGMENTS

As editors of this Second Edition focusing on embedded systems, it is remarkable how many features of embedded systems have changed since the First Edition. What is also remarkable is how may features have stayed the same. The principles in the First Edition, written 4 years ago, are still applicable but the direction the industry is presently headed in could not have been predicted. Many sections in that First Edition focused on bringing a larger ecosystem to the embedded space, with an assumption that progression in power-saving devices and systems would bring in more Linux-based products. Though true, the advent of the entire Internet of Things (IoT) caused a refocus on becoming extremely efficient in terms of hardware resources for battery life and minimizing costs.

The authors of this Second Edition were selected based on their expertise in a specific subject area. In the same spirit as the authors of the First Edition, the authors of this edition were given guidelines and parameters for their chapters, with editing undertaken in the latter stages of production, bringing chapters together into a cohesive book. Their hard work and dedication are hopefully reflected in this Second Edition.

Rob would like to thank Susan his wife and Sam and Noah for their support as well as those in the embedded industry who work with him on a regular basis. The ideas, concepts, and approaches in this book largely come from the working relationships within the embedded area of the industry.

In addition to the authors, Mark would like to thank this family, especially his wife Shannon. Additional thanks go to Wes, RJ, Garret, Shelly, Mike, Todd, Dan, Glen, Dave, Mike, Brad, Spencer, Theoni, and the embedded engineering staff who work with Mark on a regular basis. Change is good.

We hope you find the book useful and that you learn about various aspects of embedded systems and apply them immediately to your designs and developments.

The authors have been supported in this work by NXP Semiconductors. Select images in this text were reprinted with the permission of NXP Semiconductors.

Thanks, Rob Oshana and Mark Kraeling.

1

SOFTWARE ENGINEERING FOR EMBEDDED AND REAL-TIME SYSTEMS

Rob Oshana

Vice President of Software Engineering R&D, NXP Semiconductors, Austin, TX, United States

CHAPTER OUTLINE

Software Engineering for Embedded Systems. https://doi.org/10.1016/B978-0-12-809448-8.00001-1

1   Software Engineering

Over the past 10 years or so, the world of computing has moved from large, static, desk-top machines to small, mobile, and embedded devices. The methods, techniques, and tools for developing software systems that were successfully applied in the former scenario are not as readily applicable in the latter. Software systems running on networks of mobile, embedded devices must exhibit properties that are not always required of more traditional systems:

- Near-optimal performance
- Robustness
- Distribution
- Dynamism
- Mobility

This book will examine the key properties of software systems in the embedded, resource-constrained, mobile, and highly distributed world. We will assess the applicability of mainstream software engineering methods and techniques (e.g., software design, component-based development, software architecture, system integration, and testing) to this domain.

One of the differences in software engineering for embedded systems is the additional knowledge the engineer has of electrical power and electronics; physical interfacing of digital and analog electronics with the computer; and, software design for embedded systems and digital signal processors (DSPs).

Over 95% of software systems are embedded. Consider the devices you use at home daily:

- Cell phone
- iPod
- Microwave
- Satellite receiver
- Cable box
- Car motor controller
- DVD player

So what do we mean by software engineering for embedded systems? Let's look at this in the context of engineering in general. Engineering is defined as the application of scientific principles and methods to the construction of useful structures and machines. This includes disciplines such as:

- Mechanical engineering
- Civil engineering
- Chemical engineering
- Electrical engineering
- Nuclear engineering
- Aeronautical engineering

Software engineering is a term that is 35 years old, originating at a NATO conference in Garmisch, Germany, October 7–11, 1968. Computer science is its scientific basis with many aspects having been made systematic in software engineering:

- Methods/methodologies/techniques
- Languages
- Tools
- Processes

We will explore all these in this book.

The basic tenets of software engineering include:

- Development of software systems whose size/complexity warrants team(s) of engineers (or as David Parnas puts it, "multi-person construction of multi-version software").
- Scope—we will focus on the study of software processes, development principles, techniques, and notations.
- Goal, in our case the production of quality software, delivered on time, within budget, satisfying the customers' requirements and the users' needs.

With this comes the ever-present difficulties of software engineering that still exist today:

- There are relatively few guiding scientific principles.
- There are few universally applicable methods.
- Software engineering is as much managerial/psychological/sociological as it is technological.

These difficulties exist because software engineering is a unique form of engineering:

- Software is malleable
- Software construction is human-intensive
- Software is intangible
- Software problems are unprecedentedly complex
- Software directly depends upon the hardware
- Software solutions require unusual rigor
- Software has a discontinuous operational nature

Software engineering is not the same as software programming. Software programming usually involves a single developer developing "Toy" applications and involves a relatively short life span. With programming, there is usually a single stakeholder, or perhaps a few, and projects are mostly one-of-a-kind systems built from scratch with minimal maintenance.

Software engineering on the other hand involves teams of developers with multiple roles building complex systems with an indefinite life span. There are numerous stakeholders, families of systems, a heavy emphasis on reuse to amortize costs, and a maintenance phase that accounts for over 60% of the overall development costs.

There are both economic and management aspects of software engineering. Software production includes the development and maintenance (evolution) of the system. Maintenance costs represent most of all development costs. Quicker development is not always preferable. In other words, higher up-front costs may defray downstream costs. Poorly designed and implemented software is a critical cost factor. In this book we will focus on the software engineering of embedded systems, not the programming of embedded systems.

Embedded software development uses the same software development models as other forms of software development, including the Waterfall model (Fig. 1), the Spiral model (Fig. 2), and the Agile model (Fig. 3). The benefits and limitations of each of these models is well documented so we will only review them here. We will, however, spend more time later in this book on Agile development, as this approach is well suited to the changing, dynamic nature of embedded systems.

The key software development phases for embedded systems are briefly summarized below.

1. *Problem definition.* In this phase we determine exactly what the

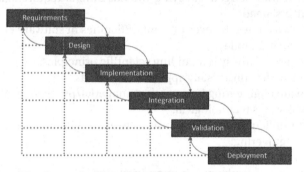

Fig. 1 Waterfall software development model.

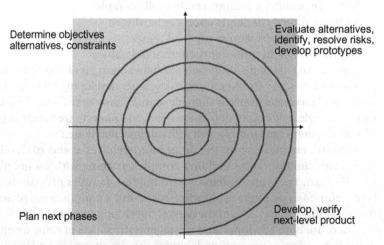

Fig. 2 Spiral software development model.

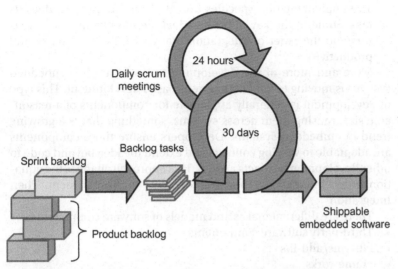

Fig. 3 Agile software development model.

customer and user want. This may include the development of a contract with the customer, depending on what type of product is being developed. The goal of this phase is to specify what the software product is to do. Difficulties include the client asking for the wrong product, the client being computer/software illiterate which limits the effectiveness of this phase, and specifications that are ambiguous, inconsistent, and incomplete.

2. *Architecture/design*. Architecture is concerned with the selection of architectural elements, their interactions, and the constraints on those elements and their interactions necessary to provide a framework with which to satisfy the requirements and serve as a basis for the design. Design is concerned with the modularization and detailed interfaces of the design elements, their algorithms and procedures, and the data types needed to support the architecture and to satisfy the requirements. During the architecture and design phases, the system is decomposed into software modules with interfaces. During design the software team develops module specifications (algorithms, data types), maintains a record of design decisions and traceability, and specifies how the software product is to do its tasks. The primary difficulties during this phase include miscommunication between module designers and developing a design that may be inconsistent, incomplete, or ambiguous.

3. *Implementation*. During this phase the develop team implements the modules and components and verifies that they meet their specifications. Modules are combined according to the design.

The implementation specifies how the software product does its task. Some of the key difficulties include module interaction errors and the order of integration that may influence quality and productivity.

More and more of the development of software for embedded systems is moving toward component-based development. This type of development is generally applicable for components of a reasonable size, reusing them across systems, something that is a growing trend in embedded systems. Developers ensure these components are adaptable to varying contexts and extend the idea beyond code to other development artifacts as well. This approach changes the equation from "Integration, Then Deployment" to "Deployment, Then Integration."

There are different makes and models of software components:

- Third-party software components
- Plug-ins/add-ins
- Frameworks
- Open systems
- Distributed object infrastructures
- Compound documents
- Legacy systems

4. *Verification and validation (V&V).* There are several forms of V&V and there is a dedicated chapter on this topic. One form is "analysis." Analysis can be in the form of static, scientific, formal verification, and informal reviews and walkthroughs. Testing is a more dynamic form of V&V. This type of testing comes in the form of white box (having access to the code) and black box (having no access to the source code). Testing can be structural as well as behavioral. There are the standard issues of test adequacy but we will defer this discussion to later when we dedicate a chapter to this topic.

As we progress through this book, we will continue to focus on foundational software engineering principles (Fig. 4):

- Rigor and formality
- Separation of concerns
  - Modularity and decomposition
  - Abstraction
- Anticipation of change
- Generality
- Incrementality
- Scalability
- Compositionality
- Heterogeneity
- Moving from principles to tools

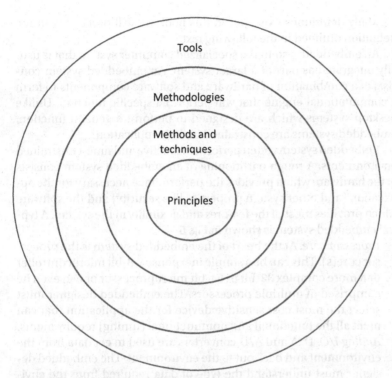

Fig. 4 Software engineering principles.

2 Embedded Systems

What is an embedded system? There are many answers to this question. Some define an embedded system simply as "a computer whose end function is not to be a computer." If we follow this definition then automobile antilock braking systems, digital cameras, household appliances, and televisions are embedded systems because they contain computers but aren't intended to be computers. Conversely, the laptop computer I'm using to write this chapter is not an embedded system because it contains a computer that is intended to be a computer (see Bill Gatliff's article "There's no such thing as an Embedded System" on www.embedded.com).

Jack Ganssle and Mike Barr, in their book Embedded Systems Dictionary, define an embedded system as "A combination of computer hardware and software, and perhaps additional mechanical or other parts, designed to perform a dedicated function. In some cases, embedded systems are part of a larger system or product, as in the case of an antilock braking system in a car."

Many definitions exist, but in this book we will proceed with the definition outlined in the following text.

An embedded system is a specialized computer system that is usually integrated as part of a larger system. An embedded system consists of a combination of hardware and software components to form a computational engine that will perform a specific function. Unlike desktop systems which are designed to perform a general function, embedded systems are constrained in their application.

Embedded systems often perform in reactive and time-constrained environments. A rough partitioning of an embedded system consists of the hardware which provides the performance necessary for the application (and other system properties like security) and the software which provides most of the features and flexibility in the system. A typical embedded system is shown in Fig. 5.

- *Processor core.* At the heart of the embedded system is the processor core(s). This can be a simple inexpensive 8-bit microcontroller or a more complex 32-bit or 64-bit microprocessor or can even be comprised of multiple processors. The embedded designer must select the most cost sensitive device for the application that can meet all the functional and nonfunctional (timing) requirements.
- *Analog I/O.* D/A and A/D converters are used to get data from the environment and back out to the environment. The embedded designer must understand the type of data required from the environment, the accuracy requirements for that data, and the input/output data rates in order to select the right converters for the application. The external environment drives the reactive nature of the embedded system. Embedded systems must be at least fast enough to keep up with the environment. This is where the analog information, such as light or sound pressure or acceleration, is sensed and input into the embedded system.
- *Sensors and actuators.* Sensors are used to sense analog information from the environment. Actuators are used to control the environment in some way.

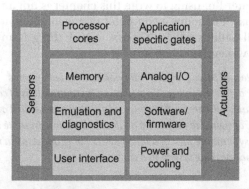

Fig. 5 Components of a typical embedded system.

- *User interfaces*. These interfaces may be as simple as a flashing LED or as sophisticated as a cell phone or digital still camera interface.
- *Application specific gates*. Hardware acceleration like ASIC or FPGA is used for accelerating specific functions in the application that have high-performance requirements. The embedded designer must be able to map or partition the application appropriately using the available accelerators to gain maximum application performance.
- *Software*. Software is a significant part of embedded system development. Over the last few years the amount of embedded software has grown faster than Moore's law, with the amount doubling approximately every 10 months. Embedded software is usually optimized in some way (performance, memory, or power). More and more embedded software is written in a high-level language like C/C++ with some of the more performance-critical pieces of code still written in assembly language.
- Memory is an important part of an embedded system and embedded applications can either run out of RAM or ROM depending on the application. There are many types of volatile and nonvolatile memory used for embedded systems and we will talk more about this later.
- *Emulation and diagnostics*. Many embedded systems are hard to see or get to. There needs to be a way to interface to embedded systems to debug them. Diagnostic ports such as a JTAG (Joint Test Action Group) are used to debug embedded systems. On-chip emulation is used to provide visibility for the behavior of the application. These emulation modules provide sophisticated visibility for the runtime behavior and performance, in effect replacing external logic analyzer functions with onboard diagnostic capability.

2.1 Embedded Systems Are Reactive Systems

A typical embedded system responds to the environment via sensors and controls the environment using actuators (Fig. 6). This imposes a requirement on embedded systems to achieve performance consistent with that of the environment. This is why embedded system are often referred to as reactive systems. A reactive system must use a combination of hardware and software to respond to events in the environment within defined constraints. Complicating the matter is the fact that these external events can be periodic and predictable or aperiodic and hard to predict. When scheduling events for processing in an embedded system, both periodic and aperiodic events must be considered, and performance must be guaranteed for worst-case rates of execution.

An example of an embedded sensor system is a tire-pressure monitoring system (TPMS). This is a sensor chipset designed to enable a

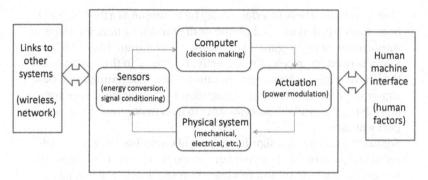

Fig. 6 A model of sensors and actuators in embedded systems.

timely warning to the driver in the case of underinflated or overinflated tires on cars, trucks, or buses—even while in motion. These sensor systems are a full integration of a pressure sensor, an 8-bit microcontroller (MCU), a radio frequency (RF) transmitter, and X-axis and Z-axis accelerometers in one package. A key to this sensor technology is the acquisition of acceleration in the X and Z directions (Fig. 7). The purpose of X-axis and Z-axis g-cells are to allow tire recognition with the appropriate embedded algorithms analyzing the rotating signal caused by the Earth's gravitational field. Motion will use either the Z-axis g-cell to detect acceleration level or the X-axis g-cell to detect a ±1-g signal caused by the Earth's gravitational field.

There are several key characteristics of embedded systems:

(a) *Monitoring and reacting to the environment.* Embedded systems typically get input by reading data from input sensors. There are many different types of sensors that monitor various analog signals in the environment including temperature, sound pressure,

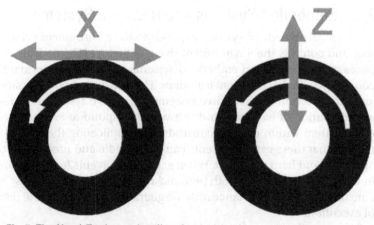

Fig. 7 The X and Z-axis sensing directions.

and vibration. This data is processed using embedded system algorithms. The results may be displayed in some format to a user or simply used to control actuators (like deploying the airbags and calling the police).

(b) *Control the environment.* Embedded systems may generate and transmit commands that control actuators, such as airbags, motors, etc.

(c) *Processing information.* Embedded systems process the data collected from the sensors in some meaningful way, such as data compression/decompression, side impact detection, etc.

(d) *Application specific.* Embedded systems are often designed for applications, such as airbag deployment, digital still cameras, or cell phones. Embedded systems may also be designed for processing control laws, finite-state machines, and signal-processing algorithms. Embedded systems must also be able to detect and react appropriately to faults in both the internal computing environment as well as the surrounding systems.

(e) *Optimized for the application.* Embedded systems are all about performing the desired computations with as few resources as possible in order to reduce cost, power, size, etc. This means that embedded systems need to be optimized for the application. This requires software as well as hardware optimization. Hardware needs to be able to perform operations in as few gates as possible, and software must be optimized to perform operations in the least number of cycles, amount of memory, or power as possible depending on the application.

(f) *Resource constrained.* Embedded systems are optimized for the application which means that many of the precious resources of an embedded system, such as processor cycles, memory, and power, are in scarce supply in a relative sense in order to reduce cost, size, weight, etc.

(g) *Real time.* Embedded systems must react to the real-time changing nature of the environment in which they operate. More on real-time systems below.

(h) *Multirate.* Embedded systems must be able to handle multiple rates of processing requirements simultaneously, for example video processing at 30 frames per second (30 Hz) and audio processing at 20-kHz rates.

Fig. 8 shows a simple embedded system that demonstrates these key characteristics:

1. *Monitoring and controlling the environment.* The embedded system monitors a fluid-flow sensor in the environment and then controls the value (actuator) in that same environment.

2. *Performing meaningful operations.* The computation task computes the desired algorithms to control the value in a safe way.

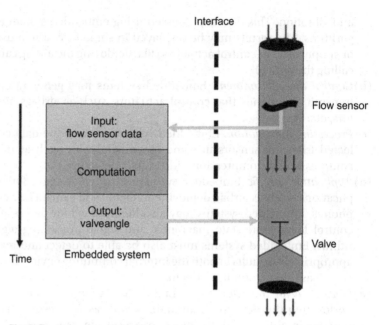

Fig. 8 Example of an embedded system.

3. *Application specific.* The embedded system is designed for a particular application.
4. *Optimized for application.* The embedded system's computation and algorithms are designed for a particular system.
5. *Resource constrained.* The embedded system executes on a small inexpensive microcontroller with a small amount of memory, operating at lower power for cost savings.
6. *Real time.* The system has to be able to respond to the flow sensor in real time, any delays in processing could lead to failure of the system.
7. *Multirate.* There may be the need to respond to the flow sensor as well as a user interface, so multiinput rates to the embedded system should be possible.

3   Real-Time Systems

A real-time system is any information-processing activity or system which must respond to externally generated input stimuli within a finite and specified period. Real-time systems must process information and produce a response within a specified time. Failure to do so will risk severe consequences, including failure. In a system with a

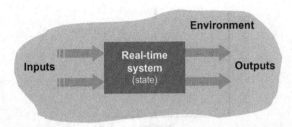

Fig. 9 A real-time system reacts to inputs from the environment and produces outputs that affect the environment.

real-time constraint, it is unacceptable to have the correct action or the correct answer *after* a certain deadline: the result must be produced by the deadline or the system will degrade or fail completely. Generally, real-time systems maintain a *continuous, timely* interaction with the environment (Fig. 9).

3.1   Types of Real-Time Systems—Soft and Hard

In real-time systems, the correctness of the computation depends not only upon its results but also the time at which its outputs are generated. A real-time system must satisfy response time constraints or suffer significant system consequences. If the consequences consist of a degradation of performance, but not failure, the system is referred to as a soft real-time system. If the consequences are system failure, the system is referred to as a hard real-time system (e.g., an antilock braking system in an automobile) (Fig. 10).

We can also think of this in terms of the real-time interval, which is defined as how quickly the system must respond. In this context, the Windows operating system is soft real-time because it is relatively slow and cannot handle shorter time constraints. In this case, the system does not "fail" but is degraded.

The objective of an embedded system is to execute as fast as necessary in an asynchronous world using the smallest amount of code with the highest level of predictability. (Note: predictability is the embedded world's term for reliability.)

Fig. 11 shows some examples of hard and soft real-time systems. As shown in this list of examples, many embedded systems also have a criticality to the computation in the sense that a failure to meet real-time deadlines can have disastrous consequences. For example, the real-time determination of a driver's intentions and driving conditions (Fig. 12) is an example of a hard real-time safety critical application.

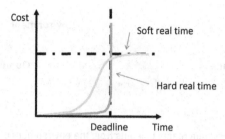

Fig. 10 A comparison between hard and soft real-time systems.

System type	Hard or soft real time?
Traffic light control	Hard RT—critical
Automated teller machine	Soft RT—noncritical
Controller for radiation therapy machine	Hard RT—critical
Car simulator for driver training	Hard RT—noncritical
Highway car counter	Soft RT—noncritical
Missile control	Hard RT—critical
Video games	Hard RT—noncritical
Network chat	Soft RT—noncritical

Fig. 11 Examples of hard and soft real-time systems.

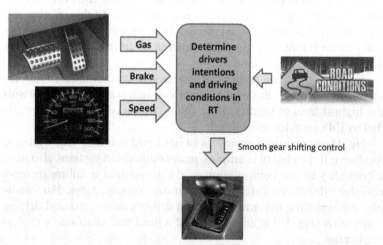

Fig. 12 An automobile shift control system is an example of a hard real-time safety-critical system.

Table 1 Real-Time Systems Are Fundamentally Different From Time-Shared Systems

Characteristic	Time-Shared Systems	Real-Time Systems
System capacity	High throughput	Schedulability and the ability of system tasks to meet all deadlines
Responsiveness	Fast average response time	Ensured worst-case latency which is the worst-case response time to events
Overload	Fairness to all	Stability; when the system is overloaded important tasks must meet deadlines while others may be starved

3.2 Differences Between Real-Time and Time-Shared Systems

Real-time systems are different from time-shared systems in three fundamental areas (Table 1):

- *High degree of schedulability*. Timing requirements of the system must be satisfied at high degrees of resource usage and offer predictably fast responses to urgent events.
- *Worst case latency*. Ensuring the system still operates under worst-case response times to events.
- *Stability under transient overload*. When the system is overloaded by events and it is impossible to meet all deadlines, the deadlines of selected critical tasks must still be guaranteed.

4 Example of a Hard Real-Time System

Many embedded systems are real-time systems. As an example, assume that an analog signal is to be processed digitally. The first question to consider is how often to *sample* or measure the analog signal in order to represent that signal accurately in the digital domain. The sample rate is the number of samples of an analog event (like sound) that are taken each second to represent the event in the digital domain. Based on a signal processing rule called Nyquist, the signal must be sampled at a rate at least equal to twice the highest frequency that we wish to preserve. For example, if the signal contains important components at 4 kHz, then the sampling frequency would need to be at least 8 kHz. The sampling period would then be:

$$T = 1/8000 = 125\ \mu s = 0.000125\ s$$

4.1 Based on Signal Sample, Time to Perform Actions Before Next Sample Arrives

This tells us that, for this signal being sampled at this rate, we would have 0.000125 s to perform *all* the processing necessary before the next sample arrived. Samples are arriving on a continuous basis and if the system falls behind in processing these samples, the system will degrade. This is an example of a soft real-time embedded system.

4.2 Hard Real-Time Systems

The collective timeliness of the hard real-time tasks is binary—i.e., either they all will always meet their deadlines (in a correctly functioning system) or they will not (the system is infeasible). In all hard real-time systems, collective timeliness is deterministic. This determinism does not imply that the actual individual task completion times, or the task execution ordering, are necessarily known in advance.

A computing system being a hard real-time system says nothing about the magnitudes of the deadlines. They may be microseconds or weeks. There is a bit of confusion with regards to the usage of the term "hard real-time." Some relate hard real-time to response time magnitudes below some arbitrary threshold, such as 1 ms. This is not the case. Many of these systems actually happen to be soft real-time systems. These systems would be more accurately termed "real fast" or perhaps "real predictable." But certainly not hard real-time systems.

The feasibility and costs (e.g., in terms of system resources) of hard real-time computing depend on how well known á priori are the relevant future behavioral characteristics of the tasks and execution environment. These task characteristics include:

- Timeliness parameters, such as arrival periods or upper bounds
- Deadlines
- Worst-case execution times
- Ready and suspension times
- Resource utilization profiles
- Precedence and exclusion constraints
- Relative importance, etc.

There are also important characteristics relating to the system itself, including:

- System loading
- Resource interactions
- Queuing disciplines
- Arbitration mechanisms
- Service latencies
- Interrupt priorities and timing
- Caching

Deterministic collective task timeliness in hard (and soft) real-time computing requires that the future characteristics of the relevant tasks and execution environment be deterministic—i.e., known absolutely in advance. Knowledge of these characteristics must then be used to preallocate resources so that hard deadlines, like motor control, will be met and soft deadlines, like responding to a key press, can be delayed.

A real-time system task and execution environment must be adjusted to enable a schedule and resource allocation which meets all deadlines. Different algorithms or schedules which meet all deadlines are evaluated with respect to other factors. In many real-time computing applications getting the job done at the lowest cost is usually more important than simply maximizing the processor utilization (if this was true, we would all still be writing assembly language). Time to market, for example, may be more important than maximizing utilization due to the cost of squeezing the last 5% of efficiency out of a processor.

Allocation for hard real-time computing has been performed using various techniques. Some of these techniques involve conducting an offline enumerative search for a static schedule which will deterministically always meet all deadlines. Scheduling algorithms include the use of priorities that are assigned to the various system tasks. These priorities can be assigned either offline by application programmers or online by the application or operating system software. The task priority assignments may either be static (fixed), as with rate monotonic algorithms or dynamic (changeable), as with the earliest deadline first algorithm.

5   Real-Time Event Characteristics

5.1   Real-Time Event Categories

Real-time events fall into one of three categories: asynchronous, synchronous, or isochronous:

- *Asynchronous events* are entirely unpredictable. An example of this is a cell phone call arriving at a cellular base station. As far as the base station is concerned, the action of making a phone call cannot be predicted.
- *Synchronous events* are predictable events and occur with precise regularity. For example, the audio and video in a camcorder take place in synchronous fashion.
- *Isochronous events* occur with regularity within a given window of time. For example, audio data in a networked multimedia application must appear within a window of time when the corresponding video stream arrives. Isochronous is a subclass of asynchronous.

In many real-time systems, task and execution environment characteristics may be hard to predict. This makes true, hard real-time scheduling infeasible. In hard real-time computing, deterministic satisfaction of the collective timeliness criterion is the driving requirement. The necessary approach to meeting that requirement is static (i.e., á priori) scheduling of deterministic tasks and execution environment characteristic cases. The requirement for advanced knowledge about each of the system tasks and their future execution environment to enable offline scheduling and resource allocation significantly restricts the applicability of hard real-time computing.

5.2 Efficient Execution and the Execution Environment

5.2.1 Efficiency Overview

Real-time systems are time critical and the efficiency of their implementation is more important than in other systems. Efficiency can be categorized in terms of processor cycles, memory, or power. This constraint may drive everything from the choice of processor to the choice of programming language. One of the main benefits of using a higher level language is to allow the programmer to abstract away implementation details and concentrate on solving the problem. This is not always true in the world of the embedded system. Some higher level languages have instructions that are an order of magnitude slower than assembly language. However, higher level languages can be used in real-time systems effectively using the right techniques. We will be discussing much more about this topic in the chapter on optimizing source code for DSPs.

5.2.2 Resource Management

A system operates in real time as long as it completes its time-critical processes with acceptable timeliness. "Acceptable timeliness" is defined as part of the behavioral or "nonfunctional" requirements for the system. These requirements must be objectively quantifiable and measureable (stating that the system must be "fast," for example, is not quantifiable). A system is said to be a real-time system if it contains some model of real-time resource management (these resources must be explicitly managed for the purpose of operating in real time). As mentioned earlier, resource management may be performed statically offline or dynamically online.

Real-time resource management comes at a cost. The degree to which a system is required to operate in real time cannot necessarily be attained solely by hardware overcapacity (e.g., high processor performance using a faster CPU).

There must exist some form of real-time resource management to be cost effective. Systems which must operate in real time consist of both real-time resource management and hardware resource capacity. Systems which have interactions with physical devices may require higher degrees of real-time resource management. One resource management approach that is used is static and requires analysis of the system prior to it executing in its environment. In a real-time system, physical time (as opposed to logical time) is necessary for real-time resource management in order to relate events to the precise moments of occurrence. Physical time is also important for action time constraints as well as measuring costs incurred as processes progress to completion. Physical time can also be used for logging history data.

All real-time systems make trade-offs of scheduling costs versus performance in order to reach an appropriate balance for attaining acceptable timeliness between the real-time portion of the scheduling optimization rules and the offline scheduling performance evaluation and analysis.

6 Challenges in Real-Time System Design

Designing real-time systems poses significant challenges to the designer. One of the significant challenges comes from the fact that real-time systems must interact with the environment. The environment is complex and changing and these interactions can become very complex. Many real-time systems don't just interact with one entity but instead interact with many different entities in the environment, with different characteristics and rates of interaction. A cell phone base station, for example, must be able to handle calls from literally thousands of cell phone subscribers at the same time. Each call may have different requirements for processing as well as different sequences of processing. All this complexity must be managed and coordinated.

6.1 Response Time

Real-time systems must respond to external interactions in the environment within a predetermined amount of time. Real-time systems must produce the correct result and produce it in a timely way. The response time is as important as producing correct results. Real-time systems must be engineered to meet these response times. Hardware and software must be designed to support response time requirements for these systems. Optimal partitioning of system requirements into hardware and software is also important.

Real-time systems must be architected to meet system response time requirements. Using combinations of hardware and software components, it is engineering that makes the architecture decisions, such as interconnectivity of system processors, system link speeds, processor speeds, memory size, I/O bandwidth, etc. Key questions to be answered include:

- *Is the architecture suitable?* To meet system response time requirements, the system can be architected using one powerful processor or several smaller processors. Can the application be partitioned among the several smaller processors without imposing large communication bottlenecks throughout the system? If the designer decides to use one powerful processor, will the system meet its power requirements? Sometimes a simpler architecture may be the better approach—more complexity can lead to unnecessary bottlenecks which cause response time issues.

- *Are the processing elements powerful enough?* A processing element with high utilization (greater than 90%) will lead to unpredictable runtime behavior. At this utilization level lower priority tasks in the system may be starved. As a general rule, real-time systems that are loaded at 90% take approximately twice as long to develop due to cycles of optimization and integration issues with the system at these utilization rates. At 95% utilization, systems can take three times longer to develop due to these same issues. Using multiple processors will help but interprocessor communication must be managed.

- *Are the communication speeds adequate?* Communication and I/O is a common bottleneck in real-time embedded systems. Many response time problems come not from the processor being overloaded but in latencies in getting data into and out of the system. In other cases, overloading a communication port (greater than 75%) can cause unnecessary queuing in different system nodes, causing delays in message passing throughout the rest of the system.

- *Is the right scheduling system available?* In real-time systems tasks that are processing real-time events must take higher priority. But how do you schedule multiple tasks that are all processing real-time events. There are several scheduling approaches available and the engineer must design the scheduling algorithm to accommodate system priorities in order to meet all real-time deadlines. Because external events may occur at any time, the scheduling system must be able to preempt currently running tasks to allow higher priority tasks to run. The scheduling system (or real-time operating system) must not introduce a significant amount of overhead into the real-time system.

6.2 Recovering From Failures

Real-time systems interact with the environment, which is inherently unreliable. Therefore real-time systems must be able to detect and overcome failures in the environment. In addition, since real-time systems are also embedded into other systems and may be hard to get at (such as a spacecraft or satellite) these systems must also be able to detect and overcome internal failures as well (there is no "reset" button in easy reach of the user!). Also, since events in the environment are unpredictable, it is almost impossible to test for every possible combination and sequence of events in the environment. This is a characteristic of real-time software that makes it somewhat nondeterministic in the sense that it is almost impossible in some real-time systems to predict the multiple paths of execution based on the nondeterministic behavior of the environment. Examples of internal and external failures that must be detected and managed by real-time systems include:

- Processor failures
- Board failures
- Link failures
- Invalid behavior of the external environment
- Interconnectivity failure

Many real-time systems are embedded systems with multiple inputs and outputs and multiple events occurring independently. Separating these tasks simplifies programming but requires switching back and forth among the multiple tasks. This is referred to as multitasking. Concurrency in embedded systems is the appearance of multiple tasks executing simultaneously. For example, the three tasks listed in Fig. 13 will execute on a single embedded processor and the scheduling algorithm is responsible for defining the priority of execution of these three tasks.

```
/* Monitor Room_Temperature */
do forever {
        measure temperature ;
        if (temperature < temperature_ setting)
                start furnace_heater ;
        else if (temperature > temperature_setting + delta)
                stop furnace_heater ;
}

/* Monitor Time of Day */
do forever {
        measure time_of_day ;
             if (7:00am)
                        setting = 72_degrees_F ;
             else if (10:00pm)
                        setting = 60_degrees_F ;
}

/* Monitor Thermostat Keypad */
do forever {
        check thermostat_keypad ;
        if (raise temperature)
                setting++ ;
        else if (lower temperature)
                setting-- ;
}
```

Fig. 13 Multiple tasks execute simultaneously on embedded systems.

7 The Embedded System's Software Build Process

Another difference in embedded systems is the software system build process, as shown in Fig. 14.

Embedded system programming is not substantially different from ordinary programming. The main difference is that each target hardware platform is unique. The process of converting the source code representation of embedded software into an executable binary image involves several distinct steps:

- Compiling/assembling using an optimizing compiler.
- Linking using a linker.
- Relocating using a locator.

In the first step, each of the source files must be compiled or assembled into object code. The job of a compiler is mainly to translate programs written in some human readable format into an equivalent set of opcodes for a particular processor. The use of the cross compiler is one of the defining features of embedded software development.

In the second step, all the object files that result from the first step must be linked together to produce a single object file, called the relocatable program. Finally, physical memory addresses must be assigned to the relative offsets within the relocatable program in a process called relocation. The tool that performs the conversion from relocatable to executable binary image is called a locator. The result of the final step of the build process is an absolute binary image that can be directly programmed into a ROM or flash device.

We have covered several areas where embedded systems differ from other desktop-like systems. Some other differences that make embedded systems unique include:

1. Energy efficiency (embedded systems, in general, consume the minimum power for their purpose).
2. Custom voltage/power requirements.
3. Security (need to be hacker proof, for example, a Femto basestation needs IP security when sending phone calls over an internet backhaul).
4. Reliability (embedded systems need to work without failure for days, months, and years).
5. Environment (embedded systems need to operate within a broad temperature range, be sealed from chemicals, and be radiation tolerant).
6. Efficient interaction with user (fewer buttons, touchscreen, etc.).
7. Designed in parallel with embedded hardware.

The chapters in this book will touch on many of these topics as they relate to software engineering for embedded systems.

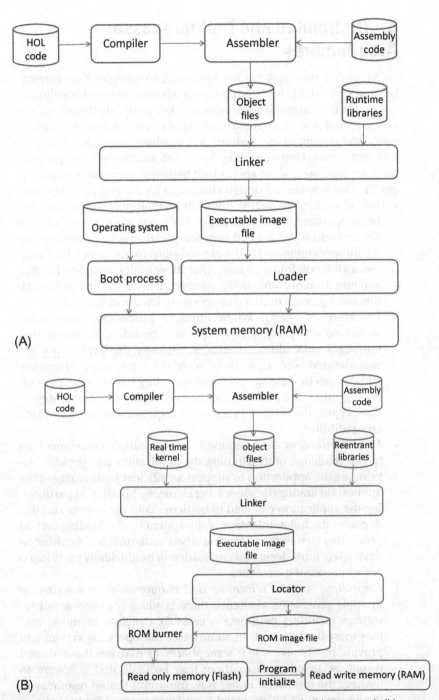

Fig. 14 Embedded system software build process and nonembedded system build process. (A) Build process for a desktop system and (B) build process for an embedded system.

8 Distributed and Multiprocessor Architectures

Some real-time systems are becoming so complex that applications are executed on multiprocessor systems that are distributed across some communication system. This poses challenges to the designer that relate to the partitioning of the application in a multiprocessor system. These systems will involve processing on several different nodes. One node may be a DSP, another a more general-purpose processor, some specialized hardware processing elements, etc. This leads to several design challenges for the engineering team:

- *Initialization of the system.* Initializing a multiprocessor system can be complicated. In most multiprocessor systems the software load file resides on the general-purpose processing node. Nodes that are directly connected to the general-purpose processor, for example, a DSP, will initialize first. After these nodes complete loading and initialization, other nodes connected to it may then go through this same process until the system completes initialization.
- *Processor interfaces.* When multiple processors must communicate with each other, care must be taken to ensure that messages sent along interfaces between the processors are well-defined and consistent with the processing elements. Differences in message protocol including endianness, byte ordering, and other padding rules can complicate system integration, especially if there is a system requirement for backwards compatibility.
- *Load distribution.* As mentioned earlier, multiple processors lead to the challenge of distributing the application and possibly developing the application to support an efficient partitioning of the application among the processing elements. Mistakes in partitioning the application can lead to bottlenecks in the system and this degrades the full entitlement of the system by overloading certain processing elements and leaving others underutilized. Application developers must design an application to be efficiently partitioned across processing elements.
- *Centralized resource allocation and management.* In a system of multiple processing elements, there is still a common set of resources including peripherals, crossbar switches, memory, etc., that must be managed. In some cases the operating system can provide mechanisms like semaphores to manage these shared resources. In other cases there may be dedicated hardware to manage the resources. Either way, important shared resources in the system must be managed in order to prevent further system bottlenecks.

9 Software for Embedded Systems

This book will spend a considerable amount of time covering each phase of software development for embedded systems. Software for embedded systems is also unique from other "run to completion" or other desktop software applications. So we will introduce the concepts here and go into more detail in later chapters.

9.1 Super Loop Architecture

The most straightforward software architecture for embedded systems is the "super loop architecture." This approach is used because when programming embedded systems it is very important to meet the deadlines of the system and to complete all the key tasks of the system in a reasonable amount of time, and in the right order. Super loop architecture is a common program architecture that is very useful in fulfilling these requirements. This approach is a program structure comprised of an infinite loop, with all the tasks of the system contained in that loop structure. An example is shown in Fig. 15.

The initialization routines are completed before entering the super loop because the system only needs to be initialized once. Once the infinite loop begins, the valves are not reset because of the need to maintain a persistent state in the embedded system.

The loop is a variant of the "batch processing" control flow: read input, calculate some values, write out values. Embedded systems software is not the only type of software which uses this kind of architecture. Computer games often use a similar loop called the *(tight) (main) game loop*. The steps that are followed in this type of gaming technology are:

```
Function Main_Game_Function()
{
    Initialization();
    Do_Forever
    {
        Game_AI();
        Move_Objects();
        Scoring();
        Draw_Objects();
    }
    Cleanup();
}
```

```
Function Main_Function()
{
    Initialization();
    Do_Forever
    {
        Check_Status_of_Task();
        Perform_Calculations();
        Output_Result();
    }
}
```

Fig. 15 Super loop architecture template.

9.2 Power-Saving Super Loop

The super loop discussed previously works fine unless the scheduling requirements are not consistent with the loop execution time. For example, assume an embedded system with an average loop time of 1 ms that needs to check a certain input only once per second. Does it really make sense to continue looping the program every 1 ms? If we let the loop continue to execute, the program will loop 1000 times before it needs to read the input again. Therefore, 999 loops of the program will effectively countdown to the next read. In situations like this an expanded super loop can be used to build in a delay as shown in Fig. 16.

Let's consider a microcontroller that uses 20 mA of current in "normal mode" but only needs 5 mA of power in "low-power mode." Assume using the super loop example outlined in the earlier text, which is in "low-power mode" 99.9% of the time (a 1-ms calculation every second) and is only in normal mode 0.1% of the time. An example of this is an LCD communication protocol used in alphanumeric LCD modules. The components provides methods to wait for a specified time. The foundation to wait for a given time is to wait for a number of CPU or bus cycles. As a result, the component

```
Function Main_Function()
{
    Initialization();
    Do_Forever
    {
        Check_Status_of_Task();
        Perform_Calculations();
        Output_Result();
        Delay_Before_Starting_Next_Loop();
    }
}
```

Fig. 16 Power-saving super loop architecture template.

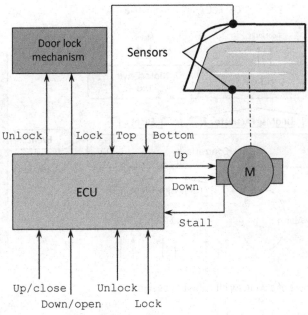

Fig. 17 Example of a window lift hardware design.

implements the two methods: Wait10Cycles() and Wait100Cycles(). Both are implemented in assembly code as they are heavily CPU dependant.

9.3 Window Lift Embedded Design

Let's look at an example of a slightly more advanced software architecture. Fig. 17 shows a simplified diagram of a window lift. In some countries it is a requirement to have mechanisms to detect fingers in window areas to prevent injury. In some cases, window cranks are now outlawed for this reason. Adding a capability like this after the system has already been deployed could result in difficult changes to the software. The two options would be to add this event and task to the control loop or add a task.

When embedded software systems become complex, we need to move away from simple looping structures and migrate to more complex tasking models. Fig. 18 is an example of what a tasking model would look like for the window lift example. As a general guideline, when the control loop gets ugly then go to multitasking and when you have too many tasks go to Linux, Windows, or some other similar type of operating system. We'll cover all these alternatives in more detail in later chapters.

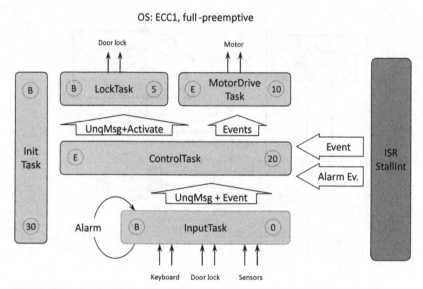

Fig. 18 Example of a window lift software design.

10 Hardware Abstraction Layers for Embedded Systems

Embedded system development is about programming at the hardware level. But hardware abstraction layers (HALs) are a way to provide an interface between hardware and software so applications can be device independent. This is becoming more common in embedded systems. Basically, embedded applications access hardware through the HAL. The HAL encapsulates the peripherals of a microcontroller, and several API implementations can be provided at different levels of abstraction. An example HAL for an automotive application is shown in Fig. 19.

There are a few problems that a HAL attempts to address:
- Complexity of peripherals and processors, this is hard for a real-time operating system (RTOS) to support out of the box, most RTOSs cover 20%–30% of the peripherals out of the box.
- Packaging of the chip-mixing function—how does the RTOS work as you move from a standard device to a custom device?
- The RTOS is basically the lowest common denominator, a HAL can support the largest number of processors. However, some peripherals, like an analog-to-digital converter (ADC) require custom support (peripherals work in either DMA mode or direct mode, and we need to support both).

The benefits of a HAL include:
- Allowing easy migration between embedded processors.
- Leveraging existing processor knowledgebase.

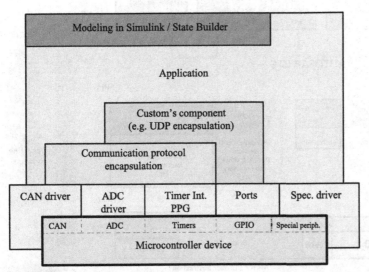

Fig. 19 Hardware abstraction layer.

- Creating code compliant with a defined programming interface, such as a standard application programming interface (API), a CAN driver source code, or an extension to a standard API, such as a higher level protocol over SCI communication (like a UDP) or even your own API.

As an example of this more advanced software architecture and a precursor to more detailed material to follow later, consider the case of an automobile "front light management" system as shown in Fig. 20. In this system, what happens if software components are running on different processors? Keep in mind that this automobile system must be a deterministic network environment. The CAN bus inside the car is not necessarily all the same CPU.

As shown in Fig. 21, we want to minimize the changes to the software architecture if we need to make a small change, like replacing a headlight. We want to be able to change the peripheral (changing the headlight or offering optional components as shown in Fig. 22) but not have to change anything else.

Finally, the embedded systems development flow follows a model similar to that shown in Fig. 23. Research is performed early in the process followed by a proof of concept and a hardware and software codesign and test. System integration follows this phase, where all of the hardware and software components are integrated together. This leads to a prototype system that is iterated until eventually a production system is deployed. We look into the details of this flow as we begin to dive deeper into the important phases of software engineering for embedded systems.

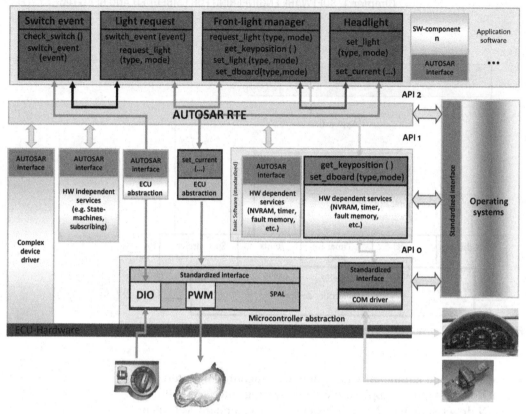

Fig. 20 Use case example of a "front light management" system.

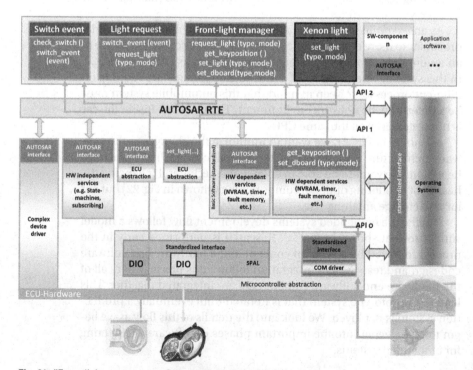

Fig. 21 "Front light management" system headlight components.

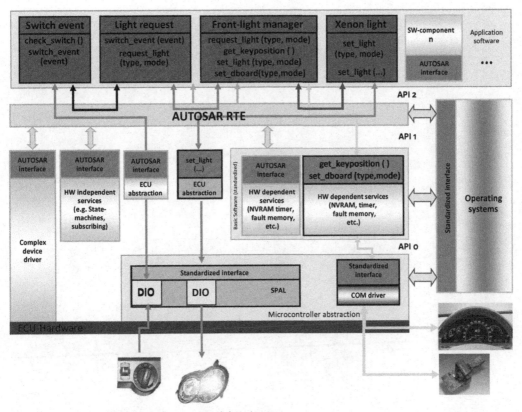

Fig. 22 "Front light management" system peripheral components.

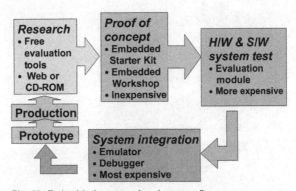

Fig. 23 Embedded system development flow.

2

SOFTWARE DEVELOPMENT PROCESS

Mark Kraeling*, Lindsley Tania†

*CTO Office, GE Transportation, Melbourne, FL, United States, †Release Train Engineer LOCOTROL® Technologies GE Transportation, a Wabtec company

Software Engineering for Embedded Systems. https://doi.org/10.1016/B978-0-12-809448-8.00002-3

1 Getting Started

The software process involves the steps necessary to produce a software product. Understanding the requirements for that software product, including regulatory, industry, and safety requirements, is fundamental before choosing a particular process. Such requirements may dictate the methods required within the software process, independence of development, or even the qualifications of the developers involved with a software product.

The software development process itself can be used for a variety of programming languages, whether it be Embedded C, larger scale languages, or scripting languages like Python. Whenever a collection of engineers exists to create a software product—the process is necessary. This can include simple modification of an existing software product—the software process is the method engineers use to create a new version.

Software processes can often be very complex due to the size of the group, the complexity of the product, or even requirements imposed (like safety) that need to be met. Computer-aided software engineering (CASE) tools often provide support for some of the activities in the software process. Far too often though, inflexibility in customization for different development processes makes these tools difficult to use, hence diminishing their value.

Depending on industry, company, or customer requirements, software processes can become very fluid. Gone are the days when all software requirements were understood up front prior to beginning work—today feature developments and additions throughout the process are the norm. To that end software development processes, like Agile, can help manage these issues and create an easier software delivery path.

The most important aspect of the software development process, is that the entire software team understands and follows the process. In many cases, the question that independent auditors of a process are evaluating is "are they doing what they said they would do?" Documenting a much more fluid or ad hoc process, and sticking to it, is arguably better than documenting a more stringent process that most people do not follow. The most important thing is to make sure that whatever process is selected that it is followed by the entire team.

1.1 Project Planning

Before any software development (especially code writing) can begin, it is necessary to get a software project plan in place. This typically involves a project manager, who will scope the project correctly before getting started. This person can either be the software manager or a person who works closely with the software manager, thereby understanding task loads and capabilities.

Software projects are often labeled as "going to be delivered late" or "going to run over budget" before they have even started. Software projects are not the same as typical engineering projects because they are so diverse and have many acceptable approaches. Software is more of a hidden type of deliverable—it is not obvious to anyone outside software coding what is being done or to what level of quality the work has been completed. Sometimes software projects are unique, where the development team or even company hasn't ever implemented anything similar before. They may represent big projects for a customer where there is no obvious possibility of using the software code elsewhere. And as mentioned, the software process itself may vary from project to project, based on software requirements. All these factors must be understood up front before the creation of a project schedule.

At the outset, when the project schedule is developed, there are a variety of inputs that need to be understood. First is understanding and translating the customer's expectations. Is the completion date a date for a preliminary field integration test or do they expect a fully tested and validated product on that date? Understanding the software resources available and their capabilities is also important. Having software engineers with experience in similar projects or even projects for the same customer is better than having a staff that is entirely inexperienced. Having inexperienced engineers working on a project is not a bad thing as long as time is allocated for experienced staff to assist and bring them up to speed in readiness for the next project. Finally, the external factors that may impact a project need to be understood. Such factors may take the form of support from other functions within the organization, such as quality or risk management, or from outside regulatory or independent assessments that are required before the software can be delivered.

After these inputs are understood, the initial project plan can be put together. It is considered "initial" because a good software project plan will undergo numerous negotiated changes throughout its development life cycle. Too often arguments can occur because the project manager holds on to the original plan instead of updating it based on the realities of the project. Various factors that can affect an initial project plan include not initially evaluating the risks correctly, having resource issues occur that are not immediately controllable, or taking on an advanced feature that has not been done before, involving a great deal of "learning."

Project plans should always have milestones in the schedule. These are points where a specific activity can be marked as complete. Avoid having milestones that are not specific to a deliverable, like a milestone stating that the design portion of the project is half done. There should be a measurable shift in activity or delivery when each milestone listed is completed. Tracking progress against the project plan and schedule can be done using a variety of tools, but typical granularity for project updates should be no less than 1 week. More often than not, the project manager will have a schedule that is constantly fluid or requiring a lot of time to status each of the pieces. Smaller software-specific plans can be done in conjunction with the larger project plan, as is done, for instance, in sprint planning for Agile.

Having a project plan that is understood and agreed upon up front, is regularly updated in terms of its status and progress, and has an easy way to assess its progress throughout the project is critical.

1.2 Risk Management

Risks will be present in any software project. Understanding those risks up front and determining mitigation plans helps in the development and then execution of the project plan. Not only are there project risks, like availability of resources, but there are business risks as well, such as having software that hits the market window. Other risks include changing the software requirements later in the software development process, hardware platform delays if developing an embedded product that the software needs to run on, or even outside influences, such as regulatory changes software needs to comply with. Risk classification, risk analysis, and risk mitigation plans need to be in place at the start of project planning and should be updated throughout development cycles.

Risk classification includes capturing all the possible risks that could influence the correct and complete delivery of the software project. They are typically categorized in groups that make sense to the project, such as resources, external groups, and even estimation risks. Each risk is preferably identified up front but risks can be added during the process of the project as well. Once they are documented and classified, a risk analysis can be performed.

A risk analysis involves taking a deeper look at each risk. This involves understanding how likely a risk is to occur, and then how much of a negative impact that risk will have if it does occur. The probability that the risk will occur is typically grouped as a percentage, with ratings for very high (maybe >90%) all the way to very low (maybe <10%). Sometimes factors are applied to each of these, with higher numbers associated with higher probabilities. Once this is complete, then the effect of that risk needs to be evaluated. This could be something severe in terms of impacting the project, to negligible. Again, a numerically weighted value can be associated with each of these assessment values. A simple way to order the risks is to multiply the two numbers for that risk together to get a score. However, a risk that has severe consequences may need special attention even if it not in the top 20% of risks—thus the suggestion of the "weighted" impact value.

Risk mitigation then involves developing a plan outlining what actions need to be taken if the risk identified occurs. This could involve early development of that feature so there is more time to address it as opposed to waiting until later in the schedule, or even developing a parallel path so that if one path for development fails the other may work. In any case, once the mitigation strategies for each of the risks is understood, then those strategies need to be executed and monitored throughout project development.

1.3   Kicking Off the Project

Ideally when the project plan including resource identification and schedule are complete, and a solid risk assessment is complete, then the project is ready to get started.

A project kickoff meeting is an effective way to get the project going in the right direction. Management can be brought in so that they understand the risks and resourcing. Developers should also participate to understand the project schedule and review the risks that were identified using the mitigation plans. The milestones (and how to evaluate whether they are complete) can also be reviewed.

It is important to get the stakeholders and the resources on the project to agree where they are, how the project will be measured, and the path that the project will take. If that is not done up front, there is little chance it will get any better after the project starts. A regular reporting rhythm for development can be agreed and getting people to buy into this idea is crucial. Once this kickoff occurs, the software development process can get started—with the software requirements.

2   Requirements

Requirements state how a system should behave or operate, without necessarily knowing the details of how it is implemented. When requirements are written, as either user requirements or system requirements, they are often written in a form to make sure that at the end of the project they are testable. The software definition for a requirement is a statement describing the functionality of the system or a constraint that the software needs to meet. Constraints could be in terms of meeting timing or mission-critical items for the system. For instance, must "turn on the A driver within 50 ms of a specific discrete input X going from low to high."

Requirements can also have varying levels of detail. For a system that is well understood, the requirements may become very detailed in the operation and functioning of the system. This is typically done when a system may be getting a technology refresh from a previous generation of product or a large amount of product definition work has already been performed to really understand the system. Traditional embedded systems tend to mix different types of requirements together, such as user requirements and system requirements. However, depending on the size of the embedded system it is often much better to separate the two and define them separately. This helps with creation, implementation, reviews, and testing of the different sets of requirements.

2.1   User Requirements

User requirements are typically written when discussing the use cases for a project. The requirements definition is done with the customer or product managers that know how the embedded system will be used by the user.

Many user requirements deal with how a user will interact with a system and what that user expects. If there is a screen or human machine interface aspect to the system, a user requirement may be based on what happens when the user selects an action on the screen. Maybe with a button press not only does a process start, but it also switches to another screen and provides an audible notification. When user requirements such as these are written down, they can often break into multiple system requirements later due to switching of screens, the maximum delays in starting the process, and finally what the next screen should look like. One pitfall is starting to try to write the system requirements during a user requirement meeting. This often detracts from gaining insight into the requirements of the user, and key functionality pieces could be missed.

In fact, as alluded to earlier, it is often better to keep user requirements and system requirements separate in their tracking and reporting. The user requirements are often more readable, understandable, and provide a better sense of how the system will operate. Even though user requirements may lack specifics on what really needs to occur in the system, they are still valuable in that they can provide the overarching system functionality expectations.

Finally, when writing user requirements, it is a good idea to have traceability in terms of where the user requirement originated.

Whether it is from a single customer or product manager, understanding where it came from is important. When capturing user requirements, there are times when separate user sessions may develop conflicting requirements. Being able to go back to the originator and understand the use case better can help facilitate the deciphering of conflicting user requirements. It could come from differing standpoints, like an operator vs. a maintenance person, so being able to go back and resolve those differences becomes important.

2.2 System Requirements

System requirements are requirements that may not necessarily be visible from a user standpoint. Many embedded systems have a much higher ratio of system requirements to user requirements, simply because embedded systems are often not seen. A braking system in an automobile, for instance, has few user inputs (the brake pedal) but many system requirements on how the brake system should function.

System requirements are often much more detailed than user requirements as well. Whether these originate from system or software engineers, they are often derived from understanding the interfaces that the embedded system must work with. So, there may be requirements to specific standards or timing requirements because of a deeper understanding of how the system must react.

One caution that should be exercised when writing system requirements is that these requirements should focus on how the embedded system works with and reacts to the external connections it has. If the embedded system must initiate or react to external signals, then this should be a primary focus for system requirement development. Too often, system requirements are written that specify the internals of the system. In some cases, this should be done, but in others it often restricts the software design unnecessarily. If it is important to include something from a product or system sense, then include it.

Once requirements are gathered up front, there are a variety of ways to manage those requirements. They should be revisited, will often change through a development process, and their management then validation is important. Because there are multiple ways to do this, one such method is described in a later section in this chapter covering Agile development.

3 Architecture

When the architecture for an embedded system is put together, there are many important factors to consider at the beginning. The following are some typical questions asked:
- Am I reusing portions of an adjacent or previous generation design?
- Do I need to deliver subsystems or components of this design to another design?

- Are there safety-critical aspects of the system that need to be addressed?
- What are the most secure pieces of the software that I need to protect?
- For the hardware that has been selected, how can my architecture take the fullest advantage of the platform?

These questions lead the development team into certain architectural decisions and its structure at the beginning of the project. Breaking an embedded system into smaller subsystems is often the best way to architect a system, so that subsystems can have a function on their own and be a separate, testable piece of the system. However, creating many subsystems may create performance degradation, especially if performance-critical dataflows get spread between multiple subsystems. The following are factors to consider when designing the architecture of an embedded system.

3.1 Safety-Critical Elements

Safety-critical elements are requirements that the system must meet either from regulatory, customer, or internal safety design practices. Safety-critical software, because of the amount of testing and rigor of the software development process itself, should try to stay within the fewest number of subsystems as possible. When this is done, that subsystem can follow more rigor and be designed in such a way that the other subsystems cannot impact its execution.

Often, this type of safety-critical architecture drives the embedded software to live on its own hardware, a safety coprocessor of sorts that further helps guarantee its isolation. Its interface to the other subsystems is then managed through serial or discrete I/O, making it much easier to apply process rigor to one part of the embedded system and have isolation to the rest of the system.

3.2 Operating Performance

When certain performance (i.e., response time) targets for an embedded system need to be met, it is good to contain these requirements in a single subsystem if possible. If simple logic exists between getting an input signal and then driving an output signal in response, this could be architected in a single subsystem.

If the architecture of the embedded system, specifically because of its complexity, doesn't allow for performance-oriented architecture items to be in an individual subsystem, then a different approach is required. There are two general types of messaging between subsystem interfaces, *dynamic* and *periodic messaging*. The most common approach is to minimize the number of interactions between subsystems when passing performance data between them, a.k.a. *dynamic messaging*. With *dynamic messaging* the response signal begins to be triggered as soon as the input stimuli is received and there is minimal delay between subsystem message transfer. This could be set up as a separate interface between subsystems.

With *periodic messaging*, a delay exits between subsystems as messages are transferred through. *Periodic messaging* could be a set of messages that is sent between subsystems every 100 ms therefore increasing the total amount of delay to receive a response output. Now imagine there are several subsystems where *periodic messaging* is used. Every time a message is transferred to a different subsystem, delay propagation continues to build. It is clear how this could affect performance-related requirements.

3.3 Security

Like safety-oriented information, it is important to try and isolate the security elements of the architecture. The goal would be to have multiple layers in the system before the secure data elements can be reached. In the security environment, embedded system layering could act as a "defense in depth" type of strategy.

In a simple example, an embedded system has layer A that interfaces externally, a layer B subsystem that it talks to, and then a layer C subsystem that contains the secure data elements. In the design, layer A receives a message, then creates a new message to layer B to request the required information. Layer B would process the request, but also check various performance and security requirements around the data being requested. This check system could also include network or formatting rules to make sure that numerous incorrect requests have been received recently. If this all checks out, then a separate message could be sent to layer C to get the data requested. Layer C could even include a check of some type of signature or encrypted part of the message that layers A and B do not know how to interpret, but the source of the request and layer C do.

Further checks, like creating an inability for layer A to talk to layer C, can provide security to the system. This could be done by using entirely different messaging checks for message formats so there exists no ability for the software to even interpret that type of message. Only layer B can interpret, check, and then reformat the message for the appropriate subsystem.

3.4 Reliability and Availability

The last of the large aspects that should be considered as part of the architectural design are the requirements for availability and reliability. These two terms are often mistaken as the same thing, but it is important to consider both for the architecture of the system.

Availability is the measurement of the uptime of the device performing a task. If a machine must spend 1 h for maintenance after every 9 h of operation, it would have an availability of 90%. Downtime of any kind, whether planned or unplanned, negatively impacts the availability of the system.

Reliability is the measurement that a system will perform its intended task when it is supposed to. Unexpected failures negatively impact a system's reliability. Measurements, like mean time between failures, are often used for a reliability assessment.

Designing an architecture to meet availability requirements often includes elements of redundancy or efficiency in the way data is handled. If there is a strict availability requirement, duplication of either hardware or software in the design may be desired. If one data path does not work, the redundant path takes over and completes the particular "high-availability" aspect of the embedded system. This could be done at a process level within a hardware processing unit or separating hardware with checks between them to make sure that one has performed the task required.

The other aspects that come into subsystem and architecture design include updating the software and configurations. Does the system need to be taken all the way down to update a single subsystem or configuration in a subsystem? This impacts the availability of the system. Instead, can the architecture be designed in such a way that subsystems or hardware elements are hot-swappable, meaning a new item can be put in place dynamically with the system continuing to operate.

After considering safety-critical, operating performance, security, and reliability/availability aspects of the architecture, then the design will start falling into place. Using interface dataflow or specifications to understand the interactions between subsystems is important. It is possible that once the list of interactions is understood, then functions or data elements may move from one subsystem to another. All of this should be done before the embedded system design is started.

4 Design

The design phase of the software process takes the requirements and architecture and produces logical data or object flows that can be implemented. At this point in the process, it is important to assess which type of software design to use.

Object-oriented software design is one popular option for embedded systems. It treats sets of data as objects, with specific data access methods for those objects. In this design, it is very understandable how data is being manipulated and how sets of objects manage their data, state, and operations. A design of this type is ideal for reusability and maintainability.

Real-time software design is another popular method when the overhead of object-oriented software cannot be tolerated. It is more focused on the set of operations that take place once a stimulus comes into the embedded system, until the desired system state is achieved. A design of this type focuses on minimizing delays in interactions, or a design that supports a smaller hardware scale (smaller than average 32-bit processors).

4.1   Object-Oriented Programming

Object-oriented programming starts with an *object*. An *object* maintains its local state and includes its own set of attributes that are specific to that object. Access to making changes in that object is handled by a set of the objects access functions, so its data can be tightly controlled. Object-oriented programming also deals with the term "class" that creates a new type, and objects are instances of a class. This would be similar in C programming, where the class could be unsigned long, and an object is each variable that is declared as an unsigned long.

Objects can store data in fields if they belong to that object. Objects can also have functionality that is specific to that object, if they belong to a class, called methods. All the fields and methods are referred to as attributes of that class.

There are two types of fields in objects. The first is a class variable which is shared data among all of the objects of that class. A change in a class variable in one object will also change in other objects of that same type. An object variable is specific to that object, so changes in an object variable do not change values in other instances of objects of that type. Consider a type called "student," where we have an embedded system for a local school. A class variable could be the address of the school. Each student would have the same address, and in the event the address changes, it is desired that a change in address for one student would alter all the others. An object variable for the type "student" could be the student's name. This would be unique to each object of the student type.

One of the major benefits to using object-oriented design is being able to reuse code in the system. This is typically gained through a term called inheritance. Consider inheritance as being able to add a subtype to any type that is specified in the system. As objects contain more and more data, they can become more difficult to manage. If object types have things in common, it is better to split this common data into a new class, and then have the appropriate classes inherit this new class as one of their attributes.

Consider a student and a teacher. There are attributes that are common to both, such as school address and possibly some type of

ID number. Students have grades and teachers do not, and teachers have salaries where students do not. A "person" class could be created, where everything common to both students and teachers can be placed. Then, when the teacher class is created, it calls out to inherit all the attributes of the person class as well. The same occurs when the student class is created. The "student" class is known as the base class, and the student/teacher classes are known as subclasses.

With this basic understanding of fields, methods, classes, and objects, the object-oriented design process can proceed. Three of the key steps in object-oriented design include system context, class identification, and design modeling.

4.1.1 System Context

This first step in object-oriented design is helpful to understand the context and interactions of the system. Use case models are used to understand the interactions with the system and help identify objects and operations. The following is an example diagram to understand which functions an operator would perform with a machine (Fig. 1).

After putting together a diagram, specific use cases descriptions can then be written. The use case descriptions describe the data, the stimulus that enables the use case, and the response of the action. These help before taking on the next step in object-oriented design.

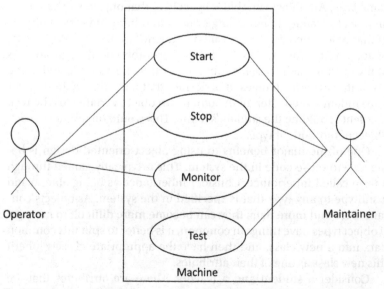

Fig. 1 Use case modeling.

4.1.2 Class Identification

This step identifies the classes and subclasses of the system. By first considering the stimuli of the system one can infer which class data should exist. Each class can also include methods or functions that use the stimulus data either directly or indirectly in subclasses. The data and how the data should be manipulated should be understood, so an initial cut at classes can be created. From the stimuli in the system, each class can also include the methods or functions that perform using the data in that type or included from the subclasses. From the system context step the use cases should be understood, so it is clear what types of methods need to be included and which data those methods operate on.

4.1.3 Design Modeling

During this step the relationships of the object classes are documented most typically using a documentation method called Unified Modeling Language™, or UML. This step bridges the requirements to the implementation of the system. UML diagrams need to show the relationships between the classes, and at the same time provide enough information for software programmers to implement. Consider the UML diagram in Fig. 2.

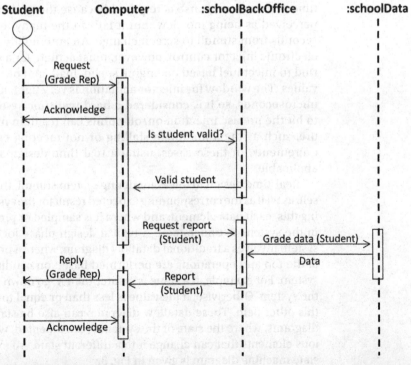

Fig. 2 UML sequence diagram.

The object classes are listed along the top with a vertical line under each. For the sequence, time starts at the top of the diagram and moves down vertically. *Arrows* then show the interactions between the classes, with rectangles showing when an object of that class is in control of the system. If the object must wait for a response from another before continuing, then that rectangle is stretched down the line until the response is received.

Each major interaction of the system, especially those that were derived from the original use cases, should be documented. Other more detailed interactions that are not necessarily part of the use cases can be done later by the programmers once they understand more of the design. Once this last step is performed then object-oriented implementation can begin.

4.2 Real-Time Design

Real-time design is when there are specific system constraints that must be met or if a response must be generated within a very tight window from when stimuli are received. Object-oriented programming could certainly be used for some "soft" real-time systems, where a response has a window of time as opposed to being very tight or "hard." A user touch screen is an example of a soft real-time system. A response is required because the screen cannot be perceived as being too slow, but it falls in the many tens of milliseconds from stimuli to screen change. An automobile motor with electronic injector control, however, must actuate for a specific period to inject fuel based on engine speed, load, and other dynamic values. The window for injector actuation is very tight, measured in microseconds, so it is considered a hard real-time system. Failure to hit the precise injection on/off points can result in mission failure, such as the automobile stalling or not meeting emission requirements. In these cases, using a real-time design approach is applicable.

Real-time design involves modeling system stimuli, the system itself, as well as the corresponding expected result for the system. By doing this, each data element, and where it is sampled or provided from in the system, is understood. A typical design phase for a real-time system involves a traditional dataflow diagram, where a process starts at the top and operations are performed based on conditions in the system. For example, if a value is greater than X, go down this path in the system. Otherwise, if the value is less than or equal to X, go down this other path. These dataflow diagrams can also be state machine diagrams, where the state of the system is documented, with the various elements that can change it to a different state. An example of a state machine diagram is given in Fig. 3.

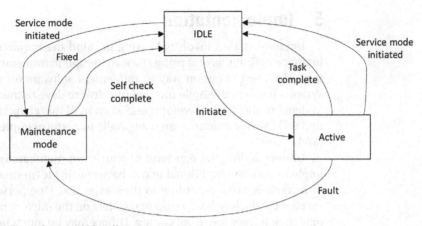

Fig. 3 State machine diagram.

Once dataflow and state machine diagrams are understood, then the design process shifts to looking specifically at the resources of the system. At this point, a real-time operating system can be selected, as this will have a bearing on the overall design. Real-time operating systems contain elements called signals, which can be sent from one process to another to initiate an action. That allows control from one process to another higher priority process to accomplish a task. For a smaller re-sourced processing system, a "roll your own" operating system can be created where an RTOS is not used. Instead, a rather simple scheduler for tasks can be used, with interrupts that wait for external stimuli be-fore kicking off a specific sequence of tasks and then quickly returning to normal operation. Microcontrollers have built-in mechanisms that assist in both commercially available or open source RTOS products, and in systems where a simple execution method is created and used.

Real-time design focuses on speed of execution and minimizing de-lays. C++ can be used for real-time programming, but certain aspects of C++ should not be used, such as dynamic garbage collection, op-erator overloading, and dynamic creation/deletion of objects during operation. Aspects like these in the C++ programming language can make execution timing inconsistent, which could lead to not meeting the real-time requirements. Typical programming languages for hard real-time systems include utilization of FPGAs for fast handling of data inputs, assembly language for a very specific sequence of timed events, and the C programming language which doesn't have much of an overhead and has easily interpreted output when looking at its generated assembly language output when compiled.

Once either object-oriented, real-time, or possibly some other design method is complete, then the team can move on to implementation.

5   Implementation

Implementation involves bearing in mind the requirements and interface definitions and using software design to implement the system. The most common way to implement software for embedded systems involves the Agile method of software development, covered in detail in the "Agile development" section of this chapter. This section will provide guidance on using Agile to perform implementation (and more).

Understanding the expertise of your team members is critical to implementation. Traditional teams have specific team members perform various tasks according to their expertise. One person may be excellent at the low-level setup of registers on the microcontroller and may stick to very low-level coding. Others may be much better at the upper level programming in C or C++ and may not have a very good hardware understanding at all. Software teams have shifted to more of a pair programming approach for implementation, where a junior software engineer may be teamed with the hardware low-level expert to learn, understand the choices being made, and ask questions or provide input to create a better low-level layer. This also helps provide backup to ensure that the project does not come to a halt due to an issue that nobody understands if an expert is unavailable.

Another aspect to consider for implementation is team member location. Traditional teams may have all the team members in one location. Teams today are becoming more global and may not be in the same location or time zone. Clear communication among team members is critically important for embedded systems whether colocated or not. Defining a communication method for the entire team must be done up front so that the project can proceed effectively. Emails sent from one team member to another without copying in other team members may be ineffective compared with a periodic meeting where all members can get together and discuss status or critical issues.

Part of implementation is the software phase where the product and project management tend to become more nervous. As with the previous phases of software development regular reporting can occur, or review meetings happen to see and review designs or architectures. During implementation, the only reporting typically provided is a measurement of percentage completion. Managers use the team's percentage of completion to forecast the expected date of completion. For software projects, a lot of time is spent in the 90th percentile because of last minute changes or issues that cause system failures that are elusive. For this reason, using an Agile process is more effective because it provides additional metrics and regular reporting, making it much easier to understand and project progress toward completion.

6 Testing

Testing of an embedded system is done throughout the system's life cycle. Unit testing, or testing of individual software elements, can be done throughout the implementation of the software itself by the software engineer. It can involve varying levels of automation, but it is done to make sure the unit of software performs as expected and that each path in the software is complete. Systems testing is done when the software itself is complete, where the system is tested against the original requirements. Test cases are set up to assess whether the original requirements have been met. This section considers the various types of testing and provides specifics for each.

6.1 Validation and Verification

During and after the implementation phase of the project, verification and validation are performed on the embedded system. Verification and validation include testing the system behavior as specified by the original requirements of the system using elements of the use cases and any performance aspects that were added. Validation answers the question "are we building the correct system?" Verification answers the question "are we building the system correctly?"

Verification is more straightforward to perform. The embedded system requirements can be translated into test cases, and then a verification plan and procedure can be put in place. For each requirement, the system is tested to make sure it is met. Pass/fail or similar criteria can be used to assess the system.

Validation is a little more complicated. It attempts to truly answer whether the embedded system can be sold into the marketplace and might be required by customers. It takes use cases and customer feedback into consideration, as opposed to hard requirements, and includes user expectations, which may be a bit more subjective than hard requirements.

Software inspection and validation testing are performed to check each of these. Such processes may involve peer reviews of software in the deepest sense or higher level activities like performance testing and user reactions to using product prototypes. Verification and validation are the processes used to identify areas in the system that need improvement, with defect tracking and reporting used to understand stages in the process. Once a defect is identified, it can go to the appropriate stage in the process (even back to architecture or requirements) so that the change can be applied through the entire software development process. Once the issue is fixed, it can be revalidated or reverified before it is closed.

Automatic static analysis can also be a part of verification. Software code is fed into an automated tool that analyzes the software and points out potential defects or latent defects. This can include branches of code that cannot normally be executed, such as a branch of software that is only executed when $X < 0$, except that X is an unsigned integer. Static analysis can also point out if a software function is too complex, by applying a McCabe™ complexity factor which is dependent on the number of paths and exit points within a function. Finally, even if the software code compiles without error, the static analysis tool can find and flag pointers that are used before assignment or typecasting being performed that may not make sense for the situation.

6.2   Integration Testing

Integration testing involves testing the system after the software components are put together. It is the first phase of testing the complete system. It involves identifying which components are necessary for a given system function and putting those together to test them to make sure functional requirements are met. An issue that arises with integration testing is attempting to understand which software component is causing the error—often there may be complex interactions between components so that it is not clear which components are problematic.

Deciding the order of components to be integrated is an important part of this process. It is possible there are mature components in the system that were developed and used for a previous project. In this case, it may be a good idea to integrate these components first since there may be more confidence with this set of software. Another method, especially if the software components are all new, is to integrate the most commonly used components in the system first. As they are the most used, it will allow for more and more testing as other components are added.

When new components are added it is important to also retest use cases that affect the components already integrated. A simple addition may end up affecting the operation of a component underneath, due to interactions or timing execution issues. Once the components have been integrated and tested the next phase of testing can occur.

6.3   Release Testing

Release testing can begin once all the components have been integrated and it is a complete set of software. During release testing, the system requirements that have been translated into testing requirements are executed. Release testing is normally called "black box" testing, as the internals of the software components are not readily understood, rather the system is tested using varying inputs while looking at the outputs to understand how the system responds.

For a given set of inputs, a given set of outputs is expected. However, combinations of inputs that may not have been specifically stated should also be tested, to make sure the system does not respond in an adverse way. As a simple example, consider that a requirement specifies if A and B are true then output X should be on. Release testing should then create test cases where A and B are both off, in addition to only A on and only B on. During this phase different combinations should be performed to see if the system can be "broken."

Testing with invalid inputs is just as important as testing with valid inputs. If a system login exists, then different combinations of usernames and passwords should be used. Even if the input is expecting alphanumeric values for a username, other types of characters should be used to make sure the system can still process those without malfunctioning.

Release testing is important to really understand how a system operates under proper, as well as adverse, inputs and conditions. This testing should be done before performance or stress testing.

6.4   Performance Testing

Performance testing involves putting the embedded system under stress to see if it can process input sets. A system may be designed to accept a serial message of 100 bytes, once per second. Performance testing takes this expectation but alters it to send 10,000 bytes per second, or maybe increases the frequency from once per second to once per every 10 ms. Even though the software design took the original typical state into consideration for the design, increasing the rate or the amount of data should not "crash" the system or cause undesirable behavior.

The first instance for performance testing is to evaluate the fault behavior of the system. Unexpected sequences of data, or sending corrupt data, should not cause additional failures to occur. Even though normal operation is a set of data that is properly formatted and sent at a precise periodic rate, the system must be able to handle nontypical conditions. This can include making sure the system throws away incorrect data, or is able to rate limit the data coming in so that it does not adversely affect the system.

The second case for doing performance testing is to make sure another defect is not caused somewhere else in the system. Even though the system may have a separate operation that is seemingly totally unrelated, understanding if there are defects because of timing becomes critical. Maybe time is being spent processing more data, so that a critical signal from one process to another cannot meet specific timing requirements and so goes to a fault state.

All these types of stress testing are particularly important for embedded systems where timing and resource constraints are more typical vs. an enterprise-oriented software architecture.

7 Rolling It Together: Agile Development

The Agile development process has taken over many forms of software design and development including the manufacture of embedded systems. Many companies are benefitting from having teams self-organize and collaborate closely. As we think about using this methodology, a few questions come to mind: How do we work to develop large complex systems using this methodology? How does the architecture for embedded devices come together if the team is Agile? How can we consider the standard system design cycle along with a well-defined operational context prior to coding when using Agile methods? If our system is a subsystem in a larger system of systems, we have many upstream and downstream functional flows, data streams, and user scenarios to consider ... how does this work when working in an Agile team? This part of the chapter will address these concerns as well as provide some guidance with special focus on roles, meetings, documentation, and flow. In this part of the chapter, I define a system-of-systems approach as an approach in which teams must:

> ... *develop a software component that is integrated with other components locally and rolled up for delivery as part of a global integration which must be installed and run as part of an overall product [1].*

When we say we are going *Agile* what does it really mean?
- Does it mean we are using a new tool? Metaphorically yes, physically no, although there are many new toolsets that foster Agile practices.
- Does it mean we no longer have documentation? No!
- Does it mean we are to be too flexible to plan? No!
- Does it mean we will never look through the lens of a phased approach? No!!

So, what exactly does going *Agile* mean?

Agile Alliance [2] defines *Agile* as:

> *The ability to create and respond to change in order to succeed in an uncertain and turbulent environment.*

And defines *Agile software development* as:

> ... *an umbrella term for a set of methods and practices based on the values and principles expressed in the Agile Manifesto.*

In my experience, going Agile means you are responding to change by simply paying attention, abandoning waste, and communicating with other people. It means you are inspecting what you are doing and regularly experimenting on how to make it better. Solutions evolve through collaboration between self-organizing, cross-functional teams utilizing the appropriate practices for their context. The meaning of

Agile for one team may be entirely different to another. Agile is not a one size fits all, it is up to the team to decide how they best want to use it. This is where we should consider the true meaning of Agile ... does it mean you no longer use common sense? No way!

This part of the chapter is not focused on regurgitating the 12 principles that capture the core Agile Manifesto values (even though there is some of that too) but is more focused more on how to balance the Agile principles to work with the natural energy your team already has, and to present solutions for those who like the *idea* of Agile to discover the *reality* of implementing Agile. This part of the chapter is a mere drop in the ocean of the content that exists on Agile and its best practices. Through training, reading, and personal conversations I have benefitted enormously from the ideas of W. Edwards Deming, Jeff Sutherland, Kent Beck, Mike Cohn, Dean Leffingwell, coaches from CA Technologies and Scaled Agile Inc., colleagues who apply Agile in their own way ... the list goes on. Rather than putting a footnote at every point where the ideas in this part of the chapter coincide with theirs, I simply want to highlight the substantial contributions and innovative thinking these scholars have given forth up front.

7.1 Scaling for Complexity/Organization

I work in a big organization therefore dataflows and infrastructure must be defined when thinking about "process." Agile can still be applied when working in a company with 500 people ... and no this doesn't mean your Scrum Teams have hundreds of people on them with one Scrum Master. It means those 500 people organize into Scrum Teams with separate Scrum Masters (think multi-"core" processing here). Different Scrum Teams have different levels of expertise. Arbitrary to the principle of "all work can flow to any team," the bottom line is: management is required to strategize about which work flows to which teams based on each team member's strengths and weaknesses (as mentioned in the "Implementation" section earlier in the chapter). The *Agile* recommendation is team members should self-organize into teams. In larger organizations, silos of personnel with varied experience levels and various functional expertise may make self-organization an unrealistic goal. There must be a spread of expertise and skill sets across each team.

Another challenge that teams encounter when attempting to be purely Agile is that by becoming Agile, they assume they must become anti-Waterfall. Don't be afraid to be the Agile team deemed as "too Waterfall-ish." A team I once worked with was so adhered to the Agile principles that they refused to conceptualize and design an initial architecture because doing so would be "too Waterfall-ish." Months and months passed, and the product was built vertical slice by vertical slice; minimal viable product (MVP) by MVP. It was not recognized until an undesired feedback cycle produced an enormous wave of defects (that were not initially caught because an integration test team would be "too Waterfall-ish"). This led me to the fact that it does not matter if you are Agile or not, complex systems must absolutely have forethought of architecture and conceptual design like the standard Waterfall practice of establishing requirements up front. We must not ignore all the system life cycle design concepts we have learned and benefited from simply to claim we operate in a particular way. We must use common sense (Fig. 4) !!!

That's not to say some of the architecture and architectural documentation requirements may emerge as the team members design, develop, and test vertical slices of value in time-boxed increments. In Agile programs, parts of the architecture may absolutely be defined incrementally along the way, however, the team must carefully assess areas where an emerging architecture makes sense and other areas where it does not. As the picture shows below, prescriptive *Waterfall* design can be done in parallel with emergent *Agile* design and development. Teams working in each direction on each side of the boundary, must work together to ensure the emergent design jives with the prescriptive design. Agile teams are completing US abiding by constraints defined in the prescriptive design. Before beginning a program, ask yourself: What part of the product architecture and high-level system design should be defined up front?

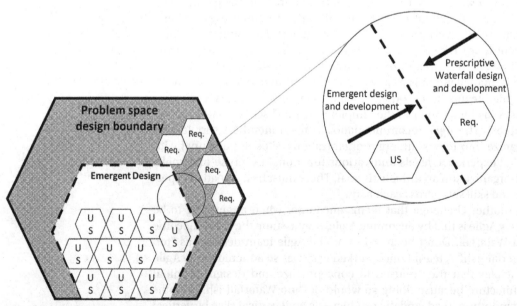

Fig. 4 Emergent meets prescriptive design and development.

Makes sense for architecture to emerge incrementally (emergent design):
- Software application architecture
- Internal interfaces and timing
- Human machine interface screen design
- Low-level configuration parameters

Does not make sense for architecture to emerge incrementally (prescriptive design defined up front):
- Platform and framework architecture
- External interfaces and timing (capture outer time boundaries)
- Software application configuration schema
- Key system dynamics
- Software application system modes
- Major software application state machines
- Nonfunctional requirements:
  - Performance characteristics
  - Availability
  - Maintainability

The meetings we have become accustomed to executing as part of the Waterfall process may not go away as we transition to a more Agile like execution. If the practices you have in place continue to serve you, do not drop them all to be Agile. Continue to hold a kickoff meeting to begin a new project. Continue to hold change control board meetings if you need them. Continue test readiness reviews. These meetings can be facilitated in addition to the Agile standups, sprint plannings, and retros (these are Agile meetings described later in this text). When applying Agile for complex integrated (and possibly multicore) embedded programs, you should merge Waterfall, Agile, or homegrown processes to come up with something that works for your team's individual needs. Software release artifacts that capture configuration changes may now log user stories and defects instead of software change requests (SCRs). Requirements, test cases, code reviews, and test reviews can now all be linked to the user stories to provide traceability. Agile processes and practices are simply tools, the focus should always be on the product, manifesting the vision, and the individuals; not the process.

7.2 Roles

An organization's roles will change when transitioning to the Agile development process. Teams adapt to Agile by using the prescribed roles, dataflows, and infrastructure prescribed by Scrum, Large Scale Scrum, eXtreme programming, and the Scaled Agile Framework model. Regardless of the framework your team chooses; it all comes down to the type of thinkers that exist in the organization. Considering which type of thinkers your team is made up of helps with mapping traditional roles to newer Agile framework roles. As you read this section of the chapter, proposal, conceptual, and detailed *thinkers* are flavors of these roles (not intended to replace traditional or Agile roles). This flexibility allows us to abstract the role from the individual to utilize individuals on our teams regardless of their traditional role or Agile role. For example, you could have a detailed thinker that is a Product Owner or a proposal phase thinker as a Scrum Master.

Proposal thinkers are often members of the commercial team, project or product managers, business analysts, architects, and/or principle engineers. These are often the people who know of the first news of a new program or system concept. Systems engineers, product owners, safety engineers, and quality assurance engineers are often well versed in this space, but it is not typically their primary focus. Proposal phase thinkers may be external facing or internal facing, meaning they collaborate with people either inside or outside of the organization. Externally facing product managers have the brunt of the work prior to program kickoff. Internally facing product managers have the brunt of the work after program kickoff. Proposal thinkers may support either side of a program kickoff. Proposal thinkers have a more intimate knowledge of a system's use cases and user scenarios, as well as organizational details like team and supplier interfaces.

Regardless of the medium, proposal thinkers typically:
- Communicate a transparent pipeline at the opportunity level
- Participate in backlog grooming
- Help to prioritize the backlog
- Communicate a regularly cadenced vision
- Negotiate what to work on and when to work on it
- Decompose and refine at the portfolio level: During a weekly sync up with leadership higher level work items are groomed and sized. At this point modifications may be made to the overall strategy going forward.

Conceptual thinkers are the people who do not see value in memorizing detailed information that can be looked up online later. The thought of trudging through details without having ultimate clarity on the higher level conceptual structure drives them crazy. When solving a problem, conceptual thinkers are "systems" thinkers. They try to define and solve the web of complexity that connects the pieces of information; not the information itself. Conceptual thinkers are different from proposal thinkers in that they know the product, the system components, their interfaces, the functions, and the functional allocation to the components. The proposal thinkers may know some of this, but it is not their primary focus. Conceptual thinkers are involved in the headspace of the "what" (unlike detailed thinkers who only think about the how). The "what" may include understanding operational context, user activity flows, dependencies/dataflows with other systems, etc. Both proposal and conceptual thinkers actively drive the vision of the future product forward.

Conceptual thinkers may also be the team members who set objective learning goals for the group. Conceptual thinkers are typically focused on the high-level work items in the backlog (think capabilities or features). The conceptual thinker's meetings are often focused around what work will go into the quarter (unlike the detailed thinkers who are digging in, focused on the user stories that will be assigned to sprints, which are 2-week cycles of work). Conceptual thinkers are your best bet to invite to feature writing and grooming sessions, as well as release planning sessions, described herein.

Detailed thinkers are focused on the details of implementation. Detailed thinkers crave order in a process. They crave determinism in operations. With Agile, some detailed thinkers may get turned off by all the flexibility and free-thinking that occurs. If you have a team of individuals that absolutely cannot stand flexibility, have them define how they want to do things and lock it in. For some teams just having this structure can make them operate better. Wrapping structure around Agile practices can help detailed thinkers feel more stability. Leaders of detailed thinkers should ensure the work to do is visible, the state of the work is visible, the metrics are made visible, and any impediments are cleared.

Traditional roles are mapped to Agile roles along with corresponding activities below.

Traditional role: *Architect*
Type of thinker: Proposal thinker
Agile role: Product Owner or Business Analyst
Typical duties (Agile + Waterfall):

- Defines multigeneration product and technology plans (then the next three generations of the product look like this given the technology at hand)
- Defines product vision and roadmap
- Responsible for high-level system architecture
- Designs validation architecture and strategy
- Defines functional concept(s)
- Performs risk management
- Defines operational use cases
- Defines use case models
- Defines sprint entry/exit criteria
- Prioritizes high-level work items in a backlog
- Determines feature dependencies
- Accepts user stories
- Assists team by providing direction
- Attends demos and provides feedback
- Updates architectural documents

Traditional role: *Project leader or project manager*
Type of thinker: Proposal and conceptual thinker
Agile role: RTE/Scrum Master
Typical duties (Agile + Waterfall):

- Facilitates creation of handoff package
- Facilitates team coordination
- Facilitates Agile meetings
- Performs risk assessment/monitors risk log
- Updates risk register, budget, plan, etc.
- Attends demos
- Attends technical readiness review
- Ensures line of sight
- Updates program management plan (iteratively)
- May draft and update software quality assurance plan
- May perform quality checks on user stories to ensure correct format with correct elements

Traditional role: *Systems engineer*
Type of thinker: Conceptual thinker
Agile role: Product Owner or Business Analyst
Typical duties (Agile + Waterfall):
- Defines product level model: this model shows where the product fits in the product line it is part of
- Drafts system engineering plan
- Assists with definition of cyber security considerations
- Assists with creation of architectural roadmap
- Assists with development of the system architecture definition/high-level system architecture
- Assists with development of the functional concept to enable generation of system requirements/acceptance criteria
- Defines system functional specification
- Defines system requirements/acceptance criteria for features and user stories with help from the team
- Is responsible for ensuring the team understands the user's intent
- Performs system functional allocation (maps system functions to system components)
- Assists with the identification of risks and assumptions
- Assists with designing validation architecture and strategy with the system architect
- Defines release and sprint DOR (definition of ready)
- Drafts features (high-level requirements)
- Determines feature dependencies with the system architect
- Ensures backlog priority is correct
- Assists with feature decomposition
- Determines user story dependencies
- Drafts a first release of user stories that demonstrate compliance to Safety objectives
- Produces and gives demos with Scrum or Kanban Team.

Traditional role: *Software developer*
Type of thinker: Detailed thinker
Agile role: Scrum Team member
Typical duties (Agile + Waterfall):
- Drafts software development plan
- Develops software design
- Drafts software design document
- Assists with determining priorities by describing dependencies between user stories and level of effort
- Sizes features/user stories with software tester(s)
- Ensures the software build passes
- Performs code reviews
- Develops software
- If tests fail, fixes software
- Assists with writing acceptance criteria with the help of the team
- Helps to prepare the sprint/release demo

Traditional role: *Software tester*
Type of thinker: Detailed thinker
Agile role: Scrum Team member
Typical duties (Agile + Waterfall):
- Ensures software functionality is correct
- Drafts test strategy and test plan
- Establishes test equipment requirements
- Identifies if requirement/acceptance criteria is testable
- Sizes features/user stories with the software developer(s)
- Assists with definition and decomposition of user stories
- Commits to user stories in sprint with team
- May assist with development of automated test scripts
- May run automated test suite
- Adds to regression test suite (manual or automated)
- Drafts software test procedures
- Writes software unit tests
- Executes software tests
- Reports software test results
- Helps to prepare the sprint/release demo

Traditional role: *Safety engineer*
Type of thinker: Conceptual and detailed thinker
Agile role: Supporting team member
Typical duties (Agile + Waterfall):
- Drafts system safety plan
- Partakes in team planning (speaking on behalf of safety type work scope)
- Partakes in sprint execution* (speaking on behalf of safety type work scope)
- Identifies safety objectives

- Drafts Preliminary Hazard Assessment (PHA), Functional Hazard Assessment (FHA), and System Hazard Analysis (SHA)
- Drafts multiple Subsystem Hazard Assessments (SSHA)
- Assesses if user stories have safety implications and if so tags them as such
- Safety acceptance criteria
- Attends demos

*Note: the safety team may work in a Kanban format a sprint behind the Scrum Team or may be embedded into the Scrum Team.

Traditional role: *Validation engineer*

Type of thinker: Conceptual and detailed thinker

Agile role: Supporting team member

Typical duties (Agile + Waterfall):

- Validates software: ensures software meets user's intent
- Ensures traceability
- Attends and participates in demos
- May develop automated test scripts
- May perform performance testing (verifies nonfunctional requirements/the "ilities")
- May perform sanity tests
- Performs test reviews

Regardless of the Agile framework and practices implemented, knowing which type of thinkers exist helps in terms of efficient organization. When transitioning to Agile, do not blindly map these roles. Talk to team members. Some systems engineers may not want the duties of Product Owner. Also, the mapping of *type of thinker* to *roles* here is meant to cover cases 90% of the time. You may have some team members that fit into many or all of these categories. Each type of thinker may work in a team with members like them, sprinting through 2-week cycles making deliveries to downstream consumers. For example, think of the delivery boundaries as a *"stage gate"* [3] *or serial* structure. Proposal cycle sprinters are preparing a backlog of high-level work items to the conceptual sprinters. They do not know much about the work yet so maybe the backlog just consists of Market Functional Requirements (MFRs). The conceptual sprinters are working to flush out the operational context, that is, system level requirements. Conceptual sprinters preload the backlog for the detailed sprinters team. The work item type that conceptual thinkers typically work on are typically features or epics; detailed thinkers work on user stories (as referred to in most Agile toolsets). The developers work in a team concentrating on vertically slicing code within the user scenarios, system architectural constraints, and system interfaces which are predefined by the conceptual sprinters. The proposal thinkers are sprinting and delivering to the conceptual thinkers who are delivering to the detailed thinkers (Fig. 5).

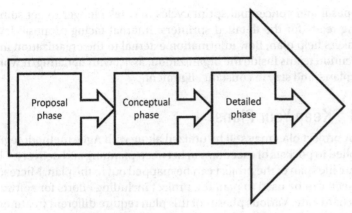

Fig. 5 Agile stage gate (serial approach).

However, this does not always have to be the case. Each type of thinker may work on a team that is "nested" (read: non-stage-gate) and has a mix of the various thinker types, sprinkling in their insight to various levels and phases of work. Each team has proposal, conceptual, and detailed thinkers either working in Scrum or Kanban configuration. For cases where the product is super mature or where bandwidth is slim the proposal, conceptual, and detailed thinkers may be the same people. Information exchange among the groups is highly interconnected and must be iterative. Quality is a by-product of how well the groups exchange information as they burn down their backlogs. The number of people on the team may influence whether your Agile teams will work in a *stage gate* or *nested* structure.

This "thinker" paradigm may also occur with development and/or testing. The test team can also work within a nested or stage gate structure. The detailed sprinters have detailed "testers" which may or may not be the developers (core Agile concepts recommend blurring developers and testers into one). Each level can have their set of tests to execute, with the proposal cycle testers focusing on user acceptance, the conceptual testers focusing on completeness and correctness of user scenarios, and the detailed testers ensuring that functionality is verified, designed, and developed correctly. Since the three thinker levels are sprinting concurrently there is a lot of knowledge sharing going on. Proposal testers are recommending conceptual functionality or maybe conceptual testers are recommending detailed functionality.

How deep the interfaces between these teams are depends on how new the technology and system is as well. For the implementation of an existing system the handoff from the proposal sprinters to the conceptual may be slim to none, information may go straight from the proposal team to the detailed team. For a brand-new system, the proposal and conceptual sprint cycles may take longer to get something ready for the detailed sprinters. Internal-facing proposal level thinkers help plan, flow information external to the organization into execution teams inside the organization, assist with updating/reworking plans, and stay in constant alignment.

7.3 Keep Your Plans!

A project plan may still be utilized along with Agile methodologies applied to pockets of execution in between planning and delivery. The entire life span of the project can be mapped out in the plan. Microsoft Project® can be used to plan the project including efforts for software and hardware. Various phases of this plan require different treatment. This allows for things like lead time on material procurement and shipments to be included in the overall vision for the program as well as aid in the understanding of the propagation of change. A preliminary project plan could be populated by the program manager who works with other members of the leadership team to define a set of preliminary target dates for each line item in the plan.

Once a draft of the overall project plan is developed, the three levels of thinkers get to work. The program manager communicates the preliminary plan to the other Agile leaders who then flow the plan to the team. The team plans out the "pockets of execution" in finer granularity. A "pocket" would correlate to a set of work items in the backlog and be represented as a line item in the project plan. JIRA® or Rally® are tools that can be used to track pockets of execution. These pockets may manifest themselves in the form of features that begin to be flushed out for each piece of functionality promised to the customer.

Planning is iterative and should change frequently as priorities are nonstatic. Involved parties should always work at reviewing and refining the plans. Once the team has filled in their part, this is a good point for the program manager to review what the team has claimed as part of the milestone deliverable and identify gaps, eliminate work that is not part of MVP, perform some prioritization, validate any assumptions, understand dependencies, etc. Changes are always flowed up to the plan to ensure everyone is in alignment. During sprint planning and refinement meetings, modifications may be made to the plan for a more granular strategy going forward. The commitment for a specified date may change as the team evaluates their story load versus velocity. This is an integral part of Agile, the team communicates to leadership what can be done but also more importantly communicates what cannot be done. It is then up to management to make the priority calls about what will be worked and what will slip. These updates can be made in the overall project plan. Once this is complete, program mangers really start to get a realistic picture of the path going forward. These steps are iterative in nature and occur with frequent cadence.

This helps to keep the plan as accurate as possible. The more transparency between management and the team equates to a smoother execution and aligned expectations between the two parties. Continuous communication about what we are planning to work on and what our capabilities are makes everyone successful.

Program management and the team continue to stay in constant alignment by ensuring everyone has visibility in terms of plan progression. By this point dashboards can be set up for each project, so you'll be able to tell if the initiatives are on track. Key views the Rally° tool supports include: initiative burnup, team commitment of initiative burnup, cumulative flow diagrams of scheduled states, team story point throughput, and scope creep. These help to forecast dates of completion and run simulations (trade out feature X for feature Y to meet completion date Z). A major part of the Agile process is that we want to make sure everyone's voice is heard. It's hard for proposal thinkers to attend every Scrum of Scrums and backlog refinement meeting so the best thing for them to do in that case is to identify another proposal thinker that can act as a proxy. During the Agile ceremonies where work is being ranked and claimed this proxy can speak up and ensure the features and/or user stories for the customer deliverables are getting boiled up to the top (higher priority) of the backlog and therefore getting executed on.

7.4 Meetings for Planning

Agile principles state that the teams (a.k.a. the *knowledge workers*) are great contributors in coming up with the plan! Teams know exactly what they need to do and can give a reasonable estimate of how long it will take. The beginning of planning for an Agile project typically entails drafting and prioritizing a backlog of work items to be accomplished by the team in a designated time frame. The work items drafted and prioritized by the proposal thinkers generally specify higher level system functionality that adds value to a *user scenario* of the system. The work items drafted and prioritized by the conceptual thinkers may be decomposed from the work items that the proposal thinkers have drafted. They typically specify mid-level functionality that adds value to a *user activity* of the system. Then the *user scenarios* and *activities* are decomposed further into user stories by the detailed thinkers. If the team is working in more of a nested structure, all types of thinkers may contribute to the user story's acceptance criteria and/or requirement collectively.

7.4.1 Quarterly Release Planning

Release planning is an Agile ceremony where team members identify what they can complete in a *release*. Some organizations define a *release* as a *software release* (*baseline of released software*) while others define a release as an arbitrary time frame, like an *annual quarter*. The SAFe Agile framework prescribes an event called PI planning every six sprints where teams plan for the next quarterly release a.k.a. *program increment* [4]. Annual quarters make sense for large organizations to use as containment mechanisms for planning because this aligns with preexisting fiscal time frames. It is up to the company and/or team to define what planning cadence makes the most sense for them.

7.4.2 Sprint Planning

Sprint planning is an Agile ceremony where team members identify what they can commit to completing in a *sprint*. Sprints are typically 2 weeks long. Typically, conceptual and detailed thinkers are involved in sprint planning. Sprint planning can begin when sprint backlog grooming is complete. Sprint planning is considered complete once the team has:

- Considered and discussed user stories with conceptual thinkers (i.e., product owners)
- Reviewed user stories with safety concerns and determined if they are ready to execute
- Ensured that they understand the intent of the user stories
- Selected several items that they forecast they can accomplish
- Created a sufficiently detailed plan to be sure they can accomplish the items[2].

7.5 Plan for Your Unplanned

Some of the details of higher level work items will most likely be unknown at the time of planning. Teams may add placeholders in their backlogs to account for the unknown. Some teams plan for a lower "loaded" capacity instead, intending unallocated capacity can be used for the unplanned. Either way, having a way for planning the unplanned scope helps the teams to be more predictable in executing their pocket of execution in the overall program plan mentioned earlier. As engineering discovery unfolds the problem space, so backlogs are rewritten and reprioritized to account for each new turn of events.

Flavors of "unplanned" work include:

- Systems engineering work
- Release preparation work
- Research spikes
- Additional validation
- Defects
- Field issues

Even though the teams plan out what they will work and plan for the unplanned, it is important that they always *embrace change*!

7.6 Documentation

The Agile Manifesto [2] states:

Working software over comprehensive documentation

In organizations producing large integrated solutions, documentation still exists and, in some cases, comprehensive documentation must still exist. Many industry's certification per safety standards still require a large set of documented artifacts to be produced. The difference when working in an incremental methodology is the documentation can be built incrementally as the program persists instead of being provided up front. Whereas with the Waterfall process documentation and meetings are tightly coupled, with the Agile process, this is not the case. In cases where it makes sense (emerging requirements) documentation becomes a by-product of the team's working cycles rather than a driver. New tooling allows for documentation to be autogenerated directly from the work items that the team creates. Many companies are coming up with solutions for integrating life cycle tooling across the board, just Google *integration of life cycle tooling* and review what comes up. Having integrated requirements, safety artifacts, design and test artifacts, code, test procedures, and automation scripts is a foundational concept in both modern DevOps as well as MBSE techniques. It is ideal to produce documentation in this way if you can.

To produce the documentation incrementally, the team should be aware of documents that they will be responsible for updating along the way as well as any material they will need for updating such documentation. Material to consider includes any information for team to learn and access document generation tooling, document templates, and configuration management program to baseline revisions. The documents that can be updated incrementally are listed below. This is not an all-inclusive list—the standards your team work under may drive modifications to this list.

- *System/software interface specifications.* Update as *interfaces* are created/modified while working user stories and features
- *System/software requirements.* Update as *requirements* are created/modified while working user stories and features
- *Software design.* Update as *software design artifacts* are created/modified while working user stories and features
- *System/software validation/verification plans.* Update with additional test cases spawning from user stories
- *System/software integration test specifications.* Update as *integration test procedures* are created/modified while working user stories and features
- *System/software architecture validation/verification report.* Update as *architecture* is created/modified while working user stories and features
- *Functional hazard analysis.* Update as user stories and features tagged with safety implications are worked

Usually conceptual and detailed thinkers are involved in the care and maintenance of these documents but in the Agile methodology the team may own this task collectively. There is no reason a software tester cannot update a requirements document. If this concept makes your team queasy, have a rule that the system engineer must review modifications made by anyone to the document. Many teams have leveraged online Wiki-type tools to capture documentation as well. Tools like Confluence® have revision tracking, data accessibility per profile permissions, searchability, and graphics editors. For most younger generations, online generation of documentation is all they know.

7.6.1 Requirements vs. Acceptance Criteria

As already mentioned, Agile teams often continue to write, review, and baseline requirements documents. The Waterfall process prescribes up-front requirements. In Agile, the requirements are a by-product, not typically a driving force. Consensus of both methodologies is achieved by using a top-down meets bottom-up approach. This covers both cases of requirements needing definition up front (Waterfall) and emerging requirements (Agile). This can leave teams striving to understand the difference between acceptance criteria and requirements for various levels of work items.

Microsoft Press [5] defines acceptance criteria as "Conditions that a software product must satisfy to be accepted by a user, customer or other stakeholder." Google defines them as "Preestablished standards or requirements a product or project must meet."

Some Agile teams believe the famous "As a ... I want ... so that ..." statement to act as the requirement. An example of the "As a ... I want ... so that ..." statement might look like this:

As a user, I want function X, so that I can action N.

The "As a user" statement works when designing embedded software if you have clear user activity and control structures defined up front and structural components mapped to use cases. In this way it is easy to see how the functionality of each bit, byte, and component adds value to the customer. If user activity is not mapped or far abstracted from control structures, functions correlating to components cannot be specified in this type of way and a standard requirement format (input in terms of an output shall statement) is used. For either format, the intended functionality should be published across the team (and ultimately to the customer). The INVEST criteria is a great way to ensure either the requirement or "As a ..." statement is accurately captured. INVEST means the requirement or acceptance criteria is independent, negotiable, valuable, estimable, small, and testable. One of the best benefits of Agile is flexibility. Teams that strive to define too much of the process on how requirements will either drive or emerge get themselves into trouble. For example, they find themselves following self-prescribed requirements processes in areas where it does not make sense, or they incur large amounts of overhead by trying to map one-to-one granular requirements to user stories, resulting in hundreds of unmanageable stories. It is up to the discretion of the team as to how they want to proceed but common sense should be applied in all cases.

7.7 Go With the Flow

The "flow" or consistent throughput of value delivery is one of the greatest benefits of using the Agile process. Control points, phase gates, formal role definition, and stringent documentation do not block the Agile team. Transparency, adaptation, and improvement through regular retrospectives and small focused teams speed up flow. When thinking about flow in this way, there are a few key points that should be considered:

1. Agile ceremonies promote and increase flow
2. In Agile, flow exists within and between sprints
3. Product vs. project teams have different flow
4. Flow of supporting teams
5. In a scaled Agile framework, flow exists in an Agile Release Train (ART)

The types of thinkers discussed previously contribute to the flow by participating in the Agile ceremonies. Proposal thinkers contribute on the front end of the value flow, conceptual thinkers in the middle, and detailed thinkers at the end. As mentioned before, different types of thinkers may contribute in two important ways: (1) thinkers work in their own concentric team delivering to downstream cycles (*stage gate*), and (2) thinkers may participate in all levels contributing their own type of knowledge to the final product in various ways *(nested)*.

7.7.1 Agile Ceremonies Promote and Increase Flow

7.7.1.1 Feature Writing, Decomposition, and Grooming

Conceptual thinkers own driving the engine of flowing features to the Scrum Teams. This engine chugs forward as features are opened in a preliminary format, enhanced with additional definition, decomposed into user stories, and prioritized. If you are using a SAFe framework this activity needs to happen prior to PI planning. After the decomposed features are made available to the team, the team is ready to capture the work to be done. A feature typically exists for each customer need (or high-level requirement in Waterfall terms). There are many resources online and in print related to feature writing. If you are in a role that requires you to draft features, I recommend you spend an afternoon Googling this topic.

Feature grooming is an activity that entails:
- Ordering feature backlog in stack-ranked priority
  - Adding or promoting features that arise or become more important
  - Removing or demoting features that no longer seem important
- Allocating safety concerns to features
- Splitting features into multiple features
- Merging features into larger ones
- Estimating a preliminary size for features

7.7.1.2 Detailed Documents and Meetings

Detailed thinkers focus on the flow and completion of lower level work items. To complete this the Agile methodology prescribes several meetings that are held with a 2-week cadence. These meetings include sprint backlog grooming, sprint planning, daily Scrum, sprint demo, release demo, and sprint retrospective. The team can decide on what day is the most sensible to hold these meetings. Once the team has decided, it is the Scrum Master's responsibility to send out a reoccurring meeting notice to the team blocking that time on each team member's calendar. The team meetings are very important and everyone on the team should make an honest effort to attend.

7.7.1.2.1 Sprint Backlog Grooming Sprint backlog grooming can begin when *feature* writing, decomposition, and grooming is complete. Typically, the conceptual and detailed thinkers get in a room for 1–2 h to groom the *sprint* backlog together. In some cases, prework can be done before this session. For example, each conceptual thinker may prepare a preliminary set of stories, then, when this meeting is held, the detailed thinkers simply point out corner cases in functionality to capture additional needed stories. Another case of prework could be where each team member is assigned a group of stories to gain special expertise in, so that when the group meets at least one person has their head wrapped around the problem space and/or intended work. Sprint backlog grooming is complete when a set of refined stories is ready to be worked for the next sprint (given all known facts at the time) and stories are prioritized and allocated to the Scrum Team that will work them. Upon leaving sprint backlog grooming each team member should understand the work items.

If the priority of work items changes in the backlog, people must change what they are working on. Many of us have seen the case where the priority changes and the team is expected to finish what they were doing while also finishing the newly deemed "hot" items. Agile frowns upon this type of work environment by pushing the agenda that all teams have a corresponding capacity and if you are going to add something to their plate, something must be removed.

I have witnessed that the hardest part about the feature and sprint backlog grooming meeting can be simply having it consistently and having people turn up to it. This meeting is key to the Agile flow, do not underestimate the criticality of this meeting. First, just like system requirements, most downstream activities depend on a well-groomed backlog including development, test environment setup, laboratory support activities, automation activities, safety activities, etc. Second, the better the planning, grooming, story definition, and sizing; the more predictable the team is and the easier it is to estimate projects. Let's explore the disfunction that could occur if the backlog grooming session was not held or was ineffective. If stories with loose definitions are brought into the sprint then stories begin to swell and eventually spawn additional user stories. If this is happening with all stories in the sprint you may arrive at the situation where nothing is completed and things are pushed to the next sprint, resulting in the inevitable and unintended "rolling wave of user stories." Or worse still, if the team has no idea what to work on next because priorities haven't been set, the roadmap is not on track, and this results in misguided execution. The team owns this activity and should understand why it is important. If the team decides not to do it, they will get through a couple of sprint cycles and the reason it needs to be done will be revealed and they will realize why they need to do it.

7.7.1.2.2 Daily Scrum A daily Scrum is held once the sprint has been planned. During Scrum each team member informs the group on:
- What I have accomplished since our last daily Scrum
- What I plan to accomplish between now and our next daily Scrum
- What is impeding my progress [2]

Typically, detailed thinkers come to a daily Scrum, but proposal and conceptual thinkers are welcome to join. If working in a systems-of-systems space then invite members from supporting teams to listen in as well. The purpose of this meeting is to touch base and get various team members aligned.

7.7.1.2.3 Sprint Demo The sprint demo is held once all user stories in the sprint are completed. The demo is a success if the Scrum Team and stakeholders reviewed the output of the sprint and found it to be satisfactory. Proposal thinkers and conceptual thinkers, regardless of their standard Agile roles, should attend demos to ensure that the customer's intent is captured correctly. In a *systems-of-systems* flow, detailed thinkers from supporting teams consuming the system output should attend the demo to ensure an alignment of interfaces and expected outcomes.

7.7.1.2.4 Release Demo The release demo can be held once all the features have been completed and user stories in each feature have been demoed in sprint demos.

During the release demo:

- The Scrum Team and stakeholders review the output of the release
- Features with safety concerns are reviewed to ensure that the concerns have been mitigated [2]

The release demo also increases the awareness of the stakeholders with respect to new functionality, how teams are doing in terms of executing the roadmap, and provides insight into where the product should and/or could go next. Invite your proposal, conceptual, and detailed thinkers to the release demo. This is a chance for all types of people to be proud of what they have collectively built together. Invite customers as well, they will be able to give you valuable feedback as to whether the objective of each feature has or has not been met.

7.7.1.2.5 Sprint Retro Sprint retrospectives are instrumental in flushing out technical issues as well as those connected to team dynamics. The sprint retro can be held once the sprint demo has been completed.

During the sprint retro:

- The team has reviewed how things went with respect to the process, the relationships among people, and the tools
- The team identified what went well and not so well and identified potential improvements
- The team came up with a plan for improving things in the future [2]

Typically, detailed thinkers attend the sprint retrospectives. This is not to say that retrospectives cannot be held among proposal and conceptual thinkers. As a matter of fact, this is encouraged by many *Agilists*. Books have been written on how to hold effective retrospectives. When working in the embedded realm, the retrospective meeting can turn into the best ceremony for the team to figure out operating practices that work for them.

As you can see, work items are considered in all the various Agile ceremonies. As these ceremonies occur with a regular cadence, so teams are meeting and discussing work with a regular cadence, and flow is therefore achieved.

7.7.2 Agile Flow Exists Within and Between Sprints

7.7.2.1 Within Sprints

7.7.2.1.1 Work Item Fields and States We will now consider the user story life cycle as it is processed through the flow. The granularity of a user story plays a factor in promoting or decreasing flow. No matter the tooling your Agile team chooses, the chances are good that the tool is going to have some representation of a work item whether it be a feature, user story, defect, etc. Chances are also good that these work items will have a ton of attribute fields.

One of the challenges a new team may face is deciding which fields are necessary and which fields to leave blank. Good practice says we give the work item a name, description, identify the state it is in, identify a size estimate, and so on. There are other fields like parent work item, release, and planned time frame that come into play when considering traceability and metrics. There should be no prescribed data field requirements for a work item unless it is decided that the information entered benefits the team in some way. For example, features should only be sized by the team if the managers assess estimated effort accordingly and assign the appropriate workload based on team estimates. If the teams continually size features and the managers continually overload … the sizing does not add value.

User stories have transition states, owners, assigned sprints, and so on. It is up to the team to decide what each user story state transition means as well as what the accepted criteria and definition of done should be. A definition of done is a standard checklist the team commits to complete for each story.

For larger integrated systems requiring certification artifacts the definition of done may look like this:

- Documentation updates
- Acceptance criteria is met (story is accepted)
- Review(s) have been performed (and documented, and location of review artifacts are accessible from user story)
- Any remaining action items have been closed out
- Safety engineering aspects are covered (including documentation or analysis)*
- Traceability has been achieved wherever necessary
- Code has been checked in (build is passing on master branch, automated test scripts checked in)

*If your safety team is running a Kanban approach this line item can be as simple as making sure the correct story exists and is closed in the safety engineering team's backlog.

7.7.2.1.1.1 User Story Cycle User stories have somewhat a life of their own. The life cycle for a story can look very similar to how SCRs are processed, except a user story's life cycle is short, owned collectively by the team, and may spawn more user stories. It involves discovery, design, development, and testing, without phase gated control point boundaries. On the first day of the sprint, testing may be done that drives the development, generates the requirements in real time, and enhances everyone's understanding of the problem space. There is no "right way" to work a story. When we had a "right way" defined in Waterfall processes and practices, everyone was *waiting to learn* or begin work, *bugs were found right before releasing the software* (instead of upstream), and *designs were often overwritten and/or abandoned* by the time coding began. Human behavior, as well as the work humans

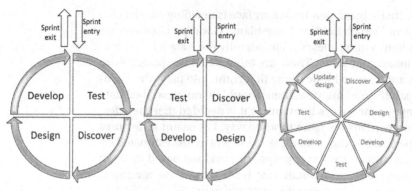

Fig. 6 Variations of a user story life cycle within a 2-week sprint.

do, is diverse, so when considering our processes, the higher the level of abstraction the better. Design, code, and test reviews can still take place as the user stories are worked on by the team but their sequence can be flipped around if necessary (Fig. 6).

One pitfall I have seen some team's fall into is to try to execute on every story consistently across the board. For example, say a team insists that every story must have a code review. At the end of the sprint no stories can be accepted because even though the software has been developed and tested, the code reviews for each of the stories were not completed. In the spirit of Agile, be flexible! It's okay to have a code review story in the next sprint that will cover this task! Heck while reviewing eight code modules at one time the task may be completed quicker and similar inferences may emerge across the code modules (hence enhancing the review!) What I am getting at is that the old process tells us to do everything consistently across the board. In the true spirit of the Agile the only thing to keep consistent is to use common sense in all scenarios!

7.7.2.1.1.2 How Do We Write User Stories and How Granular Should They Be? While it is ideal that stories are written in a way that allows developer(s) to take them and immediately go off and begin work, that is not always the case. Most of the time the Product Owner and team are scrambling to wrap up stories, get the demo prepared, and meet their sprint goals. This means, depending on the team, sometimes stories are written on the fly. Stories are initially drafted as reminders or placeholders, to capture fleeting thoughts from escaping and locking them into the backlog. Then if anyone has the time these stories may be given slightly more definition, say while a PO is facing a list of 20 of them or while the group is meeting during backlog grooming. In the embedded space, I have yet to see the case where an Agile team works off a perfectly defined backlog, and that's ok! Why? Because stories are meant to be a "placeholder for a conversation" as many Agile references say. Be careful when being too nit-picky of how to accept a user story into the sprint based on a mile-long DOR list. This can slow flow down and become a bottleneck. There is a lot of evidence supporting how having a DOR can also help flow by reducing rework.

The other tricky part about stories is that they must be completed in a 2-week time frame. Again, be careful as the definition of done described above is an Agile tool that if not used correctly can really slow down your flow. In a system-of-systems team dynamic, a 2-week delivery is a lot harder than it sounds. Teams automatically assume they can do more than they can actually do, and they do not account for all the Murphy's that come up. So, as the Agile coaches tell us, when writing your stories start small, very small. As an understanding of the problem evolves, and the work increases, add more user stories to the backlog.

7.7.2.1.2 Vertical Slices The important thing to realize when working with Agile for complex systems such as these is that we do not get too wrapped around the axle in terms of what a *vertical slice* means to us. Wikipedia describes a vertical slice, sometimes abbreviated to VS, as "a type of milestone, benchmark, or deadline, with emphasis on demonstrating progress across all components of a project." [6] For an embedded team, the scope boundary is different than the folks working with a software application. You can't respin a board every sprint or quarter. However, hardware Kanban Teams can work concurrently with Scrum Teams to complete user stories for BOM creation, schematic creation, or procurement of materials.

At the start of development and design, a web application developer may understand the user scenario. The code can then be developed, integrated, and tested all the way to the database layer for a slice of application functionality in a 2-week cycle. A win is achieved when the web application developer(s) complete many of these vertical slices and the system is built. For an embedded developer, maybe the vertical slice only goes as deep as generating a binary (Fig. 7).

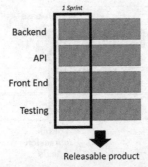

Fig. 7 Vertical slice.

7.7.2.2 Between Sprints

After your team begins to work a few sprints in a row, it will become apparent to you that it makes sense to have what is called a HIP sprint or IP sprint (SAFe). Having additional time to perform some hardening, innovation, and (additional) planning is refreshing and welcomed after the team has been sprinting for the past 6 weeks. SAFe prescribes having one HIP sprint a quarter (after five normal sprints) but other texts recommend having one whenever the team needs it. It is up to your team to decide what makes sense for them. Some teams I have seen have thresholds for the amount of technical debt they will allow themselves to incur before making it mandatory to perform some clean up in a HIP sprint (Fig. 8).

Goals Agile teams should have for their HIP sprints include:

- Documenting updates
- Reducing technical debt
- Reducing number of defects
- Developing regression test automation
- Changing impact analysis (as needed)
- The last HIP sprint in a release could also be used for release planning for the next release

7.7.3 Product vs. Project Teams Have Different Flows

Customer collaboration over contract negotiation [2].

The product vs. project topic often comes into play when strategizing on how to maximize flow. In this text, a *product* is defined as a good or service whereas a *project* is defined as a temporary endeavor that is undertaken to create a *unique* product or service [7]. Architecturally speaking, it is better to develop a *product* that becomes a *project* application than to start off a with a *project* application and try to create a *product*. As my son put it, think of a cow producing milk. Milk is available for consumption. If we buy a cow it will produce milk—an

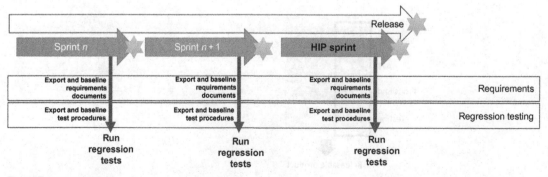

Fig. 8 Between sprints.

ongoing consumption without much additional cost! However, this does not work the other way around. The milk cannot produce the cow or more milk readily ... if it was possible or feasible at all it would require *a lot* of work, hence more cost. If you develop your product to be consumed by many consumers, sales will continually flow. If you build a product fit for one customer project application, it may not be consumable by other customers—in order to make it consumable to others would require rearchitecting, more design and development work, and further cost.

Whether the team is developing a product or executing a project, the architecture is a driver of how teams are formed, a driver of non-functional requirements, and a key contributing factor to how the product is sold ... all of which overlap with Agile practices. For example, if one formed a project team, eventually the project would finish, and the team would disperse. This is the antipattern of Agile teams. According to the Tuckman model, teams that form, storm, and norm together, perform well together. Dispersing the team would cause a loss in performance for the business therefore many times managers create product or component teams (team's that have expertise in a very specific part of the product or products in general). The disadvantage with this is that the product is not getting developed in vertical slices which may result in holes in the final design or functionality. The Scaled Agile Framework website has some great material regarding the formation of component-based or functionality-based teams to account for this [8]. It is so important to think about what you are building, what types of thinkers you have, and the optimal way to form teams, whether developing products or executing projects.

Using Agile practices for pockets of execution allows us to be flexible and work programs with customer guidance and rigor whether teams are marching toward developing a single product with various customer project configurations or developing a single project application. While companies strive to home in on a single consolidated product (think "master branch") this may be difficult or unattainable. Forethought should be given when forming teams to eliminate rework and inefficiencies that could result after the fact.

7.7.4 Supporting the Team's Flow

In the embedded world you have software teams, where the Scrum mentality fits ok. Then you have many other development life cycle teams that also need to jive in the Agile flow. These are supporting teams like safety engineering, hardware engineering, configuration management, test automation, system integration test, proposal management, and laboratory support. These types benefit from working in the Kanban format as it is often difficult for these teams to plan what they will do next because their work runs parallel or in support

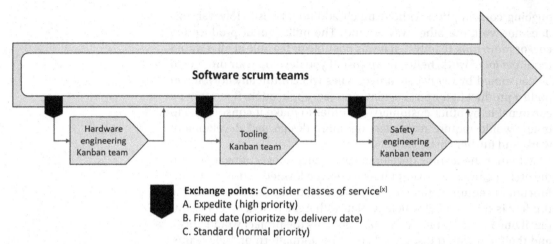

Fig. 9 Scrum ban.

of the software Scrum Teams. Supporting teams should hold their own Kanban Team meetings. At points where exchange between teams occurs, the Kanban principle classes of service [9] must be considered. These help teams to identify prioritization of supporting work items to the Scrum Teams. Kanban work in progress (WIP) limits can help a team to drive these items to completion (Fig. 9).

7.7.5 ART Flow

ART flow is the flow that exists at the Agile release train level. If your organization is following the SAFe infrastructure, an ART is a team of teams all working collectively to achieve a purpose [10]. In my opinion, SAFe is the most well-defined structure yet for implementing Agile on a large scale. In an Agile release train, internal and external dependencies impact flow. Flow can be decreased at the front end of the value stream or at the back end or point of release. The lists below highlight some internal and external dependencies to consider when thinking about how to optimize ART flow in your organization.

At the front end of your value stream, consider:

- *Time up front*: Has the contract been signed? Waiting periods during contract negotiations can eat up valuable schedule time.
- *Unclear customer expectations*: Conceptual and detailed thinkers should have calls and meetings with the customer beforehand. This will help ensure customer expectations are captured completely and correctly.
- *Unprioritized work items*: Conflicting deadlines and priorities require immediate discussion and prioritization from management. Within your value stream, consider:

- *Learning curve*: Training new team members adds time to completion.
- *Expiration dates on supporting licenses*: Can be a blocker for teams if supporting software expires.
- *Test environment lead times*: Current development and test equipment availability.
- *Availability of project's customer counterpart/point of contact*: Can impact schedule adherence.
- *Dependency on customer/supplier for work completion*: Hardware delivery, for example, can leave a team waiting for critical inputs to complete their work.
- *Additional proposal support*: Engineering time to support proposals may mean trips out of the office.
- *All other internal releases in the pipeline*: Affects current and upcoming test setups, mode of thinking of the team, release process items, and configuration manager bandwidth.
  At the back end of your value stream, consider:
- *Acceptance criteria for artifacts at release time*: Last-minute paperwork could affect the end delivery date for a project.
- *Weather, both at the delivery site and locally*: Yes, this matters ... field tests cannot be performed in snow storms or hurricanes.

8 Advanced Topics

8.1 Metrics and Transparency

Common toolsets that support the Agile process provide teams with the ability to collect many metrics as work items are defined and executed on by the teams. Common Agile metrics include things like feature completion, US point completion, sprint velocity, or length of time in each Kanban state. Metrics help us to identify bottlenecks, blockers, steer the team, and get status. Setting up a framework and data model to gather metrics for Agile projects should be done up front.

Metrics can also be helpful for determining which work should be steered to the team and when. If a project or milestone is showing a projected end date that is past its due date, we can easily increase the slope of our trendline by taking more work for this project in the next couple of sprints. Comparing accepted vs. committed lets a manager see how good the team is at estimating the work, hence giving insight into the accuracy of the estimates for all current projects.

Commonly untracked metrics which provide an enormous amount of insight to a team's flow are: amount of spanning work items, amount of unplanned work added after planning, and amount of work that is blocked.

Spanning work items: Sometimes we forget to consider items spanning our formally defined time frames. Instead, we focus on work items that can be wholly completed in a designated time frame. For example, items that started in this quarter and will finish in the next quarter. Spanning items can have a significant impact on forecast completion and should not be disregarded while coming up with a metrics schema. Ideally, work items will be split to fit into the corresponding time frames, but this cannot always be done. Think about this when setting up your tooling and designating work to time frames.

Amount of unplanned work: The amount and type of scope creep a team has from quarter to quarter can give insight into how predictable the team is. Ideally, if the team plans for the unplanned (see "Plan the unplanned" section) they will have a higher predictability than a team that does not.

Amount of work that is blocked: Whether due to an external dependency or a resource constraint all teams have things that are out of their control, that block their flow. Metrics that provide visibility into how much of their defined work is out of their control is leverage when attempting to replan or negotiate a new end date with the customer.

8.1.1 Metrics: Some is Better Than None

I once had someone say to me, "No tool can accurately capture metrics for the kind of work I do."

He felt that if he logged what he did and provided size estimates while blocking work items with corresponding lead times he would have to do so in such detail for the metrics to be accurate that he would be spending all his time doing that instead of doing real work. My advice for those who have this fear is to think about the concept of wearing a Fitbit® (a Fitbit® is a device you wear to track steps, sleep, calories, etc.) The Fitbit may not capture the exact number of steps you take a day *but* wearing it gives you a baseline number. Having a baseline number is the first step to improving. If you have a baseline with a common method of measuring it, you can try to beat your first measurement. Last week I did 8000 steps. This week I strive for 9000. In the same way an organization has a baseline of work throughput. This helps managers to understand what is coming up better than having no metrics at all.

Pure Agile recommends "relative" sizing when considering US points. Relative sizing simply means you compare each story to other user stories and ask questions like "Is this story twice or half the effort of that story?" User story points are then gauged accordingly. The point value of a user story to one team may be entirely different to that of another team. Typically, User Stories are sized using the Fibonacci series beginning at 1 and ending at 13. If the size is deemed to be greater than 13, split it into multiple User Stories. There are various splitting methods out there, just Google "user story splitting methods." In real life, operations researchers are super excited about the implications of having teams estimate their work with a granular unit of measurement. Oh, the Six Sigma potential! Data coming out of these practices could be the foundation for statistical process control and life cycle cost estimation! This does not mean a user story point must map to a specific time-based measurement and be consistently used across all teams. As teams normalize on what a point means to them, collective organizational metrics can be assessed on a level that is completely transparent to the team(s). Organizationally, asserting a common method of measuring story points can be beneficial even though some feel *Big Brother* is watching too closely. Normalizing story points to hourly estimates is a debate in Agile communities. I agree with it. We have never had a unit of measurement that touches the work we do so closely, why not normalize it and begin using it for our benefit! However, normalize and use the data in a way that results in as little impact to team productivity as possible.

If you do choose to normalize across the organization, you may find yourself with this dilemma: it is *really hard* for people to say what they are going to do and then successfully do it! Especially in software development of embedded systems for large complex integrated systems! My advice to teams that encounter the last day of the sprint and end up with nothing completed is this:

1. Claim credit for what you did! Do not roll that user story into the next sprint!*
2. Use your enhanced understanding of what you have left to do to better define stories into smaller chunks going forward.

*This is a common antipattern of the Agile process. From outside of the team, it looks like no one is working, something which could ignite micromanagement resulting in decreased flow. From inside the team, morale is low, as a large rolling wave of stories and points defeats the whole purpose of working iteratively. One of the key learning behaviors we strive for when using Agile is as we are working iteratively, we stop and take time to retrospectively assess how we did at processing small chunks of work. An expanding user story that never closes is hard to learn from.

8.1.2 Getting Comfortable With Data

In today's day and age, everyone from new hires to upper management must have the conceptual understanding of, and be comfortable with, dealing with a large amount of data on a regular basis. People must not immediately get overwhelmed by it but instead get creative with filters and queries. Every decision should be made with the screening question *can we query that*? What good is having a large amount of unorganized data that you can not draw inferences from? We must think about how we can represent certain problem spaces using the work item data attributes that we have at hand. If a portfolio management team needs a way to differentiate data set "red" from data set "green," there must be a single distinguishable attribute to key in on included in each. If the data is not structured in a way that has this distinction, the management of the data will be "death by 1000 tiny swords," as a manager of mine once said. We must not only be comfortable with having the data there, but we must also trust the data! If we find a reason to not trust it, then we must take the necessary actions to clear up discrepancies, so we can trust it.

8.1.3   The Export vs. Live Data Paradigm

In today's day and age, we also have tooling that allows us to capture our plans, what we are working on today, tomorrow, hell the whole year. Large work item repositories like JIRA and Rally allow us to collect a huge amount of inferences along the way. Inferences regarding how our teams work, how fast they work, how predictable they are, and so on. One of the paradigm shifts I have seen as we move to this type of working is a need for some individuals to export what is in the tool to a spreadsheet ... and then hold meetings with that spreadsheet ... send the spreadsheet out in an email ... modify the spreadsheet (the worst) ... create new organizing coloring schemas in the spreadsheet ... my God, make it stop! For those of you who are not yet comfortable working with the live data dynamic, let me ask this: How much of what goes into that spreadsheet is eventually translated back into the tool? If it is, how efficient is it to have to sit and then manually enter all the changes incurred in the spreadsheet into the tool after the fact? My friends, please do not fall into this trap! While it is comfortable to use an exported chunk of data, it becomes stale the second you export it. Get comfortable working with, querying, tagging, reporting on, speaking to, and emailing live data links with corresponding queries!

8.1.3.1   Inspecting and Adapting With Metrics

So, as you can tell, there's a lot of hype about metrics, but I am sure you are wondering what does inspecting and adapting around your metrics look like? Napkin scribbles become a concept of the metrics one wishes to monitor. A smart person creates a dashboard to capture these metrics. The organization starts to home in on what data should be collected (and how to normalize this data) to get meaningful, correct, and complete metrics from the dashboard. Feedback loops begin to set in. Data is investigated. Bottlenecks are revealed. Wasted throughput is recovered. Processes are changed. No archaic document of this cycle exists because the thread I have just walked through happens iteratively, many times a day/week/month/year and the adaptation of these teams is quicker to accomplish than to edit the document and publish it back to them. Teams need freedom in defined processes but enough structure for everyone to know what is going on. The best advice I have received is to just facilitate your teams to communicate and when new people come in train them. A manager of mine once told me to capture enough process to define what communicating daily cannot.

8.2 Tooling

Tooling in this context refers to the tool your Agile teams are using to track their work, you know the one that has the user stories in it, where you move the stories through a sequence of states, assign them to iterations and owners, and so on. The data model represented by the tool should be apparent such that no one needs an explanation to understand it. The data model of your work items is the process these days. We do not require training when we get a new cell phone. Based on our general understanding of how to click through menus of various contexts we figure it out. Tooling for work item tracking and how it is configured for each team should be as intuitive as that.

Product architecture can be a driver for how the tool to track the work is set up. Tool configuration created for one team may not be easy to implement for another team. For example, if a team is working on a backlog for customer X, and customer Y has a separate software baseline, a duplicate backlog of similar work items will need to be created if customer Y decides on the same new feature set as customer X. This may not need to be thought through in excruciating detail if you have a couple of teams working concurrently to develop a product. However, when you have close to 70 teams syncing to develop a product for a large integrated system, the tooling schema must absolutely be cleanly defined and have team boundaries that make sense.

8.2.1 Tooling and Metrics

Another tooling consideration is around metrics. Consider what metrics you will be collecting prior to defining the work item schema and hierarchy. At a high level, do not mix tasks for different teams under the same work item. If one team lags behind the other, you will have hundreds of work items that never reach completion. Instead, use an abstracted attribute, like a milestone, to track completion metrics. This way, from a high level, it is easy to see where the bottlenecks occur. Consider this to drive backlog data cleanup as well. Run a quick filter to see which work items are stuck at greater than 80% and have not budged for weeks. Unless there is a major blocker, chances are good that the remaining work for these items has been overcome by events and therefore these cruft stories can be deleted. One backlog organization technique you can use is to bring the cruft to the bottom of the backlog.

Agile teams will need to inspect and adapt with the tooling. I have worked with teams that have refined the data model for their work item hierarchy many times. Starting with "time-based" features that must be completed in a 3-month period, comprised of standard subtasks per review type, resulted in too much "paperwork" for the teams. Then, they changed their model to be representative of "functional" features that specified actual system functionality, comprised of subtasks defined per discipline. With this method they found the closure rate to be low because all disciplines could not finish at the same time. This led the teams to modify their model again, extracting supporting disciplines like safety and testing to the feature level (up a level of abstraction). Agile coaches recommend teams just get something started and then refine it through trial and error to meet their needs.

8.3 Agile and Alignments

Life consists not in holding good cards, but in playing those you hold well [11].

— Josh Billings.

Different teams and team members with different levels and areas of expertise require different Agile configurations. Staunch traditionalists might say that the foundations of Agile are static and should be used consistently across the board. I disagree. Different levels of system complexity and definition of architecture, software, or hardware could require different energy flows and a different focus in different areas. Hence, causing the team to use Agile practices differently. Tuning the team to create a better balance by selecting which Agile practices should be used, and when, may be necessary. Applying the Agile methods is not a one-size-fits-all process, all factors must be considered and balanced. You want to align Agile practices to work for the types of thinkers in your team (the organization) as well as to consider the phase of development (the life cycle) and lastly align Agile to the system you are building (whether a *product* or a *project instantiation* of the product).

What if we had foresight before the program started and selectively thought out the areas where we were strong and weak and then applied the Agile methodology accordingly. In Feng Shui you strive for balance in how you focus energy to enhance, produce, reduce, and control certain objectives. You let your current state speak to you and apply remedies to achieve the desired result. Consider the table below. In each cell, the maturity/knowledge level of the system construct on the x-axis is assessed for the type of thinker in the y-axis. *Strong means the concept is understood and the requirements are known whether documented formally or not. Stable means the team has an awareness and perhaps begins to generate ideas but nothing is formalized yet. Weak means the concept is not understood and the requirements are not known or documented formally.*

This snapshot shows a typical starting pattern for a product. The architecture concept may be strong to start off. At first the proposal and conceptual thinkers may be the only ones who know about the architectural concept while having a somewhat stable understanding of the software and hardware design. The detailed thinkers are left in the dark (with the purpose of keeping them focused on something already underway).

Starting off…

	Architecture	SW design	HW design
Proposal thinkers	Strong	Stable	Stable
Conceptual thinkers	Strong	Stable	Stable
Detailed thinkers	Weak	Weak	Weak

As time progresses, the contract is officially won. Now the conceptual thinkers really dig in and seek to understand what the architectural, software, and hardware design concepts will be. The detailed thinkers have probably been given a heads-up but still have limited or no knowledge of the new opportunity.

	Architecture	SW Design	HW Design
Proposal thinkers	Strong	Stable	Stable
Conceptual thinkers	Strong	Strong	Strong
Detailed thinkers	Weak	Weak	Weak

By the time the program has flowed into sprints, all the various thinkers understand the concepts. Each cell in the matrix eventually goes from weak to strong as the product is built out and delivered.

	Architecture	SW Design	HW Design
Proposal thinkers	Strong	Strong	Strong
Conceptual thinkers	Strong	Strong	Strong
Detailed thinkers	Strong	Strong	Strong

Agile best-practice recommendations for this case are:
- Use Agile ceremonies to consistently promote flow
- Vertical slices of functionality are developed incrementally to verify the functionality operates correctly on the desired platform
- The customer is involved throughout the development to ensure intent is met

There are many different scenarios for how programs may play out. For example, another case may be as the team innovates, refactors a legacy product, or completes each vertical slices of the MVP, the manifestation of the architecture emerges along the way. Let's dig into how we can balance the team's Agile execution forward in this case. They have a weak definition of what the architecture should be when the program is proposed. This happens and is surprisingly common as detailed thinkers may make a solid business case for a particular functionality to marketing and sales who sells the idea before architecture is formally defined.

Starting off...

	Architecture	SW Design	HW Design
Proposal thinkers	Weak	Weak	Weak
Conceptual thinkers	Stable	Stable	Stable
Detailed thinkers	Strong	Stable	Stable

Recommendations for this case:
- Commissioning a detailed Scrum Team on the hardware side would balance this out. Identifying the hardware includes considering operational environment, supplier constraints, and cost. As the detailed thinkers enlighten the conceptual thinkers, everyone begins to get a stronger understanding of the architectural, hardware, and software concepts. Once nailed down we know what platform we want to use, we know what our interfaces are, we know our incoming data, we know what our expected output data is, and therefore high-level requirements can begin to be generated.
- Agile practices that promote knowledge sharing are team swarming and pair programming, these may be helpful as everyone is learning together.
- Marketing and sales representatives could regularly attend system demos to ensure what is being developed is what was sold.

Let's look at the opposite case. What if the product has been in the field for 10 years, the same team has been developing it for 10 years, very little change happens—maybe a few defects here and there when it is rolled out to a new customer.

Starting off…

	Architecture	**SW Design**	**HW Design**
Proposal thinkers	Strong	Strong	Strong
Conceptual thinkers	Strong	Strong	Strong
Detailed thinkers	Strong	Strong	Strong

What Agile practices are relevant in this case? All teams can improve! This is the *perfect* setup for migrating legacy development and test practices to Agile's best friend, DevOps. In today's age continuous integration and deployment, automated testing, and remote updates are not just nice to have … these methods are necessary for survival in the market. There is a ton of wonderful knowledge on this topic. If you are in a team that has this type of setup, your primary focus should be figuring out how to automate your infrastructure.

9 Conclusion

Agile processes and practices are widespread, how to apply them is what can become the challenge. This content has touched on the basics of Agile but more importantly provided insight into some application techniques that your teams can use when applying Agile in a real world system-of-systems situation. Large integrated teams *can* work with Agile practices. Pay attention to the unseen constructs, different types of thinkers, traditional and newly prescribed roles, human interactions, documentation, levels of abstractions, energy flows, unplanned work, what is not being done that should be, what is being done that should not be, and what makes sense.

Limitations of this work include definition and guidance on the Kanban methods, Agile review processes, the many awesome SAFe principles and practices including architectural road-mapping and portfolio management, how design thinking can support Agile development, DevOps, the flipped program manager "V," and recommendations for thorough customer interaction. Further learning can and should be explored for all of the topics mentioned as well as in regard to Agile alignments and how different types of thinkers embrace Agile. Continue to embrace change and use common sense in all processes.

Exercises

Q: What are three steps for object-oriented design?

A: Identification of system context, class identification, and design modeling.

Q: *True or false?* Real-time systems respond to stimuli in milliseconds.

A: False. Real-time systems respond to stimuli in microseconds.

Q: What is the difference between *availability* and *reliability*?

A: Availability is the amount of uptime the device has, while reliability is the device's ability to perform when it's supposed to.

Q: How can large integrated teams apply Agile effectively?

A: By considering the types of thinkers in the organization and how they will contribute (whether in a supporting Kanban Team context, a nested Scrum flow, or a stage-gated Scrum flow).

Q: *True or false?* All planning and documentation goes away once a team is deemed to be an Agile team.

A: False. Planning and documentation continue once a team is deemed to be an Agile team but are iteratively updated as by-products instead of primary drivers.

Q: Name the six key Scrum meetings?

A: Release planning, sprint planning, sprint backlog grooming, daily Scrum (standup), sprint demo, and sprint retrospective.

References

[1] S. Harris, Multi-stage CI with Jenkins in an Embedded World, https://www.cloudbees.com/blog/multi-stage-ci-jenkins-embedded-world, 2014. Accessed 30 August 2018.

[2] K. Beck, Manifesto for Agile Software Development, Agile Alliance, 2001. https://www.agilealliance.org/agile101/. Accessed 30 August 2018.

[3] Innovationmanagement. Integrating Agile with Stage-Gate®—How New Agile-Scrum Methods Lead to Faster and Better Innovation. n.d. http://www.innovationmanagement.se/2016/08/09/integrating-agile-with-stage-gate/. Accessed 30 August 2018.

[4] Leffingwell, et al., Program Increment Article, Scaled Agile, Inc., 2011–2014. https://www.scaledagileframework.com/program-increment/. Accessed 30 August 2018.

[5] Segue Technologies, What Characteristics Make Good Agile Acceptance Criteria? Microsoft Press, 2015. https://www.seguetech.com/what-characteristics-make-good-agile-acceptance-criteria/. Accessed 30 August 2018.

[6] Wikipedia Vertical slice article, https://en.wikipedia.org/wiki/Vertical_slice. Accessed: 7 September 2018.

[7] Swanberg, Kate. The Difference Between Product and Project Management. Posted August 22, 2018. https://www.koombea.com/blog/the-difference-between-product-and-project-management/ . Accessed 30 August 2018.

[8] Leffingwell, et al., Features and Components Article, Scaled Agile, Inc., 2011–2014. https://www.scaledagileframework.com/features-and-components/. Accessed: 30 August 2018.

[9] Leffingwell, et al., Team Kanban Article, Scaled Agile, Inc, 2011–2014. http://v4.scaledagileframework.com/team-kanban/. Accessed 30 August 2018.

[10] Leffingwell, et al., Agile Release Train Article, Scaled Agile, Inc., 2011–2014. http://v4.scaledagileframework.com/agile-release-train/. Accessed 30 August 2018.

[11] Billings, Josh. *Josh Billings Quotes*. Brainy Quote. 2018. https://www.brainyquote.com/authors/josh_billings . Accessed 7 September 2018.

Further Reading

[1] Scrum Alliance. Core Scrum. v2014.08.15. https://www.scrumalliance.org/ScrumRedesignDEVSite/media/ScrumAllianceMedia/Files%20and%20PDFs/Learn%20About%20Scrum/Core-Scrum.pdf. Accessed 7 September 2018.

3

EMBEDDED AND MULTICORE SYSTEM ARCHITECTURE— DESIGN AND OPTIMIZATION

Michael C. Brogioli

Polymathic Consulting, Austin, TX, United States

CHAPTER OUTLINE

1 Introduction

When implementing a given application on a specific hardware target, system architects and managers must consider several different factors ranging from hardware capabilities, application requirements, software requirements, and even the technical ability of engineering teams. This chapter explores how to take a given application that demands a specific number of channels and data rates, and the steps required to systematically decompose the application for implementation on the target architecture. By properly accounting for the compute resources available, and the timing/bandwidth requirements of the application, system architects and managers can appropriately delegate implementation and analysis to appropriate engineering resources. In addition, by formally understanding the underlying application, optimization efforts can be pragmatically applied to yield the best outcome rather than simply applying premature optimization to the application which may adversely affect numerous aspects of the resultant system.

Software Engineering for Embedded Systems. https://doi.org/10.1016/B978-0-12-809448-8.00003-5

By exploring the intersection point of hardware resources, application requirements, software tooling capabilities and limitations, as well as power requirements, system architects and managers can effectively and efficiently bring well-optimized systems to market.

2   The Right Way and the Wrong Way

Like many things, in the areas of embedded and multicore software and system design, there are often right ways and wrong ways to go about things. Programmers and developers all to often set out to optimize various aspects of the system far too prematurely, often resulting in less than acceptable results.

There is a topical quote by Donald Knuth, author of *The Art of Computer Programming*, that sums this phenomenon succinctly and is reproduced below:

> *Programmers waste enormous amounts of time thinking about, or worrying about, the speed of noncritical parts of their programs, and these attempts at efficiency actually have a strong negative impact when debugging and maintenance are considered. We should forget about small efficiencies, say about 97% of the time:* **premature optimization is the root of all evil.**
>
> **Donald Knuth**

This is by no means to say that optimization is not required in embedded and multicore design, rather quite the opposite. Optimization must, however, be performed with a disciplined and iterative approach. Regarding serial performance tuning specifically, there are several key factors to consider to ensure that optimization is applied to software in which there is a firm understanding of behavior and bottlenecks. As such, a good iterative optimization approach should include such things as performing measurements and careful analysis for a guide to *informed* decision making, making changes to only one thing at a time, and meticulous and regular remeasurement of the augmented system to confirm changes have been beneficial. These should be done as part of software development, validation, measurement, simulation, and the use of profiling tools to gain insight into runtime behavior and architectural response.

There are several common metrics that are associated with embedded system design at the hardware and software level. These include, but are not limited to, nonrecurring engineering cost, size, performance, power, flexibility, as well as time to prototype, time to market, maintainability, and system correctness. Considering these complex design challenges, domain expertise in both the hardware and software is needed to optimize design metrics. The designer must be comfortable with various technologies to choose what is best for a given application and constraints.

As such, premature optimization, in addition to excessive optimization, can consume precious system resources. These include delay of the prototype or product release and compromise of the software design, often without direct or applicable improvement of system performance. To remedy this, system modeling before optimization is required to appropriately plan and deploy system design resources. Once modeling is in place, a combination of measurement, regression testing, and tuning can be employed.

3 Understanding Requirements

It is important for system architects, managers, and engineers to spend time up front to understand the nonfunctional requirements of the system. Fig. 1 shows an example of a functional requirement and various metrics and attributes that should be associated.

An example of a functional metric could be that the embedded software *shall or must* perform a specific task. Examples of these could be monitoring a certain interface or subsystem, controlling a peripheral or subcomponent, and other metrics that mandate what the system must do. Examples of nonfunctional metrics, on the other hand, could be that the system *shall be* fast, reliable, scalable, etc. In summary, functional metrics represent what the system *should do*, whereas nonfunctional metrics represent *how well* the system should do it.

Fig. 2 shows a concrete illustration of this. Here the system dimension is IP forwarding, otherwise known as internet routing. The system

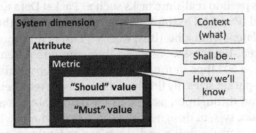

Fig. 1 Functional requirements of an application.

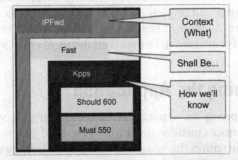

Fig. 2 System dimensions and questions.

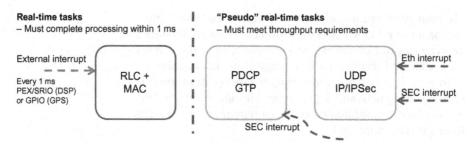

Fig. 3 eNodeB real-time and pseudo real-time tasks.

dimension has the nonfunctional requirement of being "fast." It is noted that the functional requirements are the inner block listed at kilo packets per second (kpps). Here we see that the kpps is shown as 600, however, the hard requirement is that it must be at least 550.

Following on with the above metrics, it is important to point out that there is a difference between system latency and system throughput. In general, it is not possible to design a system that provides both low latency and high throughput. However, many real-world systems have a requirement for both, such as media, wireless, eNodeB in LTE, and LTE Advanced. As such, it is a requirement for designers to be able to tune the system for the appropriate balance of latency and performance. An example is illustrated in Fig. 3 for eNodeB implementation.

Here we can see that the system has real-time tasks (to be completed in 1 ms, or the TTI interval for LTE), whereby an external interrupt is triggered for radio link control and medium access control. The system also has pseudo real-time tasks such as Packet Data Convergence Protocol and IPSec. An example set of requirements for this functionality is that latency must be 10 µs, with a 50 users maximum wake up latency for real-time tasks. Similarly, throughput requirements could be as much as 50 Mbps in the uplink, and 100 Mbps in the downlink for 512-byte packet sizes. By firmly codifying these requirements, both in latency and throughput, as well as for hard real-time and pseudo real-time tasks, system designers now have firm criteria with which to implement and focus tuning and optimization for the system.

In summary, and as touched upon previously, it is a mandate that system architects and implementers know the architecture and know the algorithms. As we will see shortly, system architects and implementers are also advised to know about the tools and compilers.

4 Mapping the Application

When mapping an application to the underlying system architecture, one must consider the various types of processing components available within the system. Some may be latency oriented, like general-purpose CPUs. Others may be throughput oriented, such as

Algorithm	Equation
Finite impulse response filter	$y(n) = \sum_{k=0}^{M} a_k x(n-k)$
Infinite impulse response filter	$y(n) = \sum_{k=0}^{M} a_k x(n-k) + \sum_{k=1}^{N} b_k y(n-k)$
Convolution	$y(n) = \sum_{k=0}^{N} x(k) h(n-k)$
Discrete Fourier transform	$X(k) = \sum_{n=0}^{N-1} x(n) \exp\left[-j(2\pi/N)nk\right]$
Discrete Cosine transform	$F(u) = \sum_{x=0}^{N-1} c(u).f(x).\cos\left[\dfrac{\pi}{2N}u(2x+1)\right]$

Fig. 4 Algorithmic breakdown of computational and memory bottlenecks.

GPU, GPGPU, FPGA, or accelerators. The system may also likely include VLIW-based DSPs. Which parts of the application map to which components is a task that must be analyzed as part of mapping the application at hand to the underlying system architecture.

Fig. 4 illustrates examples of some of the application components one might need to map to a given signal-processing or wireless system. Here we can see numerous blocks that are common in wireless and multimedia systems, such as finite impulse response, convolution, discrete Fourier transform, and so forth.

Generally, in considering these types of application blocks, the estimations for system performance should be done prior to the stage in which code is implemented. System designers will need to account for things such as:

- *Maximum CPU performance.* What is the maximum number of times the CPU can execute the algorithm per unit of time? How many channels can be supported simultaneously?
- *Maximum I/O performance.* Can the I/O system keep up with this proposed maximum number of channels?
- *High-speed memory.* Is there enough high-speed internal memory to support the desired system performance?
- *CPU load percentage.* At a given CPU load percentage, what other functions might the CPU be able to support?

4.1 Performance Calculations to Map the Application to Hardware

In this subsection, we will take the FIR algorithm component of the above table as an example of mapping the application software component to system resources. For a particular FIR benchmark, let us assume that there is a 200-tap (nh) low-pass FIR filter. Let's also assume that the frame size is 256 (nx) 16-bit elements. Lastly, let's assume that the sampling frequency is 48 kHz.

There are two main questions that this exercise will aim to answer that are listed below, each of which include a table of calculations showing the mathematics that is used to compute the final answer.

Question 1: How Many Channels Can the Core Handle Given the Complexity of the Algorithm?

Question 2: Are the I/O and Memory Capable of Handling This Many Channels?

4.1.1 How Many Channels Can the Core Handle?

Referring to the computations in Figs. 5 and 6 in the earlier sections, the goal here is to determine the maximum number of channels that this processor can handle given a specific algorithm. To do this, we must first determine the benchmark of the chosen algorithm. Again, in this case, we chose a 200-tap FIR filter. The DSPLIB documentation

CPU Mapping	FIR Benchmark	(nx/2) (nh+7) = 128 * 207 = 26496 cycles/frame
	# Times Frame Full / sec	(sampling frequency / frame size) = 48000/256 = 187.5 frames/sec
	MIP Calculation	(frame/second) (cycles/frame) = 187.5 * 26496 = 4.97M cycles/sec
	Conclusion	FIR takes ~5MIPs on Embedded Core XYZ
	Max # Channels:	60 @ 300MHz
	Max # channels does not include overhead for interrupts, control code, RTOS, etc.	

Fig. 5 CPU mapping of compute per channel.

I/O	Required I/O rate	48Ksamp/s * #channels = 48000 * 16 * 60 =	46.08 Mbps
	DSP SP rate	serial port is full duplex	50.00 Mbps
	DMA rate	(2x16-bit transfers/cycle) * 300MHz =	9600 Mbps
	Required Data Mem:	(60 * 200) + (60 * 4 * 256) + (60 * 2 * 199) = 97K x 16-bit	97K x 16-bit
	Available internal mem		32K x 16-bit X
	Required memory assumes: 60 different filters, 199 element delay buffer, double buffering receive/transmit	

Fig. 6 I/O and channel mappings per compute and memory resource.

gives us the benchmark with two variables: nx, which is the size of the buffer, and nh, which is the number of coefficients. In Table MCB-1, we have plugged these number in.

It turns out that this FIR routine takes about 26 K cycles/frame. Now, the sampling frequency comes into play. How many times is a frame full each second? Here, we divide the sampling frequency, which specifies how often a new data item is sampled, by the size of the buffer. After plugging in the numbers, we find that we fill about 47 frames/s. Next is one of the most important calculations, how many MIPS does this algorithm require of a processor? In other words, we need to find out how many cycles this algorithm will require per second. Here, we multiply frames per second by cycles per frame—if we plug in the numbers we get about 5 MIPs. Assuming this is the only thing you're doing on the processor, we can do a maximum number of 300/5 = 60 channels. This completes the CPU calculation. We'll use this number (60 channels) in the I/O calculations below.

4.1.2   Are the I/O and Memory Capable of This Many Channels?

The next question is whether the I/O interface can feed the CPU fast enough to handle the 60-channel goal? To determine this, we must first calculate the bit rate required of the serial port. Here, we take the required sampling rate which is 48 kHz and multiply it by the maximum channels (60) and then multiply by 16 (assuming the word size is 16 bits—which it is given the chosen algorithm). This calculation yields a requirement of 46 Mbps for 60 channels operating at 48 kHz.

Next, we must determine what the target architecture's serial port can support. For our target architecture, the maximum bit rate is 50 Mbps (1/2 the CPU clock rate up to 50 Mbps). It looks like we are OK here. Next, we must determine whether the DMA can move these samples to memory fast enough. This appears to not be an issue. Now, we come to the issue of required data memory. This calculation is somewhat confusing and is explained below.

First, we are assuming that all 60 channels are using *different* filters—i.e., 60 different sets of coefficients and 60 double buffers. In other words, the system is ping ponging on both receive and transmit sides, four total buffers per channel hence the multiplication by four in the fourth row of Table MCB-2, pertaining to the required data memory. This also needs to account for the delay buffers for each channel. In this exercise, only the receive side has delay buffers. This calculation is the number of channels * 2 * delay buffer size, which is 60 * 2 * 199. Yes, this is extremely conservative, and you could save some memory if this is not the case. But, this is a worst-case scenario. So, we'll have 60 sets of 200 coefficients, 60 double buffers (ping and pong on receive and transmit hence the * 4), and we'll need a delay buffer of #coeffs-1 which is 199 for each channel. So, the calculation is:

```
(#Ch * #coeffs)+(#Ch * 4 * frame size)+(#Ch * #delay_
buffers * delay_buffer_size)
(60 * 200)+(60 * 4 * 256)+(60 * 2 * 199)
```

This results in a requirement of 97 kb of memory. System designers must ensure that the target architecture has at least 97 kb of memory to support this configuration. If the target architecture does not, then the calculations can be performed again assuming only a single type of filter is used, perhaps reducing overhead and memory requirements.

4.2 How the Estimation Results Drive Options

Following on with the analysis detailed earlier, we can see that this quantitative analysis can now drive various system implementation options. For example, if we were analyzing a low-end, simple application that might only consume 5%–20% of the total CPU cycles, what might a system designer do with the remaining 80% of the compute cycles? Perhaps add additional functions or tasks? Perhaps increase the sampling rate which would result in increased accuracy? The system designer might also decide to add channels or perhaps decrease the voltage/clock speed to result in a lower system power.

Conversely, what about if the application analyzed were a complex, very high–end application that required a CPU load more than 100%! The system designer would need to wisely split up the tasks based on the data at hand. Perhaps use a GPP microcontroller for the user interface or migrate all signal processing to the DSP. Maybe the DSP could handle the user interface and most of the signal processing while an FPGA could handle the high-speed, heavy-lifting signal processing portions of the workload. Perhaps even more aggressive application partitioning could be used whereby a general-purpose processor handles the user interface, a DSP handles most but not all signal processing, and then an FPGA performs the high-speed, heavy-lifting portion of the signal-processing workload. By performing application mapping in a quantitative manner, and before the code implementation occurs, optimizations can be used effectively to meet key metrics.

5 Helping the Compiler and Build Tools

When it comes to finally optimizing the application after the exercises above have mapped it to the target architecture, software developers must become familiar with build tools and specifically the compiler. As was mentioned in Chapter [ ] the job of the compiler at the high level is to map high-level application code to the target platform. In doing so, it preserves the defined behavior of the high-level language. At the same time, the target architecture may provide functionality that is not directly present in the high-level language.

Examples of this may be fractional arithmetic, packed data moves to/ from memory, fused multiply accumulate operations, and various addressing modes. In addition, the application may be comprised of algorithmic concepts that are not handled by the high-level language, such as fractional arithmetic and vector operations.

Software engineers must understand how the compiler generates code, as it plays an important role in terms of writing code for a desired result. Fig. 7 illustrates a typical compilation tool chain.

While compiler optimization is discussed in detail elsewhere in this book, this will serve as a recap for the reader of this chapter. As can be seen in Fig. 7, high-level source code files are parsed by the front end, and then optimized by both a high-level and low-level optimizer. Finally, assembly files are output by the code generator which then

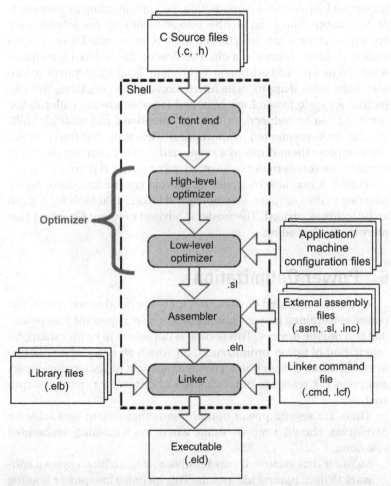

Fig. 7 Example of a modern compilation tool chain.

pass through an assembler. These assembly files are then combined with libraries, as well as various command files to produce the resulting executable. It is important to note that many build tool chains also support assembly optimization, link-time optimization, as well as various other optimizations that can be specified in the linker command file. It is important that the user refers to the built tool documentation to see which features are supported. Chapter [ ] offers additional information on compiler optimizations that are common to most tool chains.

5.1   Choosing Algorithmic Components to Work With Compilers and Architectures

Small parts of your application can often be tailored to have a big impact for loop-focused computation. By implementing these aspects of the computation in an architecture and compiler friendly manner, big improvements can be achieved in often unexpected ways. For instance, 16-bit arithmetic can often be slow on 32- and 64-bit architectures versus a packed arithmetic equivalent. Inlining of functions can also make gains if appropriate instruction cache is available, this can be true for code inside heavily nested loops where the caller/callee overhead can be reduced. Arithmetic operations, like multiply shifting, can be implemented in appropriate ways such that the compiler can compress them down to a single native instruction on the target architecture versus multiple instructions or worse! If input data types are known, it may also be advisable to avoid generic functions. Again, referring to the compiler, assembler, and linker build tools for a given architecture is advised. The reader is advised to revisit Chapter [ ] for more in-depth reading.

6   Power Optimization

As many embedded devices are battery operated or operate on low power constraints, power optimization is also important for embedded and mobile devices. This section is not meant to be an exhaustive exploration of power optimization, for which an entire text could be written. Rather, this section serves as a highlight to system developers and refers the reader to other chapters of this text for more in-depth analysis.

There are several power optimizations that system and software developers should keep in mind when implementing embedded software.

Software architecture. It may be advisable to architect system software to have natural idle points. This includes low-power booting or intelligently powering down PCI Express links and buffering transmissions on the UL and DL. Power can be conserved by only powering up these costly resources when needed by a specific application.

Interrupt driven design. Using interrupts intelligently can reduce system power consumption. By using interrupts to wake up certain functionality, rather than implementing polling loops, significant power consumption can often be saved. Use the operating system to perform blocks in in this context.

Code and data placement. By placing code and data close to the processor, one can often minimize off-chip access. Look into overlays from nonvolatile memory to fast memory. If the device has fast scratch pad memory it may be advisable to perform computations at that location.

Code size. By performing code size optimizations, the application size can be significantly reduced. These optimizations may involve using a compressed instruction set that limits functionality with a more aggressive instruction set for encoding. This will also reduce the memory required for the application and resulting leakage current.

Speed and idle modes. Often, one can optimize for speed in the computationally intensive parts of the application. While this may be unrelated to the task, it can result in increasing time during which the system can be put into idle mode, or the ability to reduce the clock rate at which the CPU and other system components operate.

Over calculation. By having a deep understanding of the application requirements, as described previously in this chapter, programmers can elect to use the minimal data widths required. This in turn can permit the use of smaller multipliers and arithmetic operations. It may also decrease the amount of bus activity and switching required during memory transfers.

Direct memory access. While it may be easier to use programmable CPU-based I/O, using the DMA engines for blocked memory transfer can be significantly more efficient in both time and resource utilization.

Coprocessors. Coprocessors are often designed to accelerate computation. By using coprocessors to efficiently handle and accelerate frequent computation, or application-specific computation, runtime can be reduced. This may increase the opportunity to put CPUs into idle mode.

Batch and buffer. By implementing the buffering of computation, and subsequent batch processing of computation, one may increase the amount of computation that can be performed during a block of time. Like the PCI Express link use case described above, this may increase the amount of time during which a device can be placed in idle/low-power mode while still meeting real-time deadlines.

Voltage and frequency. Use the operating system to your advantage, in this case by scaling voltage and frequency. Again, this requires deep knowledge of the application requirements and runtime performance, be sure to analyze and benchmark your application to achieve the right configuration.

4

BASIC PROGRAMMING TECHNIQUES

Joe Hamman
Director, Platform Software Solutions at Integrated Computer Solutions, Waltham, MA, United States

1 Introduction

The best way to learn embedded programming is through hands-on training. In this chapter, we will provide the basic techniques needed to create, cross-compile, and run programs on a common reference board. Many people learn to program natively on a PC or workstation, where the host operating system performs many actions "behind the scenes." On these host platforms, the operating system handles the

Software Engineering for Embedded Systems. https://doi.org/10.1016/B978-0-12-809448-8.00004-7

housekeeping activities performed when loading and executing programs. The operating system also restricts access to hardware registers, storage devices, and communication channels. These operations are usually performed by calling privileged operating system functions or accessed via device drivers.

To effectively develop software for embedded systems, one must understand these "behind the scenes" actions. This is a requirement when programming systems without an operating system—often referred to as "bare metal" designs. While not required to program a system running a rich operating system like Linux, understanding these concepts will help one make efficient architectural decisions and assist with low-level debugging when needed. The first place to start is with the hardware platform details.

2   Reference Platform Overview

There are many quality, low-cost platforms available to learn on. We will use the FRDM-KW41Z reference board from NXP Semiconductor (Fig. 1). This is a low-cost board using the NXP KW41Z512VHT4 System on Chip (SoC) device. This SoC contains an ARM Cortex-M0+ CPU along with an assortment of on-chip subsystems. These subsystems include serial ports, I/O ports, timers, and other devices commonly

Fig. 1 The NXP KW41Z512VHT4 development board. With permission from NXP B.V. © 2019.

used in deeply embedded devices. In addition to the SoC, the platform also contains flash and RAM memory, along with the signal conditioning circuitry and connectors needed to support the on-chip I/O devices. The design also contains an onboard JTAG debugging interface, making it easy to download and debug programs without the need for additional tools.

2.1 Understanding Hardware

To fully understand the details of an embedded platform, there are two documents that are required. The first is the platform User's Guide. This document contains the details needed to initialize the platform and access peripherals. It typically contains the platform block diagram, reset details, board schematics, and connector pin-outs. Fig. 2 shows the block diagram of the FDRM-KW41Z board we will be using.

The second document required is the processor data sheet. For this platform, it is the reference guide for the NXP KW41Z/31Z/21Z family of devices. Many times, these reference guides will refer to additional documents, like the CPU architecture data sheet. These additional documents will contain hardware-specific details, such as voltages and frequencies, bus timing, and any low-level details applicable to the system reset, system initialization, clocking, etc.

Properly initializing the system requires intimate knowledge of both hardware and software. If even one detail is overlooked or incorrect, the system may not function. Embedded hardware providers have added features to help with this phase of a project. Preconfigured initialization code is usually included in embedded software development kits (SDKs). For reference platforms like the one we are using, this is usually the easiest way to learn how to initialize a board from reset and how to program the peripherals. These documents, software development kits, and sample runtime code are available on the NXP website. For this chapter, the SDK we will use is the NXP MCUXpressoIDE.

3 SDK Installation

The coding examples given in this chapter will only run on the previously mentioned NXP reference board. The easiest way to build and run these examples is to use the NXP SDK. In addition to the required cross-development tools (C compiler, linker, debugger, etc.), the SDK provides a rich set of header files, libraries, and sample projects. These files provide utility functions to configure and use the peripherals contained in the device. The SDK also contains powerful project wizards that allow the developer to easily build simple and complex projects.

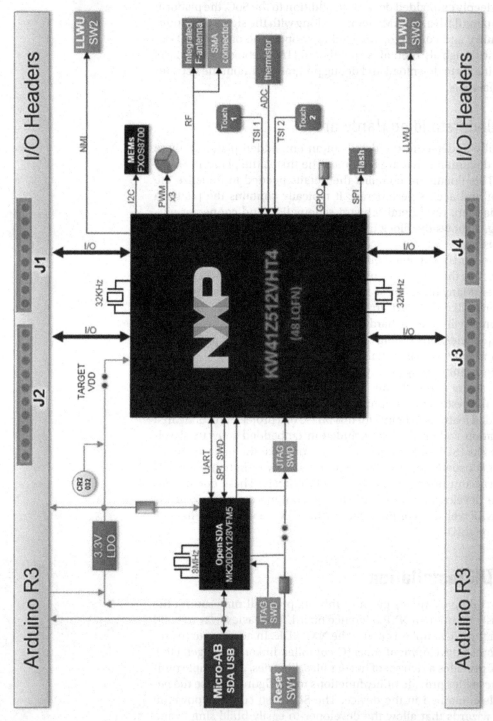

Fig. 2 The FDRM-KW41Z board block diagram. With permission from NXP B.V. © 2019.

3.1 Download and Installation

NXP's MCUXpresso Software Development Kit (SDK) can be found online at http://www.nxp.com/mcuxpresso/sdk. Click on the "Download" link and follow the steps to download and install the SDK. If prompted, the board being used is the FRDM-KW41Z.

3.2 Building a Project

Once the SDK is installed, launch the SDK and provide the location of the workspace directory you would like to use or select the default location provided (Fig. 3).

To import the example project, locate the MCUXpresso IDE–Quickstart Panel and select "Import SDK example(s) ..." When the SDK Import Wizard opens, expand the KW4x group in the list of SDK MCUs and make sure the MKW41Z512xxx4 is highlighted and click Next (Fig. 4).

On the second page of the SDK Import Project Wizard, expand the driver_examples group, then expand the gpio group, select the led_output example and click Finish (Fig. 5).

Before building the project, make sure the Console tab is selected so the build output will be visible. To build the project, right-click on the project name in the Project Explorer view and select "Build Project" from the pulldown menu. Build output should be visible in the console view (Fig. 6).

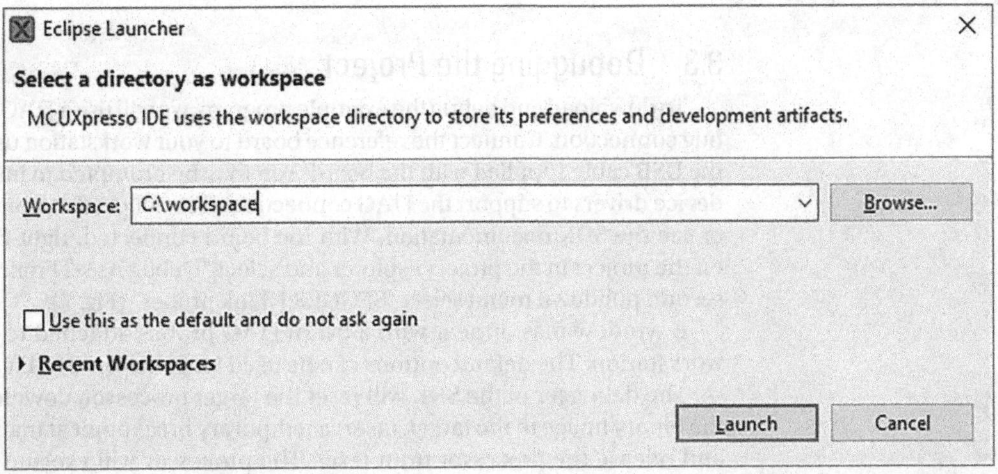

Fig. 3 The SDK workspace selection.

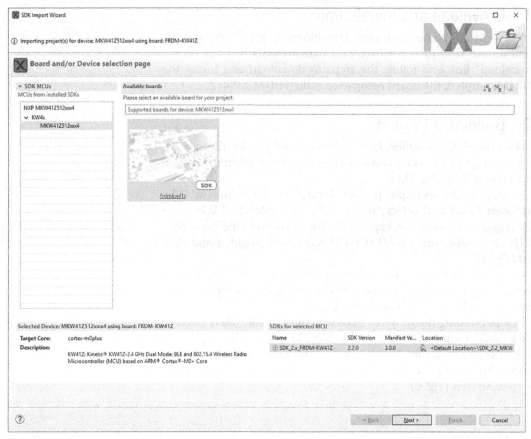

Fig. 4 Selecting the proper MCU.

3.3 Debugging the Project

To download and debug the example program, we will use a JTAG debug connection. Connect the reference board to your workstation using the USB cable supplied with the board. You may be prompted to install device drivers to support the JTAG connection. If so, follow the prompts or see the SDK documentation. With the board connected, right-click on the project in the project explorer and select "Debug As >". From the second pulldown menu select "SEGGER J-Link probes" (Fig. 7).

A window may appear with a list of JTAG probes attached to the workstation. The default options can be used and click on OK (Fig. 8).

The debugger in the SDK will reset the target processor, download the binary image to the target, insert a temporary breakpoint at main(), and release the processor from reset. The processor will execute the startup code described earlier (memory and variable initialization, etc.) and then halt execution at the temporary breakpoint at main(). At

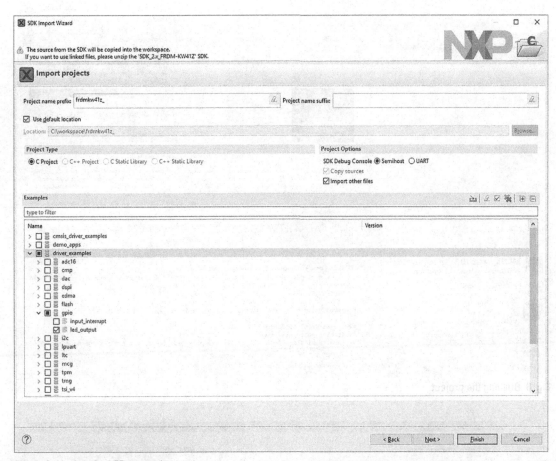

Fig. 5 Selecting the LED example.

this point the debugger IDE can be used to single step, set breakpoints, examine variables, etc. To execute the program at full speed, press the "F8" function key. The red LED on the board should begin flashing. To terminate the debug session, press the "CTRL-F2" key.

4 Target System Configuration and Initialization

4.1 System Reset

All embedded processors have a well-defined mechanism to handle the power-on reset event. However, the details of these mechanisms will vary based on many factors: CPU architecture, SoC manufacturer, and a long list of optional settings. These settings specify

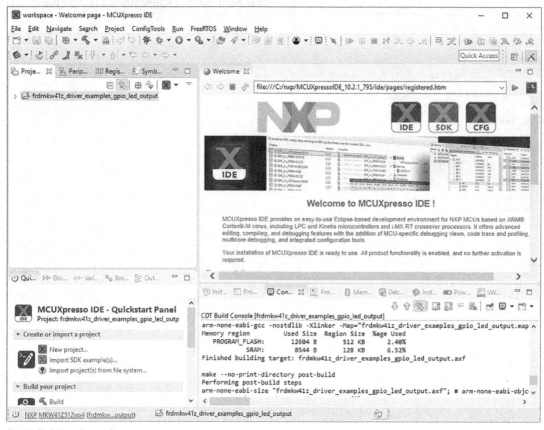

Fig. 6 Building the project.

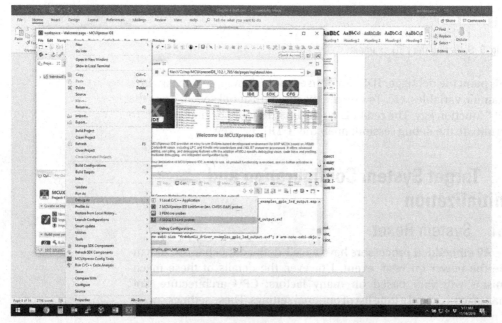

Fig. 7 Selecting the JTAG debug connection.

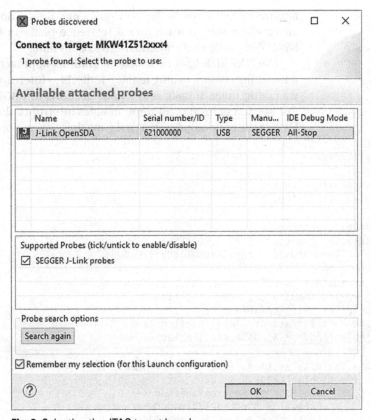

Fig. 8 Selecting the JTAG target board.

the boot device, bus settings, single-chip mode, clock sources, etc. There are also many ways these settings can be implemented. One way is to use pull-up resistors on specific I/O or bus pins. These pins are read shortly after reset and determine the optional settings for that SoC. After the settings are determined, these pins assume their primary functions. Another method employed on some SoCs is to read the settings from a specific location in flash memory.

Once the processor has come out of reset with the correct configuration, the boot software is then required to initialize many different subsystems. These initialization steps will be very specific to the SoC being used, the way the device is configured, and which peripherals will be needed by the operating system and/or application software.

There are several sources that provide working examples of boot software. These examples can often be found in the semiconductor supplier's SDKs, commercial RTOS vendor's SDKs, and open source projects like Linux (www.yoctoproject.org) and FreeRTOS (www.freertos.org). Reading sample boot code operations and identifying the appropriate sections in the previously mentioned documentation is an excellent way to learn how a reference platform like the FDRM-KW41Z is configured and initialized at boot time.

The NXP SDK boot code takes a minimal approach and only does a few things before calling main(). It disables interrupts, turns off the watchdog timer, initializes the .data sections in RAM, initializes the .bss section in RAM to all zeros, enables interrupts, then calls main(). Here is the code snippet that performs these operations:

```c
void ResetISR(void) {

    // Disable interrupts
    __asm volatile ("cpsid i");

    // Disable Watchdog
    // SIM->COPC register: COPT=0,COPCLKS=0,COPW=0
    *((volatile unsigned int *)0x40048100) = 0x00u;

    // Copy the data sections from flash to SRAM.
    unsigned int LoadAddr, ExeAddr, SectionLen;
    unsigned int *SectionTableAddr;
    // Load base address of Global Section Table
    SectionTableAddr = &__data_section_table;

    // Copy the data sections from flash to SRAM.
    while (SectionTableAddr < &__data_section_table_end) {
            LoadAddr = *SectionTableAddr++;
            ExeAddr = *SectionTableAddr++;
            SectionLen = *SectionTableAddr++;
            data_init(LoadAddr, ExeAddr, SectionLen);
    }

    // At this point, SectionTableAddr = &__bss_section_table;
    // Zero fill the bss segment
    while (SectionTableAddr < &__bss_section_table_end) {
            ExeAddr = *SectionTableAddr++;
            SectionLen = *SectionTableAddr++;
            bss_init(ExeAddr, SectionLen);
    }

    // Reenable interrupts
    __asm volatile ("cpsie i");

    main();

    // main() shouldn't return, but if it does, we'll just enter an infinite loop
    while (1) {
        ;
    }
}
```

These steps are performed for any system using the "C" programing language and they represent some of the "behind the scenes" steps mentioned in the introduction to this chapter. While this section of code performs the bare minimum to initialize the programming environment, it does nothing to initialize the I/O devices used in the examples we will present in this chapter. Many real-world implementations also include systemwide hardware initialization steps in this section. Other implementations choose to defer the hardware initialization and perform it later in a section of code dedicated to a specific hardware subsystem. It is very common to see a combination of the two—systemwide initialization being done very early in the boot code and initialization for a specific subsystem being performed later when the associated device driver is initialized. Regardless of when the hardware initialization occurs, it starts with the Clock and I/O subsystem initialization.

4.2 Clock Configuration

The clocking subsystems on modern embedded SoCs are very complex. Chapter 5 of the KW41Z Reference Manual contains the detailed information needed to properly configure the clocks. Fig. 9 shows the clocking diagram.

The clocking configuration on this device is a good example of a subsystem that uses a flash memory configuration mechanism. The clock dividers are initialized at reset based on the values in the FTFA_FOPT register (located in flash). When the flash is in its erased state (all bits are set to 1), bits are set in the FTFA_FOPT register and select the fast clocking mode. The developer has the option of programming the bits in flash to 0 and defaulting to a slower clocking mode. See section 5.5.1 in the reference manual for details.

The default clock settings do not enable clocking to the I/O subsystems. For the examples in this chapter, the clocking must be enabled by the application program. In the example program that blinks the red LED, the clocking to the GPIO module is enabled using a library call.

```
CLOCK_EnableClock(kCLOCK_PortC);          /* Port C Clock Gate Control: Clock enabled */
```

4.3 I/O Pin Configuration

Modern embedded SoCs may also contain very sophisticated I/O capabilities. To maintain flexibility and accommodate the needs of different designs, many of the I/O pins can be configured to perform one of multiple functions. These options can be general purpose input, general purpose output, or a dedicated function associated with a specific

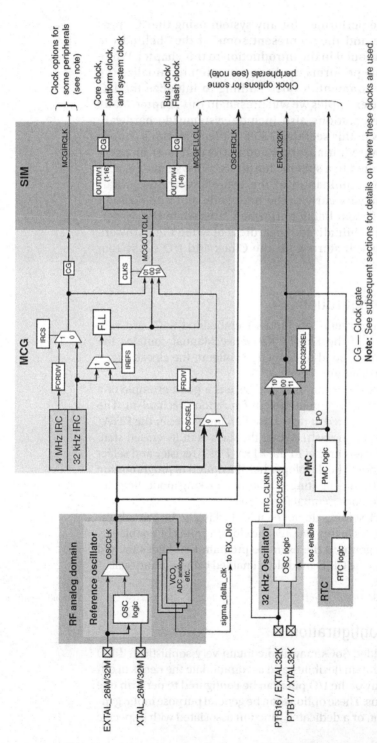

Fig. 9 The KW41Z clocking diagram. With permission from NXP B.V. © 2019.

subsystem. Two examples of dedicated functions are the transmit data signal of a serial port (UART) and a timer enable signal. If the pins associated with a specific peripheral are not used, they are often configured to function as a digital input or output. In this case, the function would be defined by the developer. Adding to the configuration complexity, many SoCs multiplex multiple dedicated functions to a single pin. It is common to see devices that have as many as six or more functions multiplexed to a single I/O pin. An example of this multiplexing can be seen in section 2.2 of the KW41Z reference manual.

There are many dependencies and restrictions on selecting specific pins for a given pin function. Sometimes a peripheral (I2C, UART, Timer, etc.) may be able to use several different sets of pins, but sometimes only a subset of the device signals can be mapped to I/O pins. A common example of this can be found in UART devices. Using one set of I/O pins, the UART exposes only the transmit and receive signals. When using an alternate set of I/O pins, the UART exposes the hardware handshake signals in addition to the transmit and receive signals.

One example from the reference manual is the CLKOUT pin for the I2C port, I2C0_SCL. This function can be found on processor pin 16 and again on pin 48 (Table 1). The first two columns identify the pin designator for two different package types. By writing specific values to the configuration registers, the physical pin can be connected internally to one of eight alternate functions (ALT0–ALT7).

To help developers determine the best pin multiplexing configuration for a given design, many of the semiconductor vendors offer configuration tools. These design tools allow the developer to select a specific device, usually by part number, and then choose the subsystem settings at the feature level. These subsystems correspond to the on-chip devices shown in the SoC block diagram. The tool allows the developer to enable/disable specific peripherals, select clock sources, enable or disable features within the subsystem, and generate the required configuration files. Some vendor tools even generate boot code, header files, and basic driver functions to match the configuration. The tool provided by NXP for the KW41Z is the MCUXpresso Config Tool. Fig. 10 shows an example of this tool.

4.4 I/O Pin Initialization

For the basic hello world example contained in the SDK, a subroutine called BOARD_InitPins() calls routines to easily setup the port pins needed to send ASCII data through the serial port (UART).

```
PORT_SetPinMux(PORTC, PIN6_IDX, kPORT_MuxAlt4);  /* PORTC6 (pin 42) is configured as UART0_RX */
PORT_SetPinMux(PORTC, PIN7_IDX, kPORT_MuxAlt4);  /* PORTC7 (pin 43) is configured as UART0_TX */
```

Table 1 Example of Processor Pin Function Assignments

KW41Z (48 LGA/ Laminate QFN)	KW41 (WLCSP)	Pin Name	DEFAULT	ALT0	ALT1	ALT2	ALT3	ALT4	ALT5	ALT6	ALT7
16	J6	PTB0	DISABLED		PTB0/LLWU_P8/ XTAL_OUT_EN		I2C0_ SCL	CMP0_OUT	TPM0_ CH1		CLKOUT
48	C8	PTC19	DISABLED	TSI0_ CH7	PTC19/LLWU_P3	SPI0_ PCS0	I2C0_ SCL	LPUART0_ CTS_b	BSM_ CLK		BLE_RF_ ACTIVE

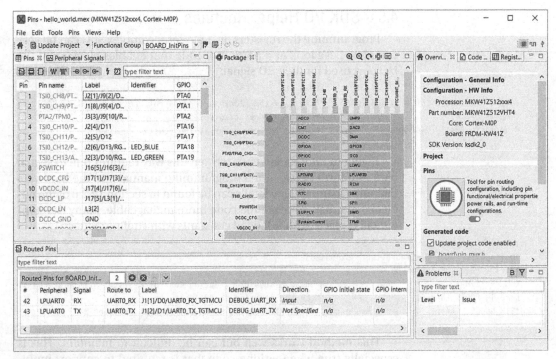

Fig. 10 Sample screenshot of the MCUXpresso pin configuration tool.

The routine PORT_SetPinMux() and the constants PORTC and PINx_IDX come predefined for the family of KW41Z devices. The constant kPORT_MuxAlt4 tells the routine to select the pin function shown in the Alt4 column of Table 2.1 in the KW41Z reference guide.

While the vendor configuration tools do a great job simplifying the details related to pin multiplexing, it is always a good idea to review all the details related to the I/O pins you will be using. Many of the options available on different pins are the pin direction (input vs. output), enable internal pull-up resistor, interrupt enable, edge vs. level activation, etc. It is also important to validate the internal connections selected while configuring a hardware subsystem, understand the dependencies and restrictions, and follow specific initialization sequences as recommended by the manufacturer.

It is also important to read any errata documentation that may be available. These documents tend to capture configuration, sequencing, limitations, and side effects that were unknown or not fully characterized when the SoC was initially released to production. The errata will identify these items as they relate to a specific silicon revision of the device. If your software is required to run on various silicon revisions of the SoC, your code may need to query the SoC registers containing the silicon version details and behave as required.

4.5 SDK I/O Helper Routines

Programming the registers on a modern, complex SoC can be very complex. Let's look at a portion of code that enables clocking to the I/O port used for the LED signal:

```
temp = *(unsigned int *)(0x40048038);  /* Read System Clock Gating Control Register 5 (SIM_SCGC5) */
temp |= 0x800;                         /* Enable clocking to PORTC */
*(unsigned int *)(0x40048038) = temp;  /* Write the new value to the control register */
```

Trying to understand code written this way requires one to switch back and forth between the programming manual and the hardware reference manual. The details can be found in section 12.2.8 of the reference manual. To make the code more manageable, many developers will replace the hardcoded constants with #define values:

```
#define SIM_SCGC5 (unsigned int *)0x40048038
#define PORTC_ENABLE 0x800
temp = *SIM_SCGC5;                     /* Read System Clock Gating Control Register 5 (SIM_SCGC5) */
temp |= PORTC_ENABLE;                  /* Enable clocking to PORTC */
*SIM_SCGC5 = temp;                     /* Write the new value to the control register */
```

This method works well but can be difficult to maintain. This is especially true when writing code that is required to support multiple variants of an SoC family that may have slightly different bitfield definitions for the same subsystems. To address this, semiconductor vendors provide files that contain various data type definitions, data structures, and routines that can be used to manage the hardware.

Here is the same portion of code using the definitions supplied by the semiconductor manufacturer:

```
CLOCK_EnableClock(kCLOCK_PortC);       /* Port C Clock Gate Control: Clock enabled */
```

For our first example, blinking an LED, we will program the hardware registers using hardcoded definitions. For the remaining examples, we will take advantage of the high-level definitions provided in the SDK.

5 Programming Examples

5.1 General Purpose I/O—Blinking LED

This example simply blinks an LED repeatedly. The initialization steps are very simple:
- Enable clocking to the GPIO port containing the LED signal.
- Configure the pin multiplexing to allow the LED signal to be a GPIO output.
- Initialize the GPIO port bit to turn off the LED.
- Enable the GPIO port bit as an output signal.

- Loop forever, toggling the LED signal.
  To build and run this example in the SDK, follow these steps:
- Close any existing projects.
- Select File | New | Project…
- Expand MCUXpresso IDE and select New C/C++ Project, click Next.
- Select the MKW41Z512xxx4 MCU and click Next.
- In the SDK Wizard, change the project name (optional), unselect all the checkbox options except baremetal, select "Empty board files," unselect both "CMSIS-Core" and "Import other files," then click Next (Fig. 11).
  Continuing with the SDK Wizard:
- Unselect "Redirect SDK 'PRINTF' to C Library 'printf."
- Unselect "Include semihost HardFault handler."
- Click Finish (Fig. 12).
  In the project explorer, expand the project and source folder and open the source file containing main(). The filename will be the same as the project name and have a .c extension. Delete all the text in the file and replace it with the text included below:

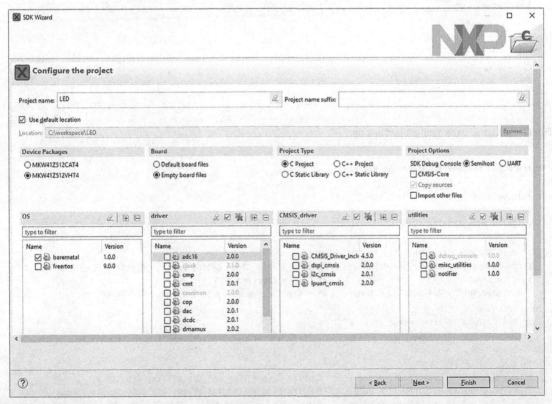

Fig. 11 Configuring a project for baremetal libraries only.

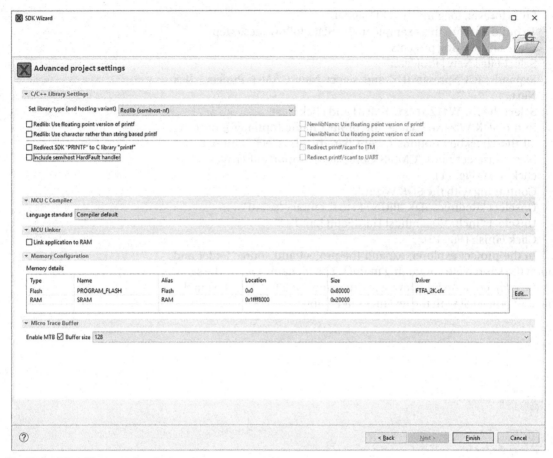

Fig. 12 Additional baremetal project setting.

```
/* Bare metal example to blink the RED LED */

int main(void) {
    unsigned int temp;
    volatile unsigned int i;

    /* Port C Clock Gate Control: Clock enabled */
    temp = *(unsigned int *)(0x40048038);      /* Read System Clock Gating Control Register 5 (SIM_SCGC5) */
    temp |= 0x800;                             /* Enable clocking to PORTC */
    *(unsigned int *)(0x40048038) = temp;      /* Write the new value to the control register */

    /* PORTC1 (pin 37) is configured as PTC1 */
    *(unsigned int *)(0x4004b004) = 0x105;     /* Setup pin mux to make RED LED I/O an output */

    /* Init output LED GPIO. */
    *(unsigned int *)(0x400ff088) = 0x02;      /* Write base + PCOR to turn off the RED LED */
    *(unsigned int *)(0x400ff094) = 0x02;      /* Write base + PDDR to set the RED LED I/O pin to output */
```

```
while (1)
{
    /* Delay for a bit */
    for (i = 0; i < 800000; ++i)
    {
        __asm("NOP"); /* delay */
    }

    /* Toggle the RED LED I/O pin */
    *(unsigned int *)(0x400ff08c) = 0x02; /* Write base + PTOR to toggle output pin for RED LED */
}
return 0;
}
```

Save the source file, then build the project by right-clicking on the project name in the Project Explorer and selecting Build Project. Monitor the build output in the console view. When finished, debug and run the program and observe the blinking LED action. To debug the program, follow the steps outlined in Section 5.3.

5.2   Basic Serial I/O—Polled UART Send/Receive

This example sends and receives characters using the serial port typically used to communicate using the RS-232 standard. Semiconductor vendors use various names for these devices, but the most common name is a UART (Universal Asynchronous Receiver Transmitter). In its simplest mode of operation, a UART will take an 8-bit byte (typically representing an ASCII character) and transmit each bit using the UART's TX signal. It also receives individual bits on the UART'S RX signal and converts them to an 8-bit byte. Modern implementations of these devices support a wide variety of configuration options—bits per word, bit rate (BAUD), hardware flow control, etc. The device can be used in polled mode where the application polls the various status bits to determine when and how to access the UART registers. The device can be configured to run in an interrupt-driven mode where an interrupt is triggered when a certain condition is true. The features and programming modes of the UART are beyond the scope of this chapter. Details can be found in the programmer's reference guide. For this example, the program initializes the configuration registers, then loops forever checking for input characters. When a character is received on the RX signal, that character is then transmitted out the TX signal.

The transmit and receive signals for UART0 can be found on the J1 header on the reference board. In its default configuration, the transmit signal, UART0_TX, is located on pin J1-2. The receive signal, UART0_RX, is located on pin J1-1. It is important to note that many reference boards are designed so the UART external I/O pins operate at "logic levels" and do not meet the RS-232 specification. These "logic level" UART signals are often referred to as "TTL level" signals. To meet the specification, additional hardware circuitry is needed to provide the required level shifting. This additional circuitry can take several forms. One way is to use an adapter board containing the required circuitry. When this type of board is used, it will typically contain the industry standard 9-pin D-style serial connector. Another way is to use a USB to TTL Serial cable. This type of cable contains a USB type-A connector on one end and several single pin connectors on the other. These single pin connectors are often called "flying leads." The USB type-A end plugs into a host computer and appears as a USB Serial port to the host operating system. The individual pin connectors on the other end are then pressed onto the appropriate pins on the reference board. For our example, the cable pins labeled Tx, Rx, and GND should be connected to J1–1 (UART0_RX), J1–2 (UART0_TX), and J2–7 (GND).

If using an adapter cable, follow the vendor's installation instructions, being sure to install any device drivers, if required. You can then use any terminal program to connect to the host's serial port corresponding to the cable (i.e., COMx:). The serial port settings should be set to 115,200 baud, 8-bit word length, no parity, 1 stop bit.

For this example, we will use the sample driver program included in the SDK. Here are the steps to build the project:
- Select File | New | Project…
- Expand the MCUXpresso IDE group and select Import SDK Examples. Click Next (Fig. 13).
- Click on the frdmkw41z board, then click Next.
- Expand the driver_examples group, expand the lpuart group, then check the box for polling. Click Finish (Fig. 14).
- Build the project by right-clicking on the project name in the project explorer (frdmkw41z_driver_examples_lpuart_polling) and left-click on Build Project.

Expand the project in the project explorer, then expand the source group. The application can be found in lpuart_polling.c. This example uses the high-level library routines described earlier to initialize the system. The application is shown below. The routine BOARD_InitPins() configures the I/O pins, routine BOARD_BootClock_RUN() configures the clocks, and routine CLOCK_SetLpuartClock(0x1U) sets up the clocking needed for the UART. The program then configures the UART communications settings using the high-level structures and routines provided by the SDK. It initializes the configuration structure to a default set of values, then sets the desired baud rate and enables both transmit and receive functions in the structure. Calling

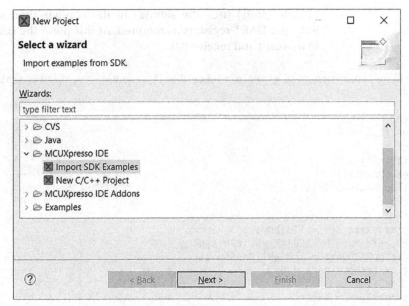

Fig. 13 Importing an SDK example project.

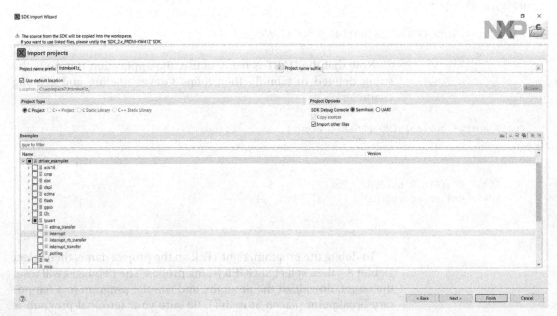

Fig. 14 Selecting the polled lpuart example.

LPUART_Init() uses the settings in the configuration structure to write the UART registers as required. At this point the UART is ready to transmit and receive data.

```
uint8_t txbuff[] = "Lpuart polling example\r\nBoard will send back received characters\r\n";
int main(void)
{
    uint8_t ch;
    lpuart_config_t config;

    BOARD_InitPins();
    BOARD_BootClockRUN();
    CLOCK_SetLpuartClock(0x1U);

    /*
     * config.baudRate_Bps = 115200U;
     * config.parityMode = kLPUART_ParityDisabled;
     * config.stopBitCount = kLPUART_OneStopBit;
     * config.txFifoWatermark = 0;
     * config.rxFifoWatermark = 0;
     * config.enableTx = false;
     * config.enableRx = false;
     */
    LPUART_GetDefaultConfig(&config);
    config.baudRate_Bps = BOARD_DEBUG_UART_BAUDRATE;
    config.enableTx = true;
    config.enableRx = true;

    LPUART_Init(DEMO_LPUART, &config, DEMO_LPUART_CLK_FREQ);
```

Now that the UART is ready to use, the application transmits the string defined in txbuff[], then loops forever reading and echoing characters:

```
    LPUART_WriteBlocking(DEMO_LPUART, txbuff, sizeof(txbuff) - 1);

    while (1)
    {
        LPUART_ReadBlocking(DEMO_LPUART, &ch, 1);
        LPUART_WriteBlocking(DEMO_LPUART, &ch, 1);
    }
}
```

To debug the program, right-click on the project name, then select Debug As, then select SEGGER J-Link probes. The debugger will reset the target, download the program, and run the program to a temporary breakpoint placed at main(). Be sure your terminal program is running on the host PC and is configured properly. To run the example program, press the F8 function key. The string in txbuff[] will be displayed in the terminal window. Typed characters will be echoed back to the terminal. Note that the characters may be displayed twice if your terminal program has its "local echo" option enabled.

5.3   Overview of Interrupt Handlers

One of the peripherals most commonly used in embedded designs is the timer. Many of the larger, highly integrated SoCs will contain multiple types of timers, and often support different operating modes and capabilities. Some are designed as periodic interrupt timers (PIT) and can be used to generate interrupts at a constant rate, and others generate periodic waveforms (PWM) or time the rise and fall of input signals (input capture). Other types of timer models can function as counters, using internal or external signals as the event to be counted. On our reference board, the SoC provides three types of timers: a Timer/PWM Module (TPM), a Periodic Interrupt Timer (PIT), and a Low-Power Timer (LPTMR). For this example, we will use the LPTMR module to count the transitions of an internal clock source.

This example also illustrates the use of an interrupt. The LPTMR will be configured to count clock cycles and then generate an interrupt when enough clocks have been counted to equal one second. This will cause our interrupt handler to be called, toggling the LED and signaling that the time interval elapsed. This process will then repeat forever.

Before we go into the details of the example, we will provide some details on interrupts and how they are handled in this example. Using Table 3-6 in the KW41Z reference manual, we can see that the LPTMR module is assigned to interrupt (IRQ) number 28 (gray shading below) and it corresponds to vector number 44. See Table 2.

The startup code described earlier takes care of initializing the processor's interrupt vector table. For the LPTMR, the entry can be found

Table 2 Interrupt assignments

Address	Vector	IRQ	Source Module	Source Description
0x0000_00AC	43	27	MCG	
0x0000_00B0	44	28	LPTMR0	
0x0000_00B4	45	29	SPI1	Single interrupt vector for all sources
0x0000_00B8	46	30	Port Control Module	Pin detect (Port A)
0x0000_00Bc	47	31	Port Control Module	Pin detect (single interrupt vector for Port B and Port C)

in startup_mkw41z4.c. Here are parts of the file that that apply to our example:

```
WEAK void LPTMR0_IRQHandler(void);
void LPTMR0_DriverIRQHandler(void) ALIAS(IntDefaultHandler);

MCG_IRQHandler,              // 43: MCG interrupt
LPTMR0_IRQHandler,          // 44: LPTMR0 interrupt
SPI1_IRQHandler,            // 45: SPI1 single interrupt vector for all sources
PORTA_IRQHandler,           // 46: PORTA Pin detect
PORTB_PORTC_IRQHandler,     // 47: PORTB and PORTC Pin detect

WEAK_AV void LPTMR0_IRQHandler(void)
{   LPTMR0_DriverIRQHandler();
}

WEAK_AV void IntDefaultHandler(void)
{ while(1) {}
}
```

Note the two functions listed in the code snippet, LPTMR0 and IntDefaultHandler, are defined using the WEAK designation. This is a feature that tells the compiler to use the locally-defined function, LPTMR0_IRQHandler(void), unless another definition of the function is encountered during the linking process. This allows a developer to define their own version of LPTMR0_IRQHandler(void) in their application and use it in the build process, with the one in the startup code being ignored.

It is also common to initialize all unused interrupt vectors to point to an infinite loop that does nothing. This allows the developer to "catch" unexpected interrupts. These are interrupts that may have been enabled in the peripheral control registers without first defining an interrupt service routine to handle the interrupt. The startup code does this by defining all the interrupt handlers to be IntDefaultHandler(). This function only contains a while(1) {} statement. Using a debugger, the developer can halt the processor. If it is executing this infinite loop, it tells the developer an unassigned interrupt occurred, and the system and peripheral registers can be examined to determine the cause.

5.4   Basic Timer Operation—Low-Power Timer (LPTMR)

For this example, we will use the sample driver program included in the SDK. Here are the steps to build the project:
-    Select File | New | Project...
-    Expand the MCUXpresso IDE group and select Import SDK Examples. Click Next (Fig. 15).
-    Click on the frdmkw41z board, then click Next.

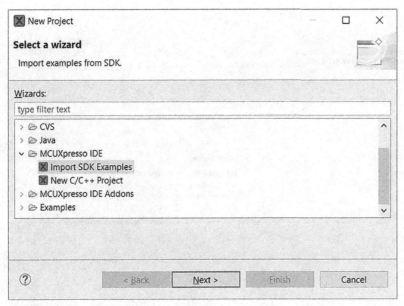

Fig. 15 Importing an SDK example project.

- Expand the driver_examples group, then check the box for lptmr. Click Finish (Fig. 16).
- Build the project by right-clicking on the project name in the project explorer (frdmkw41z_driver_examples_lptmr) and left-click on Build Project.

Expand the project in the project explorer, then expand the source group. The application can be found in lptmr.c. This example uses the high-level definitions described earlier and defines the interrupt handler needed for the timer. This is the interrupt handler routine:

```
#define LPTMR_LED_HANDLER LPTMR0_IRQHandler
volatile uint32_t lptmrCounter = 0U;

void LPTMR_LED_HANDLER(void)
{
    LPTMR_ClearStatusFlags(LPTMR0, kLPTMR_TimerCompareFlag);
    lptmrCounter++;
    LED_TOGGLE();
    /*
     * Workaround for TWR-KV58: because write buffer is enabled, adding
     * memory barrier instructions to make sure clearing interrupt flag completed
     * before go out ISR
     */
    __DSB();
    __ISB();
}
```

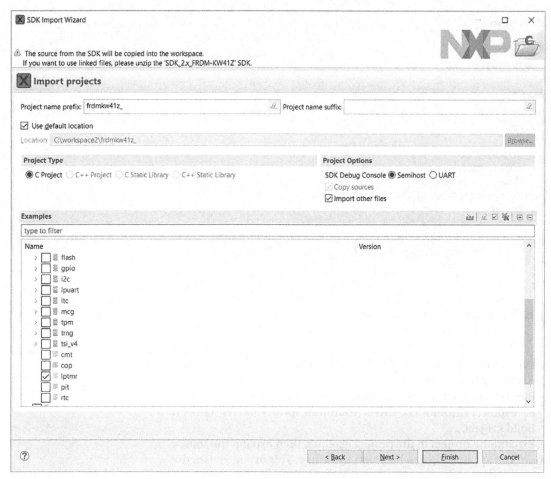

Fig. 16 Selecting the polled lptimer example.

The interrupt handler routine performs three operations:

- Resets the LPTMR status. This clears the interrupt signal the LPTMR module uses internally to alert the processor. The module is then ready to interrupt the next time the count reaches 1 s.
- Increments the variable shared between the interrupt handler (LPTMR_LED_HANDLER) and the main application. Note that this variable, lptmrCounter, is defined using the keyword "volatile." This keyword prevents the compiler from removing the code that accesses the variable when optimization is turned on. After this variable is initialized to zero in main(), it is only read repeatedly. The keyword tells the compiler that the variable is being written in another context and that it should be handled as if it may change value over time.
- Toggles the LED.

As mentioned earlier, the LPTMR0_IRQHandler is defined in the startup code as WEAK. This means the address in the interrupt vector table corresponding to vector 44 will be replaced with the address of the new interrupt handler from the example application. When the LPTMR interrupt occurs, the processor will fetch the address from the vector table (for vector 44) and jump to that location, LPTMR_LED_HANDLER in the example application.

The main application is shown below. The routine BOARD_InitPins() configures the I/O pins, routine BOARD_BootClock_RUN() configures the clocks, and routine BOARD_InitDebugConsole() sets up a debug console to redirect PRINTF() statements through a virtual UART contained in the JTAG debugger connection. This debug console allows the developer to add a console connection without using the LPUART peripheral and a physical serial cable connection. The program then configures the LPTMR settings using the high-level structures and routines provided by the SDK. It initializes the configuration structure (lptmrConfig) to a default set of values, then writes these values to the timer by calling LPTMR_Init(). The function LPTMR_SetTimerPeriod() sets the timeout interval to 1 s.

At this point in the code, the timer has been configured as required in the example. The two remaining steps are to enable the timer interrupt and to tell the timer to start counting. LPTMR_EnableInterrupts() enables the interrupt in the LPTMR module and EnableIRQ() enables the LPTMR interrupt in the processor. To tell the module to start counting, LPTMR_StartTimer() is called.

The application then goes into a loop, checking to see if the variable shared with the interrupt handler, lptmrCounter, has changed. If so, it saves a copy of the counter and prints a message on the console.

```
int main(void)
{
    uint32_t currentCounter = 0U;
    lptmr_config_t lptmrConfig;

    LED_INIT();

    /* Board pin, clock, debug console init */
    BOARD_InitPins();
    BOARD_BootClockRUN();
    BOARD_InitDebugConsole();

    /* Configure LPTMR */
    /*
     * lptmrConfig.timerMode = kLPTMR_TimerModeTimeCounter;
     * lptmrConfig.pinSelect = kLPTMR_PinSelectInput_0;
     * lptmrConfig.pinPolarity = kLPTMR_PinPolarityActiveHigh;
     * lptmrConfig.enableFreeRunning = false;
     * lptmrConfig.bypassPrescaler = true;
```

```
 * lptmrConfig.prescalerClockSource = kLPTMR_PrescalerClock_1;
 * lptmrConfig.value = kLPTMR_Prescale_Glitch_0;
 */
LPTMR_GetDefaultConfig(&lptmrConfig);

/* Initialize the LPTMR */
LPTMR_Init(LPTMR0, &lptmrConfig);

/*
 * Set timer period.
 * Note : the parameter "ticks" of LPTMR_SetTimerPeriod should be equal or greater than 1.
 */
LPTMR_SetTimerPeriod(LPTMR0, USEC_TO_COUNT(LPTMR_USEC_COUNT, LPTMR_SOURCE_CLOCK));

/* Enable timer interrupt */
LPTMR_EnableInterrupts(LPTMR0, kLPTMR_TimerInterruptEnable);

/* Enable at the NVIC */
EnableIRQ(LPTMR0_IRQn);

PRINTF("Low Power Timer Example\r\n");

/* Start counting */
LPTMR_StartTimer(LPTMR0);
while (1)
{
    if (currentCounter != lptmrCounter)
    {
        currentCounter = lptmrCounter;
        PRINTF("LPTMR interrupt No.%d \r\n", currentCounter);
    }
}
}
```

To debug the program, right-click on the project name, then select Debug As, then select SEGGER J-Link probes. The debugger will reset the target, download the program, and run the program to a temporary breakpoint placed at main(). Be sure the console tab is selected so the characters printed in the debug console are visible. To run the program, press the F8 function key. The output will appear in the console:

```
SEGGER J-Link GDB Server V6.32h - Terminal output channel
Low Power Timer Example
LPTMR interrupt No.1
LPTMR interrupt No.2
LPTMR interrupt No.3
```

Placing a breakpoint in the interrupt handler allows the developer to verify the interrupt is being generated and their handler is being called.

6 Summary

Using a low-cost reference board with the manufacturer's SDK allows a developer to quickly ramp up on low-level, embedded development. Understanding how to use the peripherals covered in this chapter provides a foundation to build small, basic embedded applications. Once a base platform is created, the developer can easily leverage more advanced features of the SoC and SDK and add support for wireless communications, displays, and storage devices.

Questions and Answers

1. List some of the mechanisms used to configure modern SoCs at reset. Provide configuration examples typically controlled by these settings.
   a. *Mechanisms.* Pull-up resistors on I/O or bus signals, configuration word in flash.
   b. *Examples of settings.* Pin direction (input vs. output), enable internal pull-up resistor, interrupt enable, edge vs. level activation.
2. What are some of the "behind the scenes" initialization steps the SDK boot code performs prior to calling main()?
   a. Disable the watchdog timer.
   b. Initialize the .data sections in RAM.
   c. Initialize the .bss section in RAM to all zeros.
3. List three types of timers on the SoC used in this chapter and give an example of a function each can perform.
   a. Periodic Interrupt Timer (PIT)—generates repetitive interrupts at a fixed interval that can be used to trigger interrupts at a constant rate.
   b. Timer/PWM module (TPM)—can be used to generate periodic waveforms.
   c. Low-power Timer module (LPTMR)—used to count internal or external signals.
4. Give an example of a use case where the Processor ID register might be used.
   a. When the errata documentation indicates different software behavior is needed based on the silicon revision of the SoC.

5

PROGRAMMING AND IMPLEMENTATION GUIDELINES

Mark Kraeling
CTO Office, GE Transportation, Melbourne, FL, United States

1 Introduction

Many approaches come to mind when considering software programming and implementation. One approach might be syntax-oriented—how the code looks and is written. Another approach might be to consider the structural rules that programmers must follow—to keep the code "cleaner." The ways that software is written and how it is formatted can bring about heated arguments between developers. This chapter was not written to provide a specific way of implementing software, rather focuses on recommendations and good practices. There isn't a single answer to how software should be implemented because there are many factors involved.

The first factor is project size. Time and time again there are arguments around project structure, use of global variables, and other factors. There are a lot of implementation guidelines that are largely

Software Engineering for Embedded Systems. https://doi.org/10.1016/B978-0-12-809448-8.00005-9

dependent on how large (in terms of source lines of code, for instance) a project is. Having 30 software engineers embark on an activity to use nothing but assembly language, using cryptic variable names, all in the same 8-bit processor space is unlikely to be fruitful. Take that same project, and instead have two software engineers working on it. This seems a little more reasonable! Keeping project size in mind is important when reading over these guidelines.

The second factor is programmer experience and background. Hopefully there is a degree of freedom to tailor some of the implementation guidelines based on what the members of a team can do well, and not so well. It's quite possible that your team may be made up of people that moved over from another project on the team, another division of the same company, or even another company altogether. There may be implementation guidelines and standards that one group is comfortable doing—providing a benefit to the rest of the team. Don't fall into the trap of believing "that is the way it has always been done, keep doing it." An assessment of the way programming and implementation is being done is healthy—if it is done at the right time. Trying to change course in the middle of project delivery isn't that time—at the beginning or between major releases may be a more appropriate time.

The third factor is future maintainability and project length. The shorter the duration of the project, or if maintainability is not a key factor, the more likely there will be a lack of effort in terms of project structure and commenting. Don't misunderstand—having useful comments in code is always good for reviewers or to jog your own memory after a weekend! Upon reading the guidelines suggested here—temper some of them if your project is comprised of one programmer putting together code for a project that lasts a month.

There are other factors as well, including safety-critical code development, software that is being sold as software for others to use in their products, and industry regulations for your product or market segment. All these influence (or even dictate) the implementation of software for your product.

1.1 Principles of High-Quality Programming

The implementation guidelines in this chapter are derived to drive higher quality programming on embedded systems. Embedded systems by their very nature are products or systems where the computer processing portion isn't necessarily evident to the user. Because of this, end customer quality assessment is not directly an assessment of the software, rather the performance characteristics of the system itself. In this way, quality can be measured in a variety of different ways.

1.1.1 Readability

Readability in software programming can be defined as the ease with which the software is read and understood. Readability of software can be somewhat objective. Programmers that are "journeyman" and move from one project to another throughout their career tend to have an easier time reading a variety of software code. However, making software more readable helps in reviewing and maintaining it over the course of its life. Simplicity in logic, conditional statements, and the structure of the code all help with readability.

The following is an example of a proper "C" code segment, that isn't entirely readable:

```
// Check for stuff to proceed
if((!(((Engine_Speed!=0)||(Vehicle_Speed!=0))) || SecureTest!=FALSE ){
    // ABC…
}
```

With a little better readability, the same conditional can be written as:

```
// Check for secure testing to be running, or if vehicle is stopped
//    along with the engine not running. Then we can execute <ABC>
if (( Secure_Test == TRUE ) ||                              \
    (( Vehicle_Speed == 0 ) && ( Engine_Speed == 0 )))
{
    // ABC…
}
```

1.1.2 Maintainability

Maintaining the code after it is written is a task that can become extremely difficult. Often the code may not make sense to others that look at it. This can lead to incorrect interpretation, so even though a new feature goes into the code, the existing code around it breaks. If someone besides the author comes into the code to make a change, and if they don't understand the existing structure, then another "if" condition can be placed at the bottom of the code to avoid making any changes to its top part.

Consider using descriptive comments in the code to capture the "intent" of what is being done. Comments can help clarify its overall purpose later when the code is being updated and the person doing the updates needs a solid reference in terms of the structure of the code. For example, a comment of "Reset timer because if we are here we have received a properly formatted, CRC-checked, ping request message" is much better than "Set timer to 10 seconds."

1.1.3 Testability

One of the key components for writing good software is writing software with testability in mind. To be "testable" (either for unit testing or debugging) each executable line of code and/or each execution path of the software must have the ability to be tested. Combining executable lines within conditional statements is not a good idea. If an equate or math operation occurs within an "*if*" evaluation, portions of it will not be testable. It is better to do that operation before the evaluation. This allows a programmer to set up a unit test case or modify memory while stepping through to allow a variety of options in choosing which path to take.

Consider the following code segment:

```
if ( GetEngineSpeed() > 700 )
{
    // Execute All Speed Governor code
}
```

For high-level source code debugging, it would not be immediately clear what the engine speed was while debugging. The tester could analyze the register being used for the return value, but it certainly is not readily apparent. Rewriting the code to use a local variable allows the variable to be placed into a watch window or other source analysis window. The code could be rewritten as follows:

```
current_engine_speed = GetEngineSpeed();
if ( current_engine_speed > 700 )
{
    // Execute All Speed Governor code
}
```

One argument for this could be program efficiency. This was certainly true years ago when embedded compilers were not very efficient in taking high-level source code and translating it to machine instructions. Today, with compiler optimizers written to look for optimizations via multiple passes through the code, most of these opportunities have been taken care of.

1.2 What Sets Embedded Apart From General Programming

The easiest way to evaluate what sets embedded apart from general programming is to look at the characteristics of an embedded programmer. The better embedded programmers tend to have a good working knowledge of hardware. They also are very aware of the resources they have, where bottlenecks could be in their system, and the speed associated with the various functions they need to perform.

There are varying definitions of what an embedded system is, but my favorite definition is "a system where the presence of a processor is not readily apparent to the user." Because the processor itself is "hidden," an embedded programmer concentrates on a set of performance and system requirements to complete specific tasks. As such, the software itself is just a part of the system, and the rest of the embedded platform around it is important as well.

An embedded software programmer keeps the following items in mind:

1) *Resources*. Every line of code and module that is written is scrutinized for the processing time it takes to execute as well as the amount of other resources (such as memory) being used. It becomes more difficult writing a tight embedded system using dynamic allocation languages such as C++ and Java compared with programming languages like C and assembly.

2) *Hardware features*. Software is split between the hardware pieces of the embedded system that can execute them more efficiently as opposed to separating software by a software-only architecture. Interrupts, DMAs, and hardware coprocessors are key components in software design.

3) *Performance*. An embedded programmer has a keen sense of what the hardware can and cannot do. For processors that do not have floating-point units, mathematical equations and calculations are done using fixed-point math. The programmer also focuses on performing calculations that are native to the atomic size of the processor, so they shy away from doing 32-bit calculations on a 16-bit processor, for instance.

2 Starting the Embedded Software Project

One of the easier things to do is to start a fresh embedded project, as opposed to inheriting a project written a long time ago. Starting a new project is typically an exciting time and programmers look forward to starting something new. Promises to not repeat previous evils are recited by programmers. The software will be correctly completed first time around! Depending on how many projects exist or are being kicked off at a company, this event may not happen very often.

It is also the easiest and best time to get organized and determine how the software team should develop embedded software. No new source code has been written yet—though there may be libraries or core modules that are going to be pulled into the software baseline. This is the best time to determine how the project is going to be handled and to get buy-in from all the programmers that will be involved in the development cycle that will be followed.

It is a lot more difficult to institute new standards or development practices in the middle of a project. If faced with that situation, the best time to make any change is after some incremental delivery has been made. Changes to standards that try to take place "weeks before software delivery" typically add more confusion and make things worse. Unless there is total anarchy, or if the project can afford to have everyone stop, come together, and agree upon a new direction, then wait until after a major release delivery of some kind before making changes.

The following subsections consider software items that are discussed and agreed upon as a team (and written down!)

2.1 Hardware Platform Input

Although this chapter is dedicated to software programming and implementation guidelines, it is worth mentioning that there should have already been an opportunity to provide input to the hardware developers on software aspects. Items like hardware interrupt request lines and what they are tied to play a key role in the organization and the performance of the embedded software. Also, other resource inputs, such as memory size, on-chip vs. off-chip resources, type of processor being used, and other hardware I/O interfaces, are critical to embedded development.

Another key aspect is the debugging interface of the processor. An interface like JTAG may be perfect for hardware checking but may not have all the functionality that is available to a software programmer. Many processors (like those based on ARM™ cores) have a JTAG interface but also have a software-centric type of debugging interface using additional lines on the same chip. Bringing those out to a header for software development boards makes debugging and insight into the operation of the software much easier.

Because this chapter focuses on software programming guidelines, there won't be any further discussion of this topic. However, make sure that the connection with the hardware developers is made early, or it could be very difficult to follow software implementation guidelines!

2.2 Project Files/Organization

There are three key components that go into project file organization. The first is identifying any dependencies that the project has on the configuration management (CM) system being used. Some CM tools prefer directory structures to look a certain way to increase interoperability with other existing systems. The second component is the compiler/debugger/linker suite that is being used for the project. The directory structure for some of the files for these components (like libraries) may need to be organized a specific way. The third is project file organization. Project file organization may be determined by team preference or a file organization that is prescribed by other embedded projects done by the same group or at the same company.

To make things easier for development, there should be a separation between the following items listed here. The most common way to separate these is by using subdirectories, or separate folders depending on the development environment.

2.2.1 Source Files Written Locally

This directory should contain all the source files that have been written by your development team. Depending on the number of modules being written or the size of the overall code base, consider further subdividing this into more subdirectories and folders. For multiple processor systems, it may make sense to separate by processor (such as "1" and "2") and have another directory at the same level that contains files common to both.

An additional way to further subdivide a large source files directory is to subdivide it by functionality. Maybe dividing it into major feature groupings, such as "display," "serial comm," and "user IO," would make sense. Indicators of a good project and good directory organization is if software falls into a category easily without a lot of searching around for it or if there are no arguments whether it belongs in one place or another.

2.2.2 Source Files From Company Libraries

This directory should either contain the software or links to the general repository where your company keeps libraries of source files useable in all projects. When doing links, it is important that some sort of control be in place so that new files just don't show up every time the software is built. Version control needs to be kept tight to ensure no unexpected changes occur between the tested and released baseline. Links to specific versions of files work best. If the files must be physically copied into this directory with no links, it is very important to remember (and have written down) exactly which version was copied. To this end, periodic checking back to the library should be done in addition to checking for newer updates or bug fix releases.

The same approach applies to this directory or folder as mentioned earlier, that is, depending on the number of files being used it may make sense to break it down further into subdirectories or subfolders as well.

2.2.3 Libraries From Third Parties

There may be libraries that are used by third parties as well. There might also be source code—maybe an operating system or network stack that has been provided for you. It is critically important to have these files in a separate directory from the other source files! Programmers need to know that these files probably shouldn't be changed, but there could be a tie-off that needs to happen with the software provider. If these are mixed in with the general population of source files that are written by the software team, there is a larger risk that they could be changed inadvertently.

Typically, there are files provided by third parties that are supposed to be changed. These may include definitions or links to pieces in the embedded system. For instance, one common entry is defining the number of tasks for an RTOS. Files that are supposed to be changed should either go in their own subdirectory in this group or be pulled over into a folder in the source files that your group is writing. Then privileges like "no modify/no write" could possibly be applied to the folder, to make sure they are not changed.

2.2.4 Libraries From Compiler/Linker Toolsets

There may be restrictions on where the libraries, that the compiler and linker toolsets provide, can be located. Typically, these can just be left alone. All developers need to agree up front which libraries are going to be used. The toolset company may include a full "C stdlib" available for use, or other alternatives like a smaller "micro" library that can be used instead. Trade-offs between the various libraries should be done, like whether the library allows reentrant library use, the functionality that is available, and the size of the library when linked in your embedded system.

There also may be options to remove libraries entirely from use. A common library that we often remove is the floating-point link library. So, library functions like a floating-point multiply (fmul) cannot be linked into the system. If a programmer has a link to this library, it won't link, and the mistake can be corrected.

2.3 Team Programming Guidelines

How a team agrees to program the system and the criteria they will use for evaluating other programmers' source code is something important to decide upon up front. If a programmer holds to a higher standard of software development, only becoming clear in the first code review after that programmer has already designed and written the code, then it is too late. The criteria for how a programmer can successfully pass a code review should be understood up front, so time isn't wasted rewriting and unit-testing code multiple times.

Guidelines could include a variety of rules or recommendations. The more the guidelines are verifiable, the more successful they will be. For example, if a guideline for the programmer is that "the code is not complex," it could be hard to verify, as the definition of complexity is largely subjective within a group of programmers. One may feel it is too complex, another may not. This guideline could be made verifiable if the word complex correlated to a set of measurable criteria.

To take this example a bit further, the group could decide to use a cyclomatic complexity measurement to evaluate a software module—the software is run through a tool that produces a complexity number for the module. Higher numbers represent more complex code according to the formula, lower numbers represent simpler. With a complexity formula that measures the number of "edges" and "nodes" in your software program, the simplest complexity represented by a value of "1" is a program that contains no "if" or "for" conditions and has a single entry and exit point. As the number of conditions and flows increase, the complexity increases. So, the evaluation criteria could change to "the code is not complex, cyclomatic complexity <= 18." Thus it is no longer subjective.

What this is hinting at is a "checklist" of sorts that a programmer could use when writing and preparing his software code for review. Having the list of accepted programming guidelines up front that everyone follows makes expectations clear. The following are examples of items that could be on a "Software Guidelines Checklist" that would be evaluated for each module reviewed:

- Conformance to syntax standard.
- Cyclomatic complexity calculation.
- Number of source lines per function/file.
- Number of comments.
- Ratio of number of source lines to number of comments.
- Run through code formatter.
- Comment and design document understandability/matches code.
- Code under CM control is linked to a "change request."
- No compiler warnings.
- Rule exceptions properly documented (if warnings ignored or don't match standard).
- #pragma directives documented clearly in source code.
- Nonconstant pointers to functions are not present.
- All members of union or struct are fully specified.
- Data representation (scale, bits, bit assignments) clearly documented.
- Data defined and initialized before being used.
- Loop bounds and terminations are correct.
- Mathematical operations correct (no divide-by-zero, overflows).
- No deadlocks, priority inversions, reentrant faults.

2.4 Syntax Standard

There are a variety of ways a coding syntax standard can appear. A syntax standard defines the way code is spaced, capitalized, and formatted when written into source code. Personal preference needs to be considered when using a syntax standard for a group. There may also be a mix of syntax rules a group could incorporate on a project—there may be some rules that are not mandatory but are recommended. This section contains some ideas about how this might look. The most important thing is getting the developers to agree on a given standard and ensuring they stick to it throughout. If the project is reusing quite a bit of code, preference should be given to the standard that the existing code uses.

This section has some ideas about how the syntax standard could be developed. There isn't a right or wrong here—apart from if developers on a team are all doing something different. In such cases, this impacts the ability to review the code or go in and easily make changes. If the code is developed by all team members using the same syntax, then it is much easier to change, as well as understand, when reviewing the work.

Adopting a set of coding standards is also desirable. One such standard set is "MISRA-C" (Motor Industry Software Reliability Association), which defines rules that C and C++ source code should follow to be reliable, secure, portable, and safe. Though it was developed for the transportation industry, it has been accepted widely outside this sector, especially when safety elements come into play. Versions of the full standard exist from "MISRA C:1998" through to the latest "MISRA C:2012," incorporating various amendments that are a bit newer. The other benefit to using a standard such as this is that there are a variety of automated tools that can be run to check compliance of the code to this standard.

Subsequent sections in this chapter outline some of the syntax-oriented coding standard items that can be found.

2.4.1 Code WhiteSpace

The following are examples of how various software lines can add white space to increase the readability of the code itself. All these examples are operationally equivalent—they produce the same machine code. They are listed in order of the amount of white space they use:

```
int i;
for(i=0;i<20;i++)
{
    printf("%02u",i*2);
}
int i;
for ( i=0; i<20; i++ )
{
    printf( "%02u", i*2 );
}
```

```
int i;
for ( i = 0; i < 20; i++ )
{
    printf( "%02u", i * 2 );
}
```

The examples above concern themselves with the white space that is between the various operators and numbers on a given line of source code. Numerous studies indicate that more white space increases readability in software code. This would support using the third example outlined above. However, if the amount of white space causes the software to wrap to the next line, then too much white space has been used, because wrapping is very unreadable.

2.4.2 Tabs in Source Files

Most syntax standards indicate that tab characters should not be used in source files when writing code. This is because the tab character could be interpreted differently by source editing tools, file viewers, or when it is printed. They are also not readily visible when editing. Source code editors typically provide a way to substitute spaces with the tab character. So, while programming, when the tab key is hit, it automatically replaces the tab with X number of spaces.

This brings about an important point. How many spaces should represent a tab key press or a normal indent in source code? Most editors have a substitution for either "3" or "4" spaces per tab. Either is fine—again, this will be based on some personal preference in addition to how the rest of the code is formatted. In terms of improved alignment, the amount of indent space depends on the spacing that is used for other things, like the "for" loop spacing identified earlier.

2.4.3 Alignment Within Source

How things are aligned in source code makes an impact on readability as well. Take into consideration the following two operationally equivalent sections of code:

```
int incubator = RED_MAX; /* Setup for Red Zone */
char marker = '\0'; /* Marker code for zone */

int  incubator = RED_MAX /* Setup for Red Zone  */
char marker    = '\0';   /* Marker code for zone */
```

White space is used on the second example, lining up the variable names, initialization values, and comments on the same column for the code block.

The examples above were quick examples to demonstrate how different code syntax with white space can be used. Consistency and readability are key components to writing good embedded software source code.

2.5 Safety Requirements in Source Code

When writing safety-critical software, the implementation guidelines for software source code change. Many considerations need to be made when developing this code.

Is all the code in your system safety-critical? If a system is safety-critical, it may not actually rely on *all* the code to be safety critical. The system itself needs to have fail-safe operations in place so that things fail to the least permissive case, as defined by FMEA analysis. There may be operations like logging that are not required to be safety-critical since they cannot cause the safety-critical code in the system to act in an unsafe manner.

Documentation of safety-critical sections of code is important. Special care and consideration should be given to mark these sections differently, or even have comments that refer directly to the safety case or documentation that the code adheres to. Using all capitals such as "SAFETY-CRITICAL CODE SECTION START" in a comment section certainly alerts programmers, who might be changing code or adding new requirements, to the fact that they should tread lightly in these sections.

As discussed above, development standards, such as "MISRA C" (Motor Industry Software Reliability Association) and "MISRA C++," can also help facilitate writing code that operates in a safe manner. There are many users of the standard outside the automotive and transportation industries, including medical and defense users. There are also many tools that can check source code for MISRA compliance that can be included as part of the overall software build process. Picking up and using this standard can be helpful for implementation.

There may be special programming requirements for safety-critical sections of code. There may be a separate development guideline list, that includes things like performing a software FMEA on the safety-critical code section being implemented. There also may be additional reviewers in the code review itself, such as representatives from a safety team or a software engineer that specializes in safety-critical code development.

The following are additional factors or checklist items that could be considered a part of safety-critical code development:
- Adherence and checking to a standard, such as MISRA C or C++.
- Safety sections clearly marked to standard.
- Data that is safety-critical incorporates "safe" or similar wording in the variable name.
- All safety-critical variables are initialized to the least permissive state.
- Any safety-critical data is clearly marked as stale and/or deleted after use.
- Comparisons between safety-critical data are handled correctly.

- All paths are covered when variables are used for path decision making.
- Checks are in place to make sure safety-critical code is executed on time.
- Periodic flash and RAM checks are done to check hardware correctness.
- Safety-critical data is protected by regular CRC or data integrity checks.
- "Voting" mechanisms between software and processors is done correctly.
- Safety dependencies on functions (like a watchdog timer) are checked periodically for correct operation.

More details on safety-critical software development are outlined in Chapter 11 (Safety-critical Development).

3 Variable Structure

3.1 Variable Declarations

One of the key components for developing an embedded software system is determining how the data in the system will be declared and used. To discuss each type of variable declaration, it is probably best to break them down by type. The three primary types of variables in a system are global variables, file-scope variables, and local variables.

3.1.1 Global Variables

Global variables are variables that are visible to any linked component of the system in a single build. They could be declared at the top of a source file but could also be present in header files where the variable is declared in one spot, and then made available as an extern to any other file that includes that header file. There certainly is an entire philosophy associated with global variables—some programmers hate them, and software leads have been known to ban them.

There are differing opinions on the usage of global variables. Programmers can define a correct and "right" way to use them, if they don't help foster the creation of unorganized (spaghetti) code. There are a couple of guidelines that could be used to allow global variables into your system, as this will typically help increase the performance of the system without using access functions to modify encapsulated local data.

The first is to declare the variable in a header file. Anyone that includes the header file would then have access to the variable, but it also helps make sure that if the global was declared as an unsigned integer all the extern references would match. The header file (ip.h) would look something like this:

```
#ifdef IP_C
    #define EXT
#else
    #define EXT extern
#endif

EXT uint16_t IP_Movement_En
EXT uint16_t IP_Direction_Ctrl

#undef EXT
```

The example above would need each of the source files to declare a definition of their "filename_C" for the variable to be declared. The source file (ip.c) would look like this:

```
#define IP_C
#include "ip.h"
#undef IP_C
#include … /* Rest of the include files needed by the source file */
```

By declaring the variable in a header file, the type will be correct, and identifying who is including the header file will also provide a good indication of who might be looking at this variable. Using this type of method could also allow the team to dictate that no global variables are declared in source files—they would be declared in this manner only.

The second recommendation for using global variables is to always prefix the name with the "owner" of the variable itself. In the example above, IP stands for "Input Processing." So, any variable used with global scope of IP_xxx is a variable declared in the input processing header. This helps by not having a bunch of random names floating around for variables.

The third recommendation that would help make global variable usage easier relates to the second recommendation mentioned above. After a global variable is declared in a header file, the only program that could modify that variable would be an input processing source file, like "ip.c". Other source files would have "read" access to that variable, but not be allowed to change its value. Of course, the compiler would allow the programmer to change it—but if this was a rule the project team wanted to use it would be easy to find in a code review. Any instance of a variable prefixed with the "ownership" shouldn't be modified by another program. Consider the source lines below in output processing (op.c):

```
if ( IP_Movement_En == TRUE )
{
    if (( IP_Direction_Ctrl == IP_FORWARD ) ||
        ( IP_Direction_Ctrl == IP_REVERSE ))
```

```
{
    OP_Display_Movement = TRUE;
    IP_Display_Shown = TRUE;    /* Unacceptable… */
  }
  else
  {
    OP_Display_Movement = FALSE;
  }
}
```

In the example above, we do not want to modify an input processing variable following the third recommendation. This hopefully would be easy to see during a code inspection or review. Instead, consider having input processing figure this out by looking at the variable OP_Display_Movement. If this cannot be done, then a function call from here to an input processing function, having that function change IP_Display_Shown, may work. For debugging purposes, and to try to keep the code organized, having a rule like this in place can make global variables a lot cleaner.

The final recommendation for global variables, in addition to showing "ownership" of the variable by prefixing the source file indicator, is to capitalize each letter in the variable name. This would be a further indication that the variable is a global variable that could be read at multiple places, so changes to meaning, scaling, or size could cause a ripple effect to propagate throughout the system.

3.1.2 File-Scope Variables

File-scope variables are used to share data between multiple functions in a single source file. They are easier to use than global variables, because typically there is a single owner in a particular source file, at least when it is initially written. File-scope variables make it easy to share data between functions, without having to pass them as arguments on the stack.

A key recommendation is to keep the keyword "static" in front of each of the file-scope variables being declared. This keeps them from being used by other files and keeps things local. One issue with this is visibility on the map (or possibly even the debugger) file. For compilers, if a variable is declared in a file without visibility to other files, there isn't a need to put a reference to it for linking. Sometimes it is nice to be able to see such a variable during source line debugging or peeking at memory while the system is running.

To give such variables visibility during debugging, consider declaring the file-scope variables in the following way in a source file:

```
STATIC uint32_t IP_time_count;
STATIC uint16_t IP_direction_override;
```

This "STATIC" definition would then reside in one of the "master" header files in your system. Further discussion of this type of header file (portable.h) is provided in Section 3.2. When compiling the debugger version of your code, the programmer can define a keyword "DEBUG" for those source files—when it is time to release the code the keyword "DEBUG" is not defined. This is particularly useful if there are special setup steps that need to be completed (like turning on a debug function in the microcontroller at initialization). With this type of setup, the following lines would appear in the common header file:

```
#ifdef DEBUG
  #define STATIC
#else
  #define STATIC static
#endif
```

Another recommendation for file-scope variables is the use of capitalized and lowercase characters. Consider prefixing all the file scope variables with the same "filename" or feature set indicator at the front, then making all other letters lowercase. In this way, it will be easy to discern between a global variable and a file-scope variable.

3.1.3 Local Variables

Local variables have the easiest recommendations of all types of variable. The first recommendation is to drop the prefix mentioned in the previous two sections, because it is just a variable within the function. Second, with the use of decent comments, the variable names for local variables really do not need to be overly descriptive. In my opinion, it is alright to have variable names such as n, i, j, etc., when using them to index arrays and loop variables. Even a simple variable like "count" is OK—again if there are comments to let an observer know what the function is trying to do.

The other type of local variable in a function is one with the keyword "static" in front of it. This is used when a variable needs to retain the data through multiple calls of the function, but it is not shared by any of the other functions in a file.

For local variables, consider keeping them as all lowercase. In the case of a "static" variable declared in a function, consider capitalizing the first character. In that way, when looking through the function or maintaining it later, the variable retains its value. The following is an example of how function local variables can look:

```
static void ip_count_iterations( void )
{
    uint16_t        i, j, n;
    static uint16_t error_count_exec = 0;
    uint32_t        *reference_ptr;
```

3.2 Data Types

One of the key attributes for embedded systems is resource management. In the preceding sections, the declarations that were made were using type definitions. To keep an embedded system portable to other processors, and to keep resources in check, type definitions can be used for various data types. The following is a list of type definitions that can be declared in a master header file, that will be included by all source files.

Consider a file called "portable.h" which is included by the source files:

```
typedef unsigned char       uint8_t;
typedef unsigned short int  uint16_t;
typedef unsigned long int   uint32_t;
typedef signed char         int8_t;
typedef signed short int    int16_t;
typedef signed long int     int32_t;
```

Because an "integer" size is dependent on the microcontroller architecture size, the programmer can use the type definitions above and then only change the file if porting to a different platform. Library templates can also be written using the type definitions provided above, so that when they are pulled in and used on any platform they work correctly. Another variation of the same concept above is to shorten the type definitions, to save some white space when writing the source files. A variation of the definitions above is shown here:

```
typedef unsigned char       UINT8;
typedef unsigned short int  UINT16;
typedef unsigned long int   UINT32;
typedef signed char         INT8;
typedef signed short int    INT16;
typedef signed long int     INT32;
```

Building on the same naming convention, when structures are declared, consider adding a suffix "_t" to identify it as a type definition. An example of a structure declaration is shown here:

```
#define DIO_MEM_DATA_BLOCKS 64
typedef struct
{
    UINT16 block_write_id;
    UINT16 block_write_words;
    UINT16 data[DIO_MEM_DATA_BLOCKS];
    UINT16 block_read_id;
    UINT16 startup_sync1;
    UINT16 startup_sync2;
} DIO_Mem_Block_t;
```

Following the same convention, a union type definition is shown here:

```
typedef union
{
    UINT16 value;
    struct
    {
        UINT16  data        :15;
        UINT16  header_flag :1;
    } bits;
} DIO_FIFO_Data;
```

For the struct and union examples provided earlier, type definitions are used for data sizes as discussed previously. Spacing and white space is decided upon by the programmers—having it maintained uniformly across all source files adds to maintainability.

Another thing to notice is that type definitions mentioned earlier are prefixed with "DIO_" and contain a mix of capital and lowercase letters. This is a stylistic choice. One thought process is to have type definitions in header files declared in this manner and file-scope type definitions all lowercase, without the need for a prefix. As discussed in the "Variable Declarations" section, this can help a reviewer understand if the structure is something that may be global in scope or just local.

3.3 Definitions

3.3.1 Conditional Compilation

Another topic in developing embedded software is the use of conditional compiles in the source code. Conditional compiles allow a compiler to dictate which code is compiled and which code is skipped. There are many books written for software engineering that suggest conditional compiles should not be used in the code.

For hardware-oriented code that is written to work on multiple processors in a system, there may be conditional compiles to specify "Processor A" vs. "Processor B." For software source code, if >15% of the source code has conditional compiles in it, consideration should be given to splitting the code, keeping the common code in one file and separating the reason for the conditional compiles between two (or more) files. As the number of conditional compiles increases the readability decreases. Files with minimal conditional compiles are likely easier to maintain than a file that has been branched or separated, but again, as the number of conditional compiles increases past 15% the maintainability drops.

Consider the following source code section for a module written to run on two processors specified as PROCA and PROCB. Depending on which makefile is selected, the compiler defines one of these two values depending on the processor target.

```
    frame_idle_usec = API_Get_Time();
#ifdef PROCA
    /* Only send data when running on processor B */
    ICH_Send_Data( ICH_DATA_CHK_SIZE, (uint32_t *)&frame_idle_usec );
#else
#ifdef PROCB
    /* Nothing to send with processor B in this situation */
#else
    /* Let's make sure if we ever add a PROCC, that we get error */
    DoNotLink();
#endif /* PROCA */
#endif /* PROCB */
```

There is one additional thing to note with this code. In this example, we do not simply just look for processor A and then do nothing if we are not processor A. There is an else condition, so that if we ever run on a processor besides A or B, a made-up function "DoNotLink()" will be called, which will result in a compiler warning and a linker error (the function doesn't exist). In this way, if another processor is added in the future it will force the software engineer to look at this code to see if a special case should be added for this new processor. It is simply a defensive technique to catch various conditional compiles that may exist in the source code baseline.

3.3.2   #Define

A commonly used symbolic constant or preprocessor macro in C or C++ coding is implemented using the #define.

Symbolic constants allow the programmer to use a naming convention for values. When used as a constant, it can allow better definition as opposed to "magic numbers" that are placed throughout the code. It allows the programmer to create a common set of either frequently used definitions in a single location, or to create more singular instances to help code readability.

Consider the following code segment:

```
// Check for engine speed above 700 RPM
if ( engine_speed > 5600 )
{
```

The code segment checks for a value of 5600. But where does this come from? The following is a slightly more readable version of the same code segment:

```
// Check for engine speed above 700 RPM
if ( engine_speed > (700 * ENG_SPD_SCALE ))
{
```

This is a little better as it uses a symbolic constant for the fixed-point scaling of engine speed, which is used throughout the software code baseline. There certainly should not be multiple definitions of this value, such as having ENGINE_SPEED_SCALE and ENG_SPD_SCALE both used in the same code baseline. This can lead to confusion or incompatibility if only one of these scalar values is changed. The code segment above also has "700" being used. What if there are other places in the code where this value is used? What is this value? The following code segment is more maintainable and readable:

```
// Check for speed where we need to transition from low-speed to all-
//    speed governor
if ( engine_speed > LSG_TO_ASG_TRANS_RPM )
{
```

A #define would be placed in a header file for engine speed for visibility to multiple files, or in the header of this source file if it is only used in a file-scope scenario. The #define would appear as:

```
// Transition from low-speed to all-speed governor in RPM
#define LSG_TO_ASG_TRANS_RPM (UINT16)( 700 * ENG_SPD_SCALE )
```

This has the additional cast of UINT16 to make sure that the symbolic constant is a fixed-point value so floating-point evaluations are not made in the code. This would be important if the transition speed was defined as 700.5 RPM, or even if a value of 700.0 was used as the transitional speed. Once floating-point values appear in the code, the compiler tends to keep any comparisons or evaluation using floating-point operations.

Preprocessor macros allow the programmer to develop commonly used formulas used throughout the code and define them in a single location. Consider the following code segment:

```
Area1 = 3.14159 * radius1 * radius1;
Area2 = 3.14159 * (diameter2 / 2) * (diameter2 / 2);
```

The code listed above can be improved by creating a preprocessor macro that calculates the circular area as opposed to listing it in the code. Another improvement would be to use a symbolic constant for PI so that the code can use the same value throughout. That way if additional decimal places are used, it can be changed in one location. The following could be defined at the top of the source file, or in a common header file:

```
#define PI 3.14159
#define AREA_OF_CIRCLE(x)     PI*(x)*(x)
```

The code could then use this preprocessor macro as follows:

```
Area1 = AREA_OF_CIRCLE(radius1);
Area2 = AREA_OF_CIRCLE(diameter2 / 2);
```

The code segments shown above could be used for a higher end microcontroller that has floating-point hardware or for a processor where floating-point libraries are acceptable. Another implementation could be in fixed-point, where tables would approximate PI values so that native fixed-point code could speed up processing times.

Content Learning Exercises

1. **Q:** Why is it important to add comment to code?
   A: The comments can help clarify its overall purpose later when the code is being updated and the person doing the updates needs a solid reference in terms of the structure of the code.
2. **Q:** What items should have the ability to be tested to consider the software testable?
   A: Each executable line of code and/or each execution path of the software must be able to be tested.
3. **Q:** What are *conditional compiles* used for and why is their use recommended to be kept to a minimum?
   A: Conditional compiles allow a compiler to dictate which code is compiled and which code is skipped. As the number of conditional compiles increases the readability of the code decreases. Files with minimal conditional compiles are likely easier to maintain than files that have been branched or separated.

6

OPERATING SYSTEMS

Jean J. Labrosse
Founder and Chief Architect, Micrium LLC, Weston, FL, United States

Real-time systems are systems in which correctness and timeliness of computed values are at the forefront; there are two types—hard and soft real-time systems.

What differentiates hard and soft real-time systems is their tolerance to missing deadlines and the consequences associated with such issues. Correctly computed values supplied after a deadline has passed are often useless.

For hard real-time systems, missing deadlines is not an option. In fact, in many cases, missing a deadline often results in loss of assets or even worse, loss of life. For soft real-time systems, however, missing deadlines is generally not as critical.

Real-time applications are wide ranging, but many real-time systems are embedded. An embedded system is where a computer is built into a system and is typically dedicated to a single use. In other words, they are systems that are designed to perform a dedicated function. The following list provides just a few examples of embedded systems:

Audio
MP3 players
Amplifiers and tuners

Automotive
Antilock braking systems
Climate control
Engine controls
Navigation systems (GPS)

Avionics
Flight management systems
Jet engine controls
Weapons systems

Office Automation
FAX machines/copiers

Home Automation
Air-conditioning units
Thermostats
White goods

Communications
Routers
Switches
Cell phones

Process control
Chemical plants
Factory automation
Food processing

Agriculture
Round balers
Square balers
Windrowers
Combines

Video
Broadcasting equipment
HD televisions

Real-time systems are typically more complicated to design, debug, and deploy than nonreal-time systems.

1   Foreground/Background Systems

Small systems of low complexity are typically designed as foreground/background systems or super loops, as shown in Fig. 1. An application consists of an infinite loop that calls modules (i.e., tasks) to perform the desired operations (background). Interrupt service routines (ISRs) are designed to handle asynchronous events (foreground events). The foreground is called the interrupt level; the background is called the task level.

Critical operations that should be performed at the task level, but must unfortunately be handled by ISRs to ensure that they are dealt with in a timely fashion, cause ISRs to take longer than they should. Also, information for a background module that an ISR makes available is not processed until the background routine gets its turn to execute, called the task-level response. The worst-case task-level response time depends on how long a background loop takes to execute and, since the execution time of typical code is not constant, the time for successive passes through a portion of the loop is nondeterministic. Furthermore, if a code change is made, the timing of the loop is affected.

Most high-volume low-cost microcontroller-based applications (e.g., microwave ovens, telephones, toys) are designed as foreground/background systems.

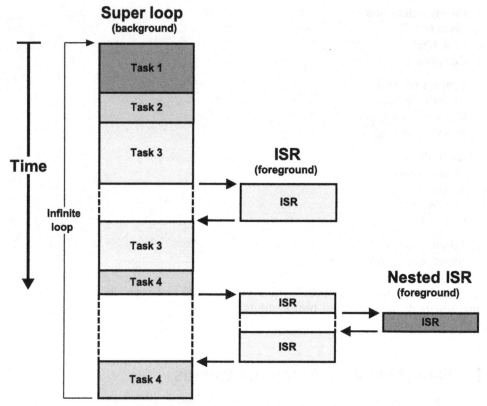

Fig. 1 Foreground/background systems.

2 Real-Time Kernels

A real-time kernel (or simply a kernel) is software that manages the time and resources of a microprocessor, microcontroller, or digital signal processor (DSP). Through functions provided by the kernel, the work of the processor is basically split into tasks, each being responsible for a portion of the job. A task (also called a thread) is a simple program that thinks it has the central processing unit (CPU) completely to itself. On a single CPU, only one task can execute at any given time.

The main function of the kernel is the management of tasks, called *multitasking*. Multitasking is the process of scheduling and switching the CPU between several tasks. The CPU switches its attention between several sequential tasks. Multitasking provides the illusion of having multiple CPUs and maximizes the use of the CPU. Multitasking also helps in the creation of modular applications. One of the most important aspects of multitasking is that it allows the application programmer to manage the complexity inherent in real-time applications. Application programs are easier to design and maintain when multitasking is used.

Most real-time kernels are preemptive, which means that the kernel always runs the most important task that is ready-to-run, as shown in Fig. 2.

(1) A low-priority task is executing.

(2) An interrupt occurs, and the CPU vectors to the ISR responsible for servicing the interrupting device.

(3) The ISR services the interrupt device, but does very little work. The ISR will typically signal or send a message to a higher priority task that will be responsible for most of the processing of the interrupting device. For example, if the interrupt comes from an Ethernet controller, the ISR simply signals a task, which will process the received packet.

(4) When the ISR finishes, the kernel notices that a more important task has been made ready-to-run by the ISR and will not return to the interrupted task, but instead context switches to the more important task.

(5) The higher priority task executes and performs the necessary processing in response to the interrupt device.

(6) When the higher priority task completes its work, it loops back to the beginning of the task code and makes a kernel function call to wait for the next interrupt from the device.

(7) The low-priority task resumes exactly at the point where it was interrupted, not knowing what happened.

Kernels are also responsible for managing communication between tasks and managing system resources (memory and I/O devices).

A kernel adds overhead to a system because the services provided by the kernel require time to execute. The amount of overhead depends on how often these services are invoked. In a well-designed

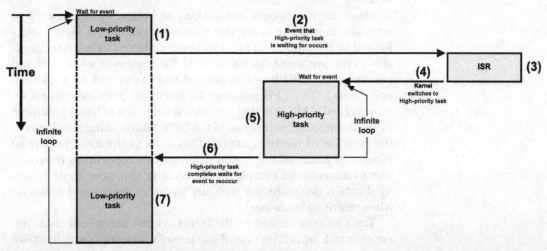

Fig. 2 Preemptive kernels.

application, a kernel uses between 2% and 4% of a CPU's time. And, since a kernel is software that is added to an application, it requires extra ROM (code space) and RAM (data space).

3 RTOS (Real-Time Operating System)

A real-time operating system generally contains a real-time kernel and other higher level services, such as file management, protocol stacks, a graphical user interface (GUI), and other components. Most of the additional services revolve around I/O devices.

3.1 Critical Sections

A critical section of code, also called a *critical region*, is code that needs to be treated indivisibly. There are many critical sections of code contained in typical kernels. If a critical section is accessible by an ISR and a task, then disabling interrupts is necessary to protect the critical region. If the critical section is only accessible by task-level code, the critical section may be protected using a *preemption lock*.

3.2 Task Management

The design process of a real-time application generally involves splitting the work to be completed into tasks, with each being responsible for a portion of the problem. Kernels make it easy for an application programmer to adopt this paradigm. A task (also called a *thread*) is a simple program that thinks it has the central processing unit (CPU) all to itself. On a single CPU, only one task can execute at any given time.

Most kernels support multitasking and allow the application to have any number of tasks. The maximum number of tasks is only limited by the amount of memory (both code and data space) available to the processor. Multitasking is the process of *scheduling* and *switching* the CPU between several tasks (this will be expanded upon later). The CPU switches its attention between several *sequential* tasks. Multitasking provides the illusion of having multiple CPUs and maximizes the use of the CPU. Multitasking also helps in the creation of modular applications. One of the most important aspects of multitasking is that it allows the application programmer to manage the complexity inherent in real-time applications. Application programs are typically easier to design and maintain when multitasking is used.

Tasks must be created for the kernel to know about your tasks. You create a task by calling one of the kernel's services (something like `OSTaskCreate()`) and you specify as arguments to the function call:

(1) The start address of the task. In C, this is the name of the function that makes up the task code.

(2) The priority of the task based on the relative importance of the task.

(3) The stack space and its size that will be used by the task. In a multitasking environment, each task requires its own stack space.

There are possibly other parameters specific to the task that could be specified. These greatly depend on kernel implementation, however, the above three elements represent the minimum.

When a task is created, it is assigned what's called a *Task Control Block* or *TCB*. The TCB is used to hold runtime information about your task. The TCB is managed by the kernel and the user of the kernel generally doesn't need to worry about this data structure.

A task can access variables, data structures, or tables that it either owns or shares with other tasks. If these are shared then the application programmer needs to ensure that the task has exclusive access to these variables, data structures, or tables. Fortunately, the kernel provides services that allow you to protect such shared resources. These are discussed later.

A task can also access I/O devices which it can own or share with other tasks. As expected, services are available from the kernel to ensure exclusive access to these I/O devices.

Fig. 3 shows elements that a task can interact with. You should note that the stack is managed by the compiler (function calls, local variables, etc.) and the TCB is managed by the kernel.

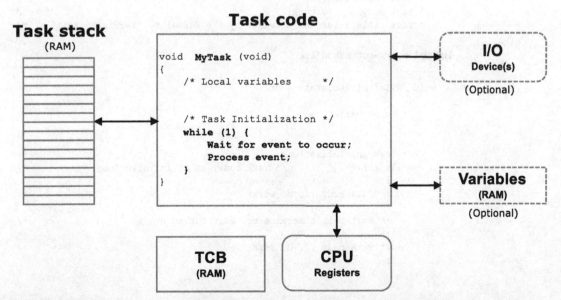

Fig. 3 Task resources.

Tasks are used for such chores as monitoring inputs, updating outputs, performing computations, controls, updating one or more displays, reading buttons and keyboards, communicating with other systems, and more. One application may contain a handful of tasks while another application may have hundreds. The number of tasks does not represent how good or effective a design may be; it really depends on what the application (or product) needs to do. The amount of work a task performs also depends on the application. One task may have a few microseconds worth of work to perform while another may require tens of milliseconds.

Tasks look like just any other C function except for a few small differences. There are typically two types of tasks: run-to-completion (Listing 1) and infinite loop (Listing 2). In most embedded systems, tasks typically take the form of an infinite loop. Also, no task is allowed to return as other C functions can. Given that a task is a regular C function, it can declare local variables.

A run-to-completion task must delete itself by calling on of the services provided by the kernel. In other words, the task starts, performs its function, and terminates. There would typically not be too many of these tasks in an embedded system because of the generally high overhead associated with "creating" and "deleting" tasks at runtime.

```
void  MyRunToCompletionTask (void)
{
    /* Local variables                                             */


    /* Task initialization                                         */
    /* Task body ... do work!                                      */
    /* Task calls a service provided by the kernel to 'terminate self'  */
}
```

Listing 1 Run-to-completion task.

```
void  MyInfiniteLoopTask (void)
{
    /* Local variables                                             */


    /* Task initialization                                         */
    while (1) {            /* Task body, as an infinite loop.      */
        :
        /* Task body ... do work!                                  */
        :
        /* Must call a service to 'wait for an event'              */
        :
        /* Task body ... do work!                                  */
        :
    }
}
```

Listing 2 Infinite loop task.

The body of the task can invoke other services provided by the kernel. Specifically, a task can create another task, suspend and resume other tasks, send signals or messages to other tasks, share resources with other tasks, and more. In other words, tasks are not limited to only make "wait for an event" function calls.

You can either call C or assembly language functions from a task. In fact, it is possible to call the same C function from different tasks if the functions are reentrant. A *reentrant* function is a function that does not use static or otherwise global variables unless they are protected (kernels provide mechanisms for this) from multiple access. If shared C functions only use local variables, they are generally reentrant (assuming that the compiler generates reentrant code). An example of a nonreentrant function is the popular strtok(), provided by most C compilers as part of the standard library. This function is used to parse an ASCII string for "tokens." The first time you call this function, you specify the ASCII string to parse and a list of token delimiters. As soon as the function finds the first token, it returns. The function "remembers" where it was last, so when it is called again it can extract additional tokens. Two tasks cannot use strtok() at the same time because strtok() can only remember one string position. Thus strtok() is nonreentrant.

The use of an infinite loop is more common in embedded systems because of the repetitive work needed in such systems (reading inputs, updating displays, performing control operations, etc.). This is one aspect that makes a task different than a regular C function. Note that you could use a "while (1)" or "for (;;)" to implement the infinite loop, since both behave the in the same manner—it is simply a matter of personal preference. The infinite loop must call a service provided by the kernel (i.e., function) that will cause the task to wait for an event to occur. It is important that each task wait for an event to occur, otherwise the task would be a true infinite loop and there would be no easy way for other, lower priority tasks to execute.

The event the task is waiting for may simply be the passage of time. Kernels provide "sleep" or "time delay" services. For example, a design may need to scan a keyboard every 50 ms as shown in the pseudocode of Listing 3. In this case, you would simply delay the task for 100 ms then see if a key was pressed on the keyboard and, possibly perform some action based on which key was pressed. Typically, however, a keyboard-scanning task should just buffer an "identifier" unique to the key pressed and use another task to decide what to do as a result of the key(s) pressed.

Similarly, the event the task is waiting for could be the arrival of a packet from an Ethernet controller. The task will have nothing to do until the packet is received. Once the packet is received, the task processes the contents of the packet and possibly moves the packet

```
void  KeyboardScanningTask (void)
{
    Setup the I/O devices needed for the keyboard scanning;
    while (1) {
        Call kernel to delay task 50 ms;          /* Suspend task execution for 50 ms */
        if (a key was pressed) {
            Determine which key it was;
            Place the scan-code of the key into a buffer;
        }
    }
}
```

Listing 3 Scanning a keyboard.

along a network stack. Kernels provide signaling and message-passing mechanisms.

It is important to note that when a task waits for an event, it does not consume any CPU time.

4   Assigning Task Priorities

Sometimes determining the priority of a task is both obvious and intuitive. For example, if the most important aspect of the embedded system is to perform some type of control, and it is known that the control algorithm must be responsive, then it is best to assign the control task a high priority while display and operator interface tasks are assigned low priority. However, most of the time, assigning task priorities is not so cut and dry because of the complex nature of real-time systems. In most systems, not all of the tasks are considered critical—noncritical tasks should obviously be given a low priority.

An interesting technique called rate monotonic scheduling (RMS) assigns task priorities based on how often tasks execute. Simply put, tasks with the highest rate of execution are given the highest priority. However, RMS makes several assumptions, including that:

- All tasks are periodic (they occur at regular intervals).
- Tasks do not synchronize with one another, share resources, or exchange data.
- The CPU must always execute the highest priority task that is ready-to-run. In other words, preemptive scheduling must be used.

Given a set of **n** tasks that are assigned RMS priorities, the basic RMS theorem states that all tasks that have hard real-time deadlines are always met if the following inequality holds true:

$$\sum_i \frac{E_i}{T_i} \le n\left(2^{\frac{1}{n}} - 1\right)$$

Table 1 Allowable CPU Usage Based on Number of Tasks

Number of Tasks	$n(2^{1/n} - 1)$
1	1.000
2	0.828
3	0.779
4	0.756
5	0.743
:	:
Infinite	0.693

where E_i corresponds to the maximum execution time of task i, and T_i corresponds to the execution period of task i. In other words, E_i/T_i corresponds to the fraction of CPU time required to execute task i.

Table 1 shows the value for size $n(2^{1/n} - 1)$, based on the number of tasks. The upper bound for an infinite number of tasks is given by $\ln(2)$, or 0.693, which means that you meet all hard real-time deadlines based on RMS, the CPU usage of all time-critical tasks should be less than 70%!

Note that you can still have nontime-critical tasks in a system and thus use close to 100% of the CPU's time. However, using 100% of your CPU's time is not a desirable goal as it does not allow for code changes and added features. As a rule of thumb, you should always design a system to use less than 60%–70% of the CPU.

RMS says that the highest rate task has the highest priority. In some cases, the highest rate task might not be the most important task. The application should dictate how to assign priorities. Also, RMS assumes that you know ahead of time the execution of your tasks, which might not be necessarily the case when you start your design. However, RMS is an interesting starting point.

5   Determining the Size of a Stack

The size of the stack required by the task is application specific. When sizing the stack, however, you must account for the nesting of all the functions called by the task, the number of local variables to be allocated by all functions called by the task, and the stack requirements for all nested ISRs (if the ISR uses the task's stack). In addition, the stack must be able to store all CPU registers and possibly floating-point unit (FPU) registers if the processor has an FPU. In addition, as a rule in embedded systems, avoid writing recursive code.

It is possible to manually figure out the stack space needed by adding all the memory required by all function call nesting (one pointer each function call for the return address), plus all the memory required by all the arguments passed in those function calls, plus storage for a full CPU context (depends on the CPU), plus another full CPU context for each nested ISR (if the CPU doesn't have a separate stack to handle ISRs), plus whatever stack space is needed by those ISRs. Adding all this up is a tedious chore and the resulting number is a minimum requirement. Most likely you would not make the stack size that precise in order to account for "surprises." The number arrived at should probably be multiplied by some safety factor, possibly 1.5 to 2.0. This calculation assumes that the exact path of the code is known at all times, which is not always possible. Specifically, when calling a function, such as `printf()` or some other library function, it might be difficult or nearly impossible to even guess just how much stack space `printf()` will require. In this case, start with a fairly large stack space and monitor the stack usage at runtime to see just how much stack space is actually used after the application runs for a while.

There are really cool and clever compilers/linkers that provide this information in a link map. For each function, the link map indicates the worst-case stack usage. This feature clearly enables you to better evaluate stack usage for each task. It is still necessary to add the stack space for a full CPU context plus another full CPU context for each nested ISR (if the CPU does not have a separate stack to handle ISRs) plus whatever stack space is needed by those ISRs. Again, allow for a safety net and multiply this value by some factor.

Always monitor stack usage at runtime while developing and testing the product as stack overflows occur often and can lead to some curious behavior. In fact, whenever someone mentions that his or her application behaves "strangely," insufficient stack size is the first thing that comes to mind.

A task can be in any one of five states as shown in Fig. 4.

(1) The *dormant* state corresponds to a task that resides in memory but has not been made available to the kernel. A task is made available to the kernel by calling a function to create the task. The task code actually resides in code space but the kernel needs to be informed about it.

(2) When the task is no longer needed, your code can call the kernel's task delete function. The code is not actually deleted, it is simply not eligible to access the CPU.

(3) A task is in the *ready* state when it is ready-to-run. There can be any number of tasks ready and the kernel keeps track of all ready tasks in a ready list (discussed later). This list is sorted by priority.

(4) The most important ready-to-run task is placed in the *running* state. On a single CPU, only one task can be running at any given time.

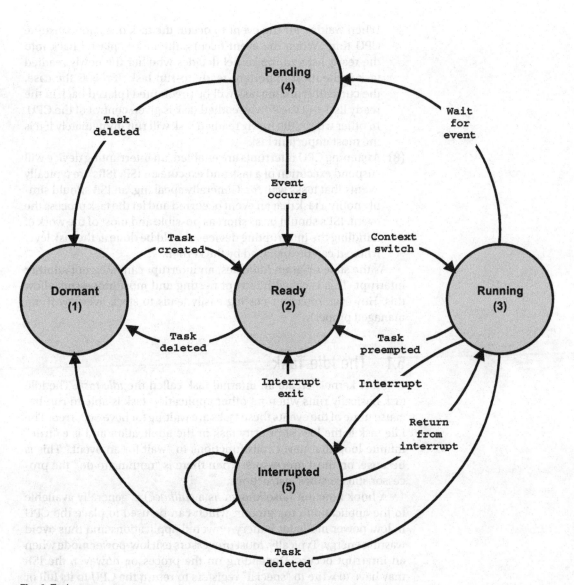

Fig. 4 Task states.

(5) The task selected to run on the CPU is *switched in* by the kernel when the it determines that it's the highest priority task that is ready-to-run.

(6) As previously discussed, tasks must wait for an event to occur. A task waits for an event by calling one of the functions that brings the task to the pending state if the event has not occurred.

(7) Tasks in the *pending* state are placed in a special list called a *pend list* (or wait list) associated with the event the task is waiting for.

When waiting for the event to occur, the task does not consume CPU time. When the event occurs, the task is placed back into the ready list and the kernel decides whether the newly readied task is the most important ready-to-run task. If this is the case, the currently running task will be preempted (placed back in the ready list) and the newly readied task is given control of the CPU. In other words, the newly readied task will run immediately if it is the most important task.

(8) Assuming CPU interrupts are enabled, an interrupting device will suspend execution of a task and execute an ISR. ISRs are typically events that tasks wait for. Generally speaking, an ISR should simply notify a task that an event occurred and let the task process the event. ISRs should be as short as possible and most of the work of handling the interrupting devices should be done at the task level where it can be managed by the kernel.

As the state diagram indicates, an interrupt can interrupt another interrupt. This is called interrupt nesting and most processors allow this. However, interrupt nesting easily leads to stack overflow if not managed properly.

5.1   The Idle Task

Most kernels create an internal task called the *idle task*. The idle task basically runs when no other application task is able to run because none of the events these tasks are waiting for have occurred. The idle task is the lowest priority task in the application and is a "true" infinite loop that never calls functions to "wait for an event." This is because, on most processors, when there is "nothing to do," the processor still executes instructions.

A hook function (also known as a *callback*) is generally available to the application programmer which can be used to place the CPU in low-power mode for battery-powered applications and thus avoid wasting energy. Typically, most processors exit low-power mode when an interrupt occurs. Depending on the processor, however, the ISR may have to write to "special" registers to return the CPU to its full or desired speed. If the ISR wakes up a high-priority task (every task is higher in priority than the idle task) then the ISR will not immediately return to the interrupted idle task, but instead switch to the higher priority task.

5.2   Priority Levels

All kernels allow you to assign priorities to tasks based on their importance in your application. Typically, a low-priority number means a high priority. In other words, "priority 1" is more important than "priority 10." The number of different priority levels greatly depends on the implementation of the kernel. It's not uncommon to have up to 256 different priority levels and thus the kernel can use an 8-bit variable to represent the priority of a task.

On most kernels, an application can have multiple tasks assigned to the same priority. When this priority becomes the highest priority, the kernel generally executes each task at that priority in a *round-robin* fashion. In other words, each task gets to execute for up to a configurable amount of time.

5.3 The Ready List

Tasks that are ready-to-run are placed in the *ready list*. The ready list is ordered by priority. The highest priority task is at the beginning of the list and the lower priority tasks are placed at the end of the list. There are techniques that allow inserting and removing tasks from the ready list. However, this is beyond the scope of this chapter.

6 Preemptive Scheduling

The *scheduler*, also called the *dispatcher*, is a part of the kernel responsible for determining which task runs next. Most kernels are implemented using a *preemptive* scheme. The word preemptive means that when an event occurs, and that event makes a more important task ready-to-run, then the kernel will immediately give control of the CPU to that task. Thus when a task signals or sends a message to a higher priority task, the current task is suspended and the higher priority task is given control of the CPU. Similarly, if an ISR signals or sends a message to a higher priority task, when the message has been sent, the interrupted task remains suspended, and the new higher priority task resumes. Preemptive scheduling is illustrated in Fig. 5.

(1) A low-priority task is executing, and an interrupt occurs.
(2) If interrupts are enabled, the CPU vectors (i.e., jumps) to the ISR that is responsible for servicing the interrupting device.
(3) The ISR services the device and signals or sends a message to a higher priority task waiting to service this device. This task is thus ready-to-run.
(4) When the ISR completes its work it makes a service call to the kernel.
(5) ☐
(6) Since there is a more important ready-to-run task, the kernel decides to not return to the interrupted task but switches to the more important task. This is called a *context switch*.

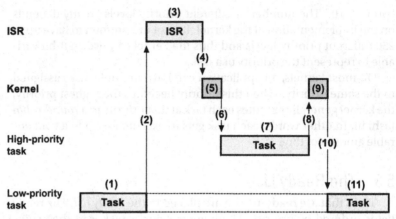

Fig. 5 Preemptive scheduling.

(7) □
(8) The higher priority task services the interrupting device and, when finished, calls the kernel asking it to wait for another interrupt from the device.
(9) The kernel blocks the high-priority task until the next time the device needs servicing. Since the device has not interrupted a second time, the kernel switches back to the original task (the one that was interrupted).
(10) □
(11) The interrupted task resumes execution, exactly at the point where it was interrupted.

7 Scheduling Points

Scheduling occurs at *scheduling points* and nothing special must be done in the application code since scheduling occurs automatically based on some conditions described below. This is a partial list for brevity.

A task signals or sends a message to another task.
This occurs when the task signals or sends a message to another task.

A task "sleeps" for a certain amount of time.
Scheduling always occurs since the calling task is placed in a list waiting for time to expire. Scheduling occurs as soon as the task is inserted in the wait list and this call will always result in a context switch to the next task that is ready-to-run at the same or lower priority than the task that is placed to sleep.

A task waits for an event to occur and the event has not yet occurred.
The task is placed in the wait list for the event and, if a nonzero timeout is specified, the task is also inserted in the list of tasks waiting to timeout. The scheduler is then called to select the next most important task to run.

If a task is created.
The newly created task may have a higher priority than the task's creator. In this case, the scheduler is called.

If a task is deleted.
When terminating a task, the scheduler is called if the current task is deleted.

A task changes the priority of itself or another task.
The scheduler is called when a task changes the priority of another task (or itself) and the new priority of that task is higher than the task that changed the priority.

At the end of all nested ISRs.
The scheduler is called at the end of all nested ISRs to determine whether a more important task is made ready-to-run by one of the ISRs.

A task gives up its time quanta by voluntarily relinquishing the CPU through a kernel call.
This assumes that the task is running alongside other tasks at the same priority and the currently running task decides that it can give up its time quanta and let another task run.

8   Round-Robin Scheduling

When two or more tasks have the same priority, most kernels allow one task to run for a predetermined amount of time (called a *time quanta*) before selecting another task. This process is called *round-robin scheduling* or *time slicing*. If a task does not need to use its full time quanta it can voluntarily give up the CPU so that the next task can execute. This is called *yielding*.

9   Context Switching

When the kernel decides to run a different task, it saves the current task's context, which typically consists of the CPU registers, onto the current task's stack and restores the context of the new task and resumes execution of that task. This process is called a *context switch*.

Context switching adds overhead and, the more registers a CPU has, the higher the overhead. The time required to perform a context switch is generally determined by how many registers must be saved and restored by the CPU.

The context switch code is generally part of a processor's *port* which adapts the kernel (typically written in C or other higher level languages) to the processor architecture. The latter is typically written in assembly language.

Here, we will discuss the context-switching process in generic terms using a fictitious CPU, as shown in Fig. 6. Our fictitious CPU contains 16 integer registers (R0 to R15), a separate ISR stack pointer, and a separate status register (SR). Every register is 32 bits wide and each of the 16 integer registers can hold either data or an address. The program counter (or instruction pointer) is R15 and there are two separate stack pointers labeled R14 and R14′. R14 represents a task stack pointer (TSP) and R14′ represents an ISR stack pointer (ISP). The CPU automatically

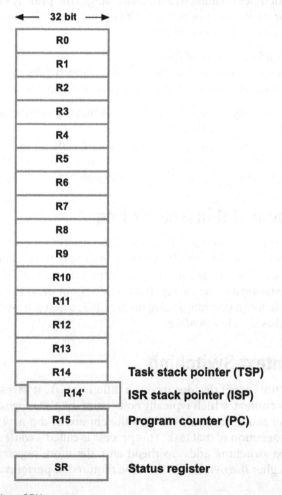

Fig. 6 Fictitious CPU.

switches to the ISR stack when servicing an exception or interrupt. The task stack is accessible from an ISR (i.e., we can push and pop elements onto the task stack when in an ISR), and the interrupt stack is also accessible from a task.

The task initialization code (i.e., the task create function) for a kernel generally sets up the stack frame for a ready task to look as if an interrupt has just occurred and all processor registers were saved onto it. Tasks enter the ready state upon creation and thus their stack frames are preinitialized by software in a similar manner. Using our fictitious CPU, we will assume that a stack frame for a task that is ready to be restored is shown in Fig. 7.

The task stack pointer points to the last register saved onto the task's stack. The program counter (PC or R15) and status register (SR) are the first registers saved onto the stack. In fact, these are saved automatically by the CPU when an exception or interrupt occurs (assuming interrupts are enabled) while the other registers are pushed onto the stack by software in the exception handler. The stack pointer (SP or R14) is not actually saved on the stack but instead is saved in the task's control block (TCB).

The interrupt stack pointer points to the current top-of-stack for the interrupt stack, which is a different memory area. When an ISR executes, the processor uses R14' as the stack pointer for function calls and local arguments.

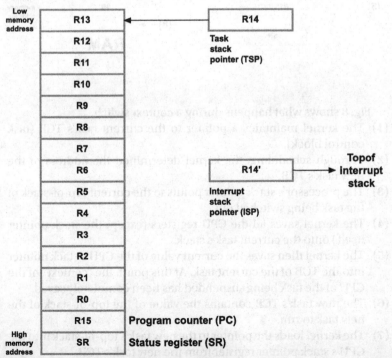

Fig. 7 CPU register stacking order for a ready task.

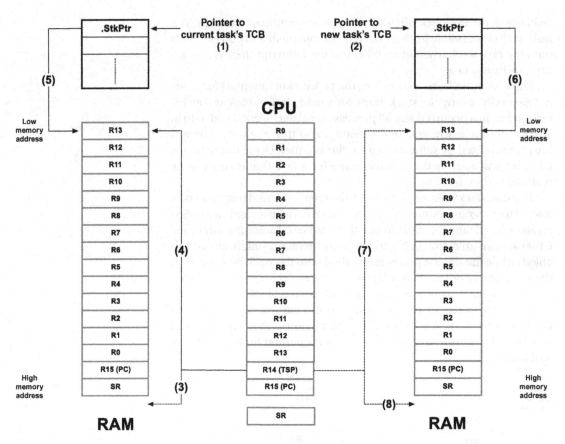

Fig. 8 Context switch.

Fig. 8 shows what happens during a context switch.

(1) The kernel maintains a pointer to the current task's TCB (task control block).

(2) Through scheduling, the kernel determined the address of the new task's TCB.

(3) The processor's stack pointer points to the current top-of-stack of the task being switched out.

(4) The kernel saves all the CPU registers (except the stack pointer itself) onto the current task's stack.

(5) The kernel then saves the current value of the CPU's stack pointer into the TCB of the current task. At this point, the "context" of the CPU of the task being suspended has been completely saved.

(6) The new task's TCB contains the value of the top-of-stack of the new task to run.

(7) The kernel loads the pointer to the new task's top-of-stack into the CPU's stack pointer register from the new task's TCB.

(8) Finally, the CPU registers are loaded from the stack frame of the new task and, once the PC is loaded into the CPU, the CPU executes the code of the new task.

The execution time of the above process greatly depends on the number of CPU registers to save and restore and, in fact, should be about the same from one kernel to another. Also, a context switch is normally performed with interrupts disabled so that the whole process is treated atomically.

10 Interrupt Management

An *interrupt* is a hardware mechanism used to inform the CPU that an asynchronous event occurred. When an interrupt is recognized, the CPU saves part (or all) of its context (i.e., registers) and jumps to a special subroutine called an ISR. The ISR processes the event, and—upon completion of the ISR—the program either returns to the interrupted task, or the highest priority task, if the ISR made a higher priority task ready-to-run.

Interrupts allow a microprocessor to process events when they occur (i.e., asynchronously), which prevents the microprocessor from continuously *polling* (looking at) an event to see if it occurred. Task-level response to events is typically better using interrupt mode as opposed to polling mode. Microprocessors allow interrupts to be ignored or recognized through the use of two special instructions: disable interrupts and enable interrupts, respectively.

In a real-time environment, interrupts should be disabled as little as possible. Disabling interrupts affects interrupt latency, possibly causing interrupts to be missed.

Processors generally allow interrupts to be nested, which means that while servicing an interrupt, the processor recognizes and services other (more important) interrupts.

All real-time systems disable interrupts to manipulate critical sections of code and reenable interrupts when critical sections are completed. The longer interrupts are disabled, the higher the interrupt latency.

Interrupt response is defined as the time between the reception of the interrupt and the start of the user code that handles the interrupt. Interrupt response time accounts for the entire overhead involved in handling an interrupt. Typically, the processor's context (CPU registers) is saved on the stack before the user code is executed.

Interrupt recovery is defined as the time required for the processor to return to the interrupted code or to a higher priority task if the ISR made such a task ready-to-run.

Task latency is defined as the time it takes from the time the interrupt occurs to the time task-level code resumes.

10.1 Handling CPU Interrupts

There are many popular CPU architectures on the market today, and most processors typically handle interrupts from a multitude of sources. For example, a UART receives a character, an Ethernet controller receives a packet, a DMA controller completes a data transfer, an analog-to-digital converter (ADC) completes an analog conversion, a timer expires, etc.

In most cases, an *interrupt controller* captures all of the different interrupts presented to the processor, as shown in Fig. 9 (note that the "CPU interrupt enable/disable" is typically part of the CPU, but is shown here separately for sake of the illustration).

Interrupting devices signal the interrupt controller, which then prioritizes the interrupts and presents the highest priority interrupt to the CPU.

Modern interrupt controllers have built-in intelligence enabling the user to prioritize interrupts, remember which interrupts are still pending and, in many cases, have the interrupt controller provide the address of the ISR (also called the vector address) directly to the CPU.

If "global" interrupts (i.e., the switch in Fig. 9) are disabled, then the CPU will ignore requests from the interrupt controller. However, interrupts will be held pending by the interrupt controller until the CPU reenables interrupts.

CPUs deal with interrupts using one of two models:

1) All interrupts vector to a single interrupt handler.
2) Each interrupt vectors directly to an interrupt handler.

Before discussing these two methods, it is important to understand how a kernel handles CPU interrupts.

In most cases, ISRs are written in assembly language. However, if a C compiler supports in-line assembly language, the ISR code can be placed directly into a C source file. The pseudocode for a typical ISR when using a kernel is shown in Listing 4.

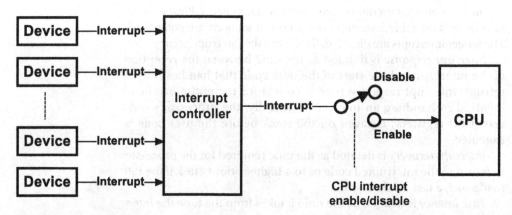

Fig. 9 Interrupt controllers.

```
MyKernelAwareISR:                                                   (1)
    ; ISR Prologue
    Disable all interrupts;                                         (2)
    Save the CPU registers;                                        (3)
    Increment ISR nesting counter;                                  (4)
    Save the CPU's stack pointer register value into the TCB of the current task; (5)

    Clear interrupting device;                                      (6)
    Re-enable interrupts (optional);                                (7)
    Call user ISR;                                                  (8)

    ; ISR Epilogue
    Notify the kernel that the ISR has completed;                   (9)
    Restore the CPU registers;                                     (10)
    Return from interrupt;                                          (11)
```

Listing 4 Kernel-aware interrupt service routine.

(1) As mentioned earlier, an ISR is typically written in assembly language. `MyKernelAwareISR()` corresponds to the name of the handler that will handle the interrupting device.

(2) It is important that all interrupts are disabled before going any further. Some processors have interrupts disabled whenever an interrupt handler starts. Others require the user to explicitly disable interrupts as shown here. This step may be tricky if a processor supports different interrupt priority levels. However, there is always a way to solve the problem.

(3) The first thing the interrupt handler must do is save the context of the CPU onto the interrupted task's stack. On some processors, this occurs automatically. However, on most processors it is important to know how to save the CPU registers onto the task's stack. You should save the full "context" of the CPU, which may also include floating-point unit (FPU) registers if the CPU used is equipped with an FPU. However, it's possible that some tasks may not do any floating-point calculations and it would be a waste of CPU cycles to save the FPU registers. Luckily, you can tell some kernels (through task create options) that a task will not require floating-point capabilities.

Certain CPUs also automatically switch to a special stack just to process interrupts (i.e., an interrupt stack). This is generally beneficial as it avoids using valuable task stack space. However, for most kernels, the context of the interrupted task needs to be saved onto that task's stack.

If the processor does not have a dedicated stack pointer to handle ISRs then it is possible to implement one in the software. Specifically, upon entering the ISR, you would simply save the current task stack, switch to a dedicated ISR stack, and when done with the ISR, switch back to the task stack. Of course, this means that there would be additional code to write, however, the benefits are enormous since it is not necessary to allocate extra space on the task stacks to accommodate for worst-case interrupt stack usage including interrupt nesting.

(4) Next, the ISR would increment a nesting counter to keep track of interrupt nesting. This is done because upon completing the ISR, the kernel needs to know whether it will return to a task or a previous ISR.

(5) If this is the first nested interrupt, you need to save the current value of the stack pointer of the interrupted task into its TCB.

The previous four steps are called the *ISR prolog*.

(6) At this point, the ISR needs to clear the interrupting device so that it does not generate the same interrupt. However, most people defer the clearing of the interrupting device within the user ISR handler which can be written in "C."

(7) If the interrupting source has been cleared, it is safe to reenable interrupts if you want to support nested interrupts. This step is optional.

(8) At this point, further processing can be deferred to a C function called from assembly language. This is especially useful if there is a large amount of processing to do in the ISR handler. However, as a rule, keep the ISRs as short as possible. In fact, it is best to simply signal or send a message to a task and let the task handle the details of servicing the interrupting device.

The ISR must call a kernel function to signal or send a message to a task that is waiting for this event. In other words, most likely you would have designed your task to wait for ISRs to notify them. If the ISR does not need to signal or send a message to a task then you might consider writing the ISR as a "nonkernel-aware interrupt service routine," as described in the next section.

(9) When the ISR completes, the kernel is notified once more. The kernel simply decrements the nesting counter and if all interrupts have nested (i.e., the counter reaches 0) then the kernel will need to determine whether the task that was signaled or sent a message is now the most important task because it has a higher priority than the interrupted task, or not.

If the task that was waiting for this signal or message has a higher priority than the interrupted task then the kernel will context switch to this higher priority task instead of returning to the interrupted task. In this latter case, the kernel doesn't return from the ISR but takes a different path.

(10) If the ISR signaled or sent a message to a lower priority task than the interrupted task, then the kernel code returns to the ISR and the ISR restores the previously saved registers.

(11) Finally, the ISR performs a return from interrupts to resume the interrupted task.

These last three steps are called the *ISR epilog*.

10.2   NonKernel-Aware Interrupt Service Routine (ISR)

The above sequence assumes that the ISR signals or sends a message to a task. However, in many cases, the ISR may not need to notify a task and can simply perform all its work within the ISR (assuming it can be done quickly). In this case, the ISR will appear as shown in Listing 5.

```
MyNonKernelAwareISR:                                     (1)
    Save enough registers as needed by the ISR;          (2)
    Clear interrupting device;                           (3)
    DO NOT re-enable interrupts;                         (4)
    Call user ISR;                                       (5)
    Restore the saved CPU registers;                     (6)
    Return from interrupt;                               (7)
```

Listing 5 Nonkernel-aware interrupt service routine.

(1) As mentioned above, an ISR is typically written in assembly language. `MyNonKernelAwareISR()` corresponds to the name of the handler that will handle the interrupting device.
(2) Here, you save sufficient registers required to handle the ISR.
(3) The user probably needs to clear the interrupting device to prevent it from generating the same interrupt once the ISR returns.
(4) You should not reenable interrupts at this point since another interrupt could be kernel aware thus forcing a context switch to a higher priority task. This means that the above ISR would complete, but at a much later time.
(5) Now you can take care of the interrupting device in assembly language or call a C function, if necessary.
(6) Once finished, simply restore the saved CPU registers.
(7) The ISR completes by performing a return from interrupt to resume the interrupted task.

10.3   Processors with Multiple Interrupt Priorities

There are some processors that actually support multiple interrupt levels, as shown in Fig. 10.
(1) Here, we are assuming that the processor supports 16 different interrupt priority levels. Priority 0 is the lowest priority while 15 is the highest. As shown, interrupts are always higher in priority than tasks (assuming interrupts are enabled).
(2) The designer of the product decided that interrupt levels 0 through 12 will be "kernel aware" and thus will be able to notify tasks that

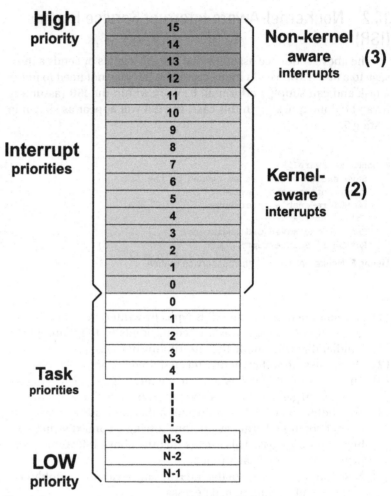

Fig. 10 Processors supporting multiple interrupt priorities.

are assigned to service these interrupts. It's important to note that disabling interrupts (when entering critical sections) for task-aware interrupts means raising the interrupt mask to level 12. In other words, interrupt levels 0 through 11 would be disabled but, levels 12 and above would be allowed.

(3) Interrupt levels 12 through 15 are "nonkernel aware" and thus are not allowed to make any kernel calls and are thus implemented as shown in Listing 5. It is important to note that since the kernel cannot disable these interrupts, interrupt latency for these interrupts is very short.

Listing 6 shows how to implement nonkernel-aware ISRs when the processor supports multiple interrupt priorities.

```
MyNonKernelAwareISR:
    Save enough registers as needed by the ISR;
    Clear interrupting device;
    Call user ISR;
    Restore the saved CPU registers;
    Return from interrupt;
```

Listing 6 Nonkernel-aware ISR with processors supporting multiple interrupt priorities.

10.4   All Interrupts Vector to a Common Location

Even though an interrupt controller is present in most designs, some CPUs still vector to a common interrupt handler, and the ISR needs to query the interrupt controller to determine the source of the interrupt. At first glance, this might seem silly since most interrupt controllers are able to force the CPU to jump directly to the proper interrupt handler. It turns out, however, that for some kernels, it is easier to have the interrupt controller vector to a single ISR handler than to vector to a unique ISR handler for each source. Listing 7 describes the sequence of events to be performed when the interrupt controller forces the CPU to vector to a single location.

```
An interrupt occurs;                                        (1)
The CPU vectors to a common location;                       (2)
ISR Prologue;                                               (3)

The C handler performs the following:                       (4)
    while (there are still interrupts to process) {         (5)
        Get vector address from interrupt controller;
        Call interrupt handler;
    }

ISR Epilogue;                                               (6)
```
Listing 7 Single interrupt vector for all interrupts.

(1) An interrupt occurs from any device. The interrupt controller activates the interrupt pin on the CPU. If there are other interrupts that occur after the first one, the interrupt controller will latch them and properly prioritize the interrupts.

(2) The CPU vectors to a single interrupt handler address. In other words, all interrupts are to be handled by this one interrupt handler.

(3) The ISR executes the "ISR prologue" (see Listing 4) code needed by the kernel.

(4) The ISR calls a special handler which is typically written in C. This handler will continue processing the ISR. This makes the code easier to write (and read). Notice that interrupts are not reenabled at this point.

(5) The kernel handler then interrogates the interrupt controller and asks: "Who caused the interrupt?" The interrupt controller will either respond with a number (0 to $N-1$) or with the address of the interrupt handler of the highest priority interrupting device. Of course, the handler will know how to handle the specific interrupt controller since the C handler is written specifically for that controller.

If the interrupt controller provides a number between 0 and $N-1$, the C handler simply uses this number as an index into a table (in ROM or RAM) containing the address of the ISR associated with the interrupting device. A RAM table is handy to change interrupt handlers at runtime. For many embedded systems, however, the table may also reside in ROM.

If the interrupt controller responds with the address of the ISR, the C handler only needs to call this function.

In both the cases above, the ISRs for all the interrupting devices need to be declared as follows:

```
void MyISRHandler (void);
```

There is one such handler for each possible interrupt source (obviously, each having a unique name).

The "while" loop terminates when there are no other interrupting devices to service.

(6) Finally, the ISR executes the "ISR epilogue" (see Listing 4) code. A couple of interesting points to note:

- If another device caused an interrupt before the C handler had a chance to query the interrupt controller, most likely the interrupt controller will capture that interrupt. In fact, if that second device happens to be a higher priority interrupting device, it will most likely be serviced first, as the interrupt controller will prioritize the interrupts.
- The loop will not terminate until all pending interrupts are serviced. This is like allowing nested interrupts, but better, since it is not necessary to redo the ISR prologue and epilogue.

The disadvantage of this method is that a high-priority interrupt that occurs after the servicing of another interrupt that has already started must wait for that interrupt to complete before it will be serviced. So, the latency of any interrupt, regardless of priority, can be as long as it takes to process the longest interrupt.

10.5 Every Interrupt Vectors to a Unique Location

If the interrupt controller vectors directly to the appropriate interrupt handler, each of the ISRs must be written in assembly language as described in Section 10.1 and as shown in Listing 4. This, of course, slightly complicates the design. However, you can copy and paste the majority of the code from one handler to the other and just change what is specific to the actual device.

If the interrupt controller allows the user to query it for the source of the interrupt, it may be possible to simulate the mode in which all interrupts vector to the same location by simply setting all vectors to point to the same location. Most interrupt controllers that vector to a unique location, however, do not allow users to query it for the source of the interrupt since, by definition, having a unique vector for all interrupting devices should not be necessary.

11   The Clock Tick (or System Tick)

Kernel-based systems generally require the presence of a periodic time source called the clock tick or system tick.

A hardware timer configured to generate an interrupt at a rate of between 10 and 1000 Hz provides the clock tick. A tick source may also be obtained by generating an interrupt from an AC power line (typically 50 or 60 Hz). In fact, you can easily derive 100 or 120 Hz by detecting zero crossings of the power line. That said, if your product is subject to use in regions that use both power line frequencies then you may need to have the user specify which frequency to use or, have the product automatically detect which region it's in.

The clock tick interrupt can be viewed as the system's heartbeat. The rate is application specific and depends on the desired resolution of this time source. However, the faster the tick rate, the higher the overhead imposed on the system.

The clock tick interrupt allows the kernel to delay (also called sleep) tasks for an integral number of clock ticks and provide timeouts when tasks are waiting for events to occur.

A common misconception is that a system tick is always needed with a kernel. In fact, many low-power applications may not implement the system tick because of the power required to maintain the tick list. In other words, it is not reasonable to continuously power down and power up the product just to maintain the system tick. Since most kernels are preemptive, an event other than a tick interrupt can wake up a system placed in low-power mode by either a keystroke from a keypad or other means. Not having a system tick means that the user is not allowed to use time delays and timeouts on system calls. This decision needs to be made by the designer of the low-power product.

11.1   Wait Lists

A task is placed in a *wait list* (also called a *pend list*) when it is waiting on a *kernel object*. A kernel object is generally a data structure that provides an abstraction of a concept, such as a semaphore, mailbox, message queue, or other. Tasks will generally be waiting on these objects to be *signaled* or *posted* by other tasks or ISRs.

A wait list is similar to the *ready list*, except that instead of keeping track of tasks that are ready-to-run, the wait list keeps track of tasks waiting for an object to be signaled or posted. In addition, the wait list is sorted by priority; the highest priority task waiting on the object is placed at the head of the list, and the lowest priority task waiting on the object is placed at the end of the list. A kernel object, along with tasks waiting for this object to be signaled or posted to, is show in Fig. 11. We will be looking at different types of kernel objects in upcoming sections.

11.2 Time Management

Kernels typically provide time-related services to the application programmer.

As previously discussed, kernels require that the user provide a periodic interrupt to keep track of time delays and timeouts. This periodic time source is called a clock tick and should occur between 10 and 1000 times per second (Hertz). The actual frequency of the clock tick depends on the desired tick resolution of the application.

A kernel provides a number of services to manage time: delay (or sleep) for "N" ticks, delay for a user-specified amount of time in seconds and milliseconds, get the current tick count, set the current tick count, and more. Example kernel APIs for these functions could be:

```
OSTimeDly() or OSTaskSleep()
OSTimeDlySecMilli() or OSTaskSleepSecMilli()
OSTimeGet() or OSTickCntGet()
OSTimeSet() or OSTickCntSet()
```

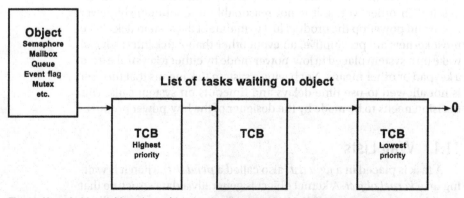

Fig. 11 Kernel object with tasks waiting for the object to be signaled or posted to.

A task can call `OSTimeDly()` to suspend execution until some amount of time expires. The calling function will not execute until the specified time expires. Listing 8 shows a typical use of this function.

```
void MyTask (void)
{
    :
    :
    while (1) {
        :
        :
        OSTimeDly(2);          /* Suspend execution of this task until 2 tick passes by */
        :
        :
    }
}
```

Listing 8 Delaying (i.e., sleeping) a task for some period of time.

The actual delay is not exact, as shown in Fig. 12.

(1) We get a tick interrupt and the kernel services the ISR.

(2) At the end of the ISR, all higher priority tasks (HPTs) execute. The execution time of HPTs is unknown and can vary.

(3) Once all HPTs have executed, the kernel runs the task that has called `OSTimeDly()`, as shown in Listing 8. For the sake of discussion, it is assumed that this task is a lower priority task (LPT).

(4) The task calls `OSTimeDly()` and specifies to delay for two ticks. At this point, the kernel places the current task in the tick list where it will wait for two ticks to expire. The delayed task consumes zero CPU time while waiting for the time to expire.

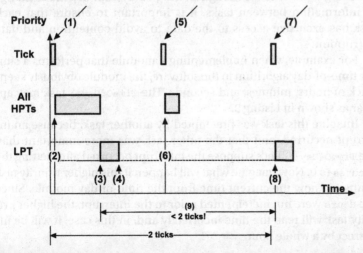

Fig. 12 Time delays are not exact.

(5) The next tick occurs. If there are HPTs waiting for this particular tick, the kernel will schedule them to run at the end of the ISR.

(6) The HPTs execute.

(7) The next tick interrupt occurs. This is the tick that the LPT was waiting for and will now be made ready-to-run by the kernel.

(8) Since there are no HPTs to execute on this tick, the kernel switches to the LPT.

Given the execution time of the HPTs, the time delay is not exactly two ticks, as requested. In fact, it is virtually impossible to obtain a delay of exactly the desired number of ticks. You might ask for a delay of two ticks, but the very next tick could occur almost immediately after calling `OSTimeDly()`! In fact, imagine what might happen if all HPTs took longer to execute and pushed (3) and (4) further to the right. In this case, the delay would actually appear as one tick instead of two.

12   Resource Management

In this section we will consider the services provided by kernels to manage shared resources. A shared resource is typically a variable (static or global), a data structure, table (in RAM), or registers in an I/O device.

When protecting a shared resource, it is preferable to use mutual exclusion semaphores, as will be described later. Other methods are also presented.

Tasks can easily share data when all tasks exist in a single address space and can reference global variables, pointers, buffers, linked lists, ring buffers, etc. Although sharing data simplifies the exchange of information between tasks, it is important to ensure that each task has exclusive access to the data to avoid contention and data corruption.

For example, when implementing a module that performs a simple time-of-day algorithm in the software, the module obviously keeps track of hours, minutes, and seconds. The `TimeOfDay()` task may appear as shown in Listing 9.

Imagine this task was preempted by another task, because an interrupt occurred, and that the other task was more important than the `TimeOfDay()`. Let's suppose the interrupt occurred after setting the `Minutes` to 0. Now imagine what will happen if this higher priority task wants to know the current time from the time-of-day module. Since the `Hours` were not incremented prior to the interrupt, the higher priority task will read the time incorrectly and, in this case, it will be incorrect by a whole hour.

```
int  Hours;
int  Minutes;
int  Seconds;


void  TimeOfDay (void)
{
    while (1) {
        OSTimeDlySecMilli(1, 0);              /* Suspend execution of task for 1 second */
        Seconds++;
        if (Seconds > 59) {
            Seconds = 0;
            Minutes++;
            if (Minutes > 59) {
                Minutes = 0;
                Hours++;
                if (Hours > 23) {
                    Hours = 0;
                }
            }
        }
    }
}
```

Listing 9 Time-of-day task.

The code that updates variables for the `TimeOfDay()` task must treat all of the variables indivisibly (or atomically) whenever there is possible preemption. Time-of-day variables are considered shared resources and any code that accesses those variables must have exclusive access through what is called a *critical section*. All kernels provide services to protect shared resources and enable the easy creation of critical sections.

The most common methods of obtaining exclusive access to shared resources and to create critical sections are:

- disabling interrupts.
- disabling the scheduler.
- using semaphores.
- using mutual exclusion semaphores (a.k.a. a mutex).

The mutual exclusion mechanism used depends on how fast the code will access a shared resource, as shown in Table 2.

12.1 Resource Management—Disable/Enable Interrupts

The easiest and fastest way to gain exclusive access to a shared resource is by disabling and enabling interrupts, as shown in the pseudocode in Listing 10.

```
Disable interrupts;
Access the resource;
Enable interrupts;
```
Listing 10 Disabling and enabling interrupts to access a shared resource.

Table 2 Mutual Exclusion Mechanisms

Resource Sharing Method	When should you use?
Disable/enable interrupts	When access to a shared resource is very quick (reading from or writing to few variables) and access is faster than the kernel's interrupt disable time. It is highly recommended to not use this method as it impacts interrupt latency
Semaphores	When all tasks that need to access a shared resource do not have deadlines. This is because semaphores may cause unbounded priority inversions (described later). However, semaphore services are slightly faster (in execution time) than mutual exclusion semaphores
Mutual exclusion semaphores	This is the preferred method for accessing shared resources, especially if the tasks that need to access a shared resource have deadlines µC/OS-III's mutual exclusion semaphores have a built-in priority inheritance mechanism, which avoids unbounded priority inversions However, mutual exclusion semaphore services are slightly slower (in execution time) than semaphores since the priority of the owner may need to be changed, which requires CPU processing

Most kernels use this technique to access certain internal variables and data structures, ensuring that these variables and data structures are manipulated atomically. Note that this is the only way that a task can share variables or data structures with an ISR. Although this method works, you should avoid disabling interrupts as it affects the responsiveness of the system to real-time events.

12.2 Resource Management—Semaphores

A semaphore was originally a mechanical signaling mechanism. The railroad industry used the device to provide a form of mutual exclusion for railroad tracks shared by more than one train. In this form, the semaphore signaled trains by closing a set of mechanical arms to block a train from a section of track that was currently in use. When the track became available, the arm would swing up and the waiting train would then proceed.

The notion of using a semaphore in software as a means of mutual exclusion was invented by the Dutch computer scientist Edgser

Dijkstra in 1959. In computer software, a semaphore is a protocol mechanism offered by most multitasking kernels. Semaphores were originally used to control access to shared resources but now are used for synchronization, as described later. However, it is useful to describe here how semaphores can be used to share resources. The pitfalls of semaphores will be discussed in a later section.

A semaphore was originally a "lock mechanism" where code acquired a key to the lock to continue execution. Acquiring a key means that the executing task has permission to enter the section of otherwise locked code. Entering a section of locked code causes the task to wait until a key becomes available.

Typically, two types of semaphores exist: binary semaphores and counting semaphores. As its name implies, a binary semaphore can only take two values: 0 or 1. A counting semaphore allows for values between 0 and 255, 65,535, or 4,294,967,295 depending on whether the semaphore mechanism is implemented using 8, 16, or 32 bits, respectively. Along with the semaphore's value, the kernel contains a list of tasks waiting for the semaphore's availability. Only tasks are allowed to use semaphores when semaphores are used for sharing resources; ISRs are not allowed.

Listing 11 shows how semaphores are typically used.

(1) A semaphore is a kernel object and an application can have any number of semaphores (limited only by the amount of RAM available). The semaphore object must be globally accessible to all tasks that will be sharing the resources guarded by the semaphore.

(2) A semaphore must be created before it can be used. Creating a semaphore is done by calling a function provided by the kernel. When you create a semaphore, you need to specify its maximum value, which represents the number of resources the semaphore is *guarding*. In other words, if you are protecting a single variable or data structure, you would create a semaphore with a count of 1. If you are protecting a pool of 100 identical buffers then you'd initialize the semaphore to 100. In the code of Listing 11, the semaphore is initialized to 1—this type of semaphore is typically called a *binary semaphore*.

Kernel objects are typically created prior to the start of multitasking.

(3) A task that wants to acquire a resource must perform a *wait* (or *pend*) operation. If the semaphore is available (the semaphore value is greater than 0) the semaphore value is decremented and the task continues execution (owning the resource). If the semaphore's value is 0 the task performing a wait on the semaphore is placed in a waiting list.

(4) A task releases a semaphore by performing a *release* (or *post*) operation. If no task is waiting for the semaphore, the semaphore

```
OS_SEM  MySem;                      (1)


void  main (void)
{
    :
   OSInit();
    :
    :
   OSSemCreate(&MySem, 1);          (2)
    :
    :
   OSStart();
}


void MyTask (void)
{
    while (1) {
            :
            :
       OSSemWait(&MySem);           (3)
       Access the resource;
       OSSemRelease(&MySem);        (4)
            :
            :
    }
}
```

Listing 11 Using a binary semaphore to access a shared resource.

value is simply incremented. If there is at least one task waiting for the semaphore, the highest priority task waiting on the semaphore is made ready-to-run, and the semaphore value is not incremented. If the readied task has a higher priority than the current task (the task releasing the semaphore), a context switch occurs and the higher priority task resumes execution. The current task is suspended until it again becomes the highest priority task that is ready-to-run.

The application must declare a semaphore as a variable of type OS_SEM. This variable will be referenced by other semaphore services.

You create a semaphore by calling OSSemCreate() and pass the address to the semaphore allocated in (1). The semaphore must be created before it can be used by other tasks. Here, the semaphore is initialized in startup code (i.e., main ()), however, it could also be initialized by a task (but it must be initialized before it is used).

You can assign an ASCII name to the semaphore, which can be used by debuggers or µC/Probe to easily identify the semaphore. Storage for the ASCII characters is usually in ROM, which is typically more plentiful than RAM. If it is necessary to change the name of the semaphore at runtime, you can store the characters in an array in RAM and simply pass the address of the array to OSSemCreate(). Of course, the array must be NUL terminated.

Semaphores are especially useful when tasks share I/O devices. Imagine what would happen if two tasks were allowed to send characters to a printer at the same time. The printer would contain interleaved data from each task. For instance, the printout from Task 1 printing "I am Task 1," and Task 2 printing "I am Task 2," could result in "I Ia amm T Tasask k1 2." In this case, you can use a semaphore and initialize it to 1. The rule is simple: to access the printer each task must first obtain the resource's semaphore. Fig. 13 shows tasks competing for a semaphore to gain exclusive access to the printer. Note that the key in this figure, indicating that each task must obtain a key to use the printer, represents the semaphore symbolically.

The above example implies that each task knows about the existence of the semaphore to access the resource. It is almost always better to encapsulate the critical section and its protection mechanism. Each task would therefore not know that it is acquiring a semaphore when accessing the resource.

A counting semaphore is used when elements of a resource can be used by more than one task at the same time. For example, a counting semaphore is used in the management of a buffer pool, as shown in Fig. 14.

Let's assume that the buffer pool initially contains 10 buffers. A task obtains a buffer from the buffer manager by calling BufReq(). When the buffer is no longer needed, the task returns the buffer to the buffer manager by calling BufRel(). The buffer manager satisfies the first 10 buffer requests because the semaphore was initialized to 10. When all buffers are used, a task requesting a buffer is suspended (placed in the semaphore wait list) until a buffer becomes available. When a task is done with the buffer it acquired, the task calls BufRel() to return the buffer to the buffer manager and the buffer is inserted into the

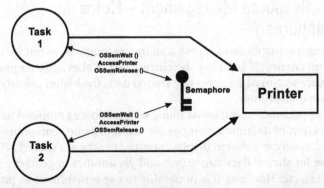

Fig. 13 Accessing a shared peripheral device.

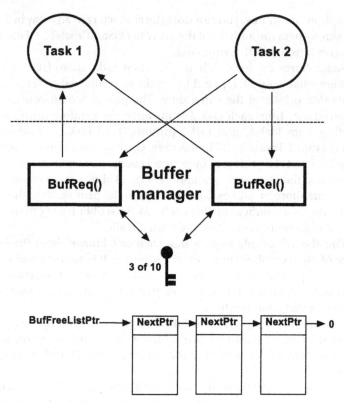

Fig. 14 Using a counting semaphore to access a pool of identical buffers.

linked list before the semaphore is signaled. If there are tasks in the wait list then the buffer is allocated to the highest priority task waiting for a buffer. By encapsulating the interface to the buffer manager in BufReq() and BufRel(), the caller does not need to be concerned with actual implementation details.

12.3 Resource Management—Notes on Semaphores

Using a semaphore to access a shared resource does not increase interrupt latency. If an ISR or the current task makes a higher priority task ready-to-run while accessing shared data, the higher priority task executes immediately.

An application may have as many semaphores as required to protect a variety of different resources. For example, one semaphore may be used to access a shared display, another to access a shared printer, another for shared data structures, and yet another to protect a pool of buffers, etc. However, it is preferable to use semaphores to protect access to I/O devices rather than memory locations.

Semaphores are often overused. The use of a semaphore to access a simple shared variable is overkill in most situations. The overhead involved in acquiring and releasing the semaphore consumes valuable CPU time. You can perform the job more efficiently by disabling and enabling interrupts, however, there is an indirect cost to disabling interrupts: even higher priority tasks that do not share the specific resource are blocked from using the CPU. Suppose, for instance, that two tasks share a 32-bit integer variable. The first task increments the variable, while the second task clears it. When considering how long a processor takes to perform either operation, it is easy to see that a semaphore is not required to gain exclusive access to the variable. Each task simply needs to disable interrupts before performing its operation on the variable and enable interrupts when the operation is complete. A semaphore should be used if the variable is a floating-point variable and the microprocessor does not support hardware floating-point operations. In this case, the time involved in processing the floating-point variable may affect interrupt latency if interrupts are disabled.

Semaphores are subject to a serious problem in real-time systems called priority inversion, which is described next.

12.4   Resource Management—Priority Inversions

Priority inversion is a problem in real-time systems and occurs only when using a priority-based preemptive kernel. Fig. 15 illustrates a priority-inversion scenario. Task H (high priority) has a higher priority than Task M (medium priority), which in turn has a higher priority than Task L (low priority).

(1) Task H and Task M are both waiting for an event to occur and Task L is executing.

(2) At some point, Task L acquires a semaphore, which it needs before it can access a shared resource.

(3) Task L performs operations on the acquired resource.

(4) The event that Task H was waiting for occurs, and the kernel suspends Task L and starts executing Task H since Task H has a higher priority.

(5) Task H performs computations based on the event it just received.

(6) Task H now wants to access the resource that Task L currently owns (i.e., it attempts to get the semaphore that Task L owns). Because Task L owns the resource, Task H is placed in a list of tasks waiting for the semaphore to be available.

(7) Task L is resumed and continues to access the shared resource.

(8) Task L is preempted by Task M since the event that Task M was waiting for occurred.

(9) Task M handles the event.

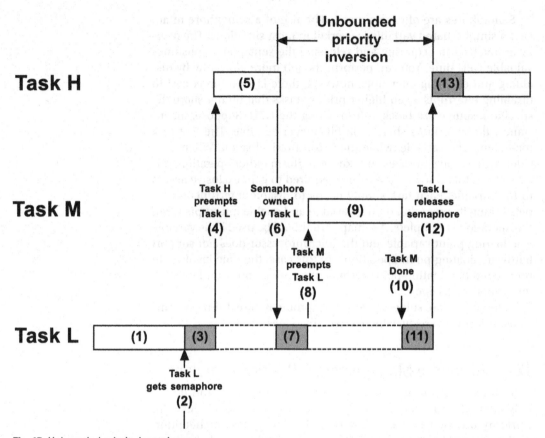

Fig. 15 Unbounded priority inversion.

(10) When Task M completes, the kernel relinquishes the CPU back to Task L.

(11) Task L continues accessing the resource.

(12) Task L finally finishes working with the resource and releases the semaphore. At this point, the kernel knows that a higher priority task is waiting for the semaphore, and a context switch takes place to resume Task H.

(13) Task H has the semaphore and can access the shared resource.

So, what has happened here is that the priority of Task H has been reduced to that of Task L since it waited for the resource that Task L owned. The trouble began when Task M preempted Task L, further delaying the execution of Task H. This is called an *unbounded priority inversion*. It is unbounded because any medium priority can extend the time Task H has to wait for the resource. Technically, if all medium-priority tasks have known worst-case periodic behavior and bounded execution times, the priority inversion time is computable.

This process, however, may be tedious and would need to be revised every time the medium-priority tasks change.

This situation can be corrected by raising the priority of Task L, for the time it takes to access the resource, and restoring its original priority level when the task is finished. The priority of Task L should be raised to the priority of Task H. In fact, many kernels contain a special type of semaphore that does just that—a *mutual exclusion semaphore*.

12.5 Resource Management—Mutual Exclusion Semaphores (Mutex)

Some kernels support a special type of binary semaphore called a mutual exclusion semaphore (also known as a *mutex*) which eliminates unbounded priority inversions. Fig. 16 shows how priority inversions are bounded using a mutex.

(1) Task H and Task M are both waiting for an event to occur and Task L is executing.

(2) At some point, Task L acquires a mutex, which it needs before it is able to access a shared resource.

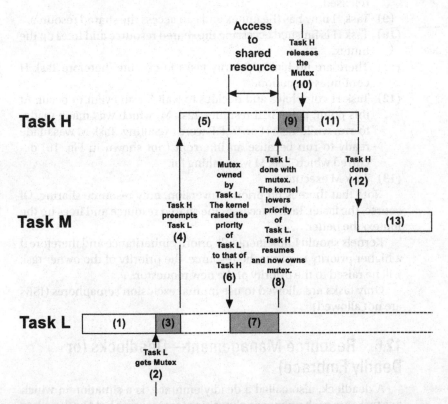

Fig. 16 Eliminating unbounded priority inversions with mutexes.

(3) Task L performs operations on the acquired resource.

(4) The event that Task H waited for occurs and the kernel suspends Task L and begins executing Task H since Task H has a higher priority.

(5) Task H performs computations based on the event it just received.

(6) Task H now wants to access the resource that Task L currently owns (i.e., it attempts to get the mutex from Task L). Given that Task L owns the resource, the kernel raises the priority of Task L to the same priority as Task H to allow Task L to finish with the resource and prevent Task L from being preempted by medium-priority tasks.

(7) Task L continues accessing the resource, however, it now does so while it is running at the same priority as Task H. Note that Task H is not actually running since it is waiting for Task L to release the mutex. In other words, Task H is in the mutex wait list.

(8) Task L finishes working with the resource and releases the mutex. The kernel notices that Task L was raised in priority and thus lowers Task L to its original priority. After doing so, the kernel gives the mutex to Task H, which was waiting for the mutex to be released.

(9) Task H now has the mutex and can access the shared resource.

(10) Task H is finished accessing the shared resource and frees up the mutex.

(11) There are no higher priority tasks to execute therefore Task H continues execution.

(12) Task H completes and decides to wait for an event to occur. At this point, µC/OS-III resumes Task M, which was made ready-to-run while Task H or Task L were executing. Task M was made ready-to-run because an interrupt (not shown in Fig. 16) occurred which Task M was waiting for.

(13) Task M executes.

Note that there is no priority inversion, only resource sharing. Of course, the faster Task L accesses the shared resource and frees up the mutex, the better.

Kernels should implement full-priority inheritance and therefore if a higher priority requests the resource, the priority of the owner task will be raised to the priority of the new requestor.

Only tasks are allowed to use mutual exclusion semaphores (ISRs are not allowed).

12.6 Resource Management—Deadlocks (or Deadly Embrace)

A deadlock, also called a deadly embrace, is a situation in which two tasks are each unknowingly waiting for resources held by the other.

Assume Task T1 has exclusive access to Resource R1 and Task T2 has exclusive access to Resource R2, as shown in the pseudocode of Listing 12.

```
void  T1 (void)
{
    while (1) {
        Wait for event to occur;          (1)
        Acquire M1;                        (2)
        Access  R1;                        (3)
        :
        :
        \--------  Interrupt!              (4)
        :
        :                                  (8)
        Acquire M2;                        (9)
        Access  R2;
    }
}


void  T2 (void)
{
    while (1) {
        Wait for event to occur;          (5)
        Acquire M2;                        (6)
        Access  R2;
        :
        :
        Acquire M1;                        (7)
        Access  R1;
    }
}
```
Listing 12 Deadlocks.

(1) Assume that the event that Task T1 is waiting for occurs and T1 is now the highest priority task that must execute.
(2) Task T1 executes and acquires Mutex M1.
(3) Resource R1 is accessed.
(4) An interrupt occurs causing the CPU to switch to Task T2 since T2 has a higher priority than Task T1.
(5) The ISR is the event that Task T2 was waiting for and therefore T2 resumes execution.
(6) Task T2 acquires Mutex M2 and is able to access Resource R2.
(7) Task T2 tries to acquire Mutex M1, but the kernel knows that Mutex M1 is owned by another task.
(8) The kernel switches back to Task T1 because Task T2 can no longer continue. It needs Mutex M1 to access Resource R1.

(9) Task T1 now tries to access Mutex M2 but, unfortunately, Mutex M2 is owned by Task T2. At this point, the two tasks are deadlocked, neither one can continue because each owns a resource that the other one wants.

Some techniques used to avoid deadlocks are for tasks to:

- Acquire all resources before proceeding.
- Always acquire resources in the same order.
- Use timeouts on wait calls (the kernel must provide timeouts on wait calls).

13 Synchronization

This section focuses on how tasks can synchronize their activities with ISRs, or other tasks.

When an ISR executes, it can signal a task telling the task that an event of interest has occurred. After signaling the task, the ISR exits and, depending on the signaled task priority, the scheduler is run.

The signaled task may then service the interrupting device, or otherwise react to the event. Servicing interrupting devices from task level is preferred whenever possible, since it reduces the amount of time interrupts are disabled and the code is easier to debug.

13.1 Synchronization—Semaphores

As previously described, a semaphore is a protocol mechanism offered by most multitasking kernels. Semaphores were originally used to control access to shared resources. However, a mutex is a better mechanism to protect access to shared resources, as previously described.

Semaphores are best used to synchronize an ISR with a task or synchronize a task with another task, as shown in Fig. 17. This is called a *unilateral rendez-vous*.

Note that the semaphore is drawn as a flag to indicate that it is used to signal the occurrence of an event. The initial value for the semaphore is typically zero (0), indicating the event has not yet occurred.

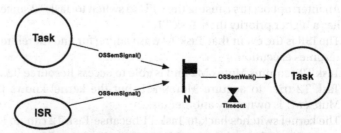

Fig. 17 Semaphore used as a signaling mechanism for synchronization.

The value "*N*" next to the flag indicates that the semaphore can accumulate events or *credits*. An ISR (or a task) can signal a semaphore multiple times and the semaphore will remember how many times it was signaled. It is possible to initialize the semaphore with a value other than zero, indicating that the semaphore initially contains that number of events.

Also, the small hourglass close to the receiving task indicates that the task has an option to specify a timeout. This timeout indicates that the task is willing to wait for the semaphore to be signaled within a certain amount of time. If the semaphore is not signaled within that time, the kernel will resume the task and return an error code indicating that the task was made ready-to-run because of a timeout and not because the semaphore was signaled.

A few interesting things are worth noting in Fig. 17. First, the task that calls OSSemWait() will not consume any CPU time until it is signaled and becomes the highest priority task ready-to-run. In other words, as far as the task is concerned, it called a function (OSSemWait()) that will return when the event it is waiting for occurs. Second, if the signal does not occur, the kernel maximizes the use of the CPU by selecting the next most important task to run. In fact, the signal may not occur for many milliseconds and, during that time, the CPU will work on other tasks.

Again, semaphores must be created before they can be signaled or waited on.

13.2 Synchronization—Credit Tracking

As previously mentioned, a semaphore "remembers" how many times it was signaled. In other words, if an ISR occurs multiple times before the task waiting for the event becomes the highest priority task, the semaphore will keep count of the number of times it was signaled. When the task becomes the highest priority ready-to-run task, it will execute without blocking as many times as there were ISR signals. This is called *credit tracking* and is illustrated in Fig. 18 and described in the following text.

(1) A high-priority task is executing.

(2) ☐

(3) An event meant for a lower priority task occurs which preempts the task (assuming interrupts are enabled). The ISR executes and posts the semaphore. At this point the semaphore count is 1.

(4) ☐

(5) ☐

(6) A kernel API is called at the end of the ISR to see if the ISR caused a higher priority task to be ready-to-run. Since the ISR was an event that a lower priority task was waiting for, the kernel will resume execution of the higher priority task at the exact point where it was interrupted.

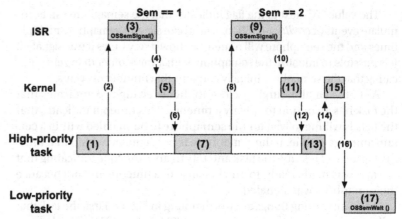

Fig. 18 Semaphore and credit tracking.

(7) The high-priority task is resumed and continues execution.

(8) ☐

(9) The interrupt occurs a second time. The ISR executes and posts the semaphore. At this point the semaphore count is 2.

(10) ☐

(11) ☐

(12) The kernel is called at the end of the ISR to see if the ISR caused a higher priority task to be ready-to-run. Since the ISR was an event that a lower priority task was waiting for, the kernel resumes execution of the higher priority task at the exact point where it was interrupted.

(13) ☐

(14) The high-priority task resumes execution and actually terminates the work it was doing. This task will then call one of the kernel services to wait for "its" event to occur.

(15) ☐

(16) The kernel will then select the next most important task, which happens to be the task waiting for the event and will context switch to that task.

(17) The new task executes and will know that the ISR occurred twice since the semaphore count is two. The task will handle this accordingly.

14   Bilateral Rendez-vous

Two tasks can synchronize their activities by using two semaphores, as shown in Fig. 19. This is called a *bilateral rendez-vous*. A bilateral rendez-vous is similar to a unilateral rendez-vous, except that both tasks must synchronize with one another before proceeding.

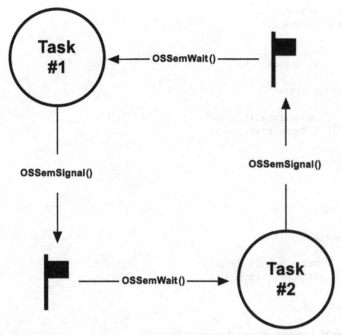

Fig. 19 Bilateral rendez-vous.

A bilateral rendez-vous cannot be performed between a task and an ISR because an ISR cannot wait on a semaphore.

The code for a bilateral rendez-vous is shown in Listing 13.

(1) Task #1 is executing and signals Semaphore #2.

(2) Task #1 waits on Semaphore #1. Because Task #2 has not executed yet, Task #1 is blocked waiting on its semaphore to be signaled. The kernel context switches to Task #2.

(3) Task #2 executes, and signals Semaphore #1.

(4) Since it has already been signaled, Task #2 is now synchronized to Task #1. If Task #1 is higher in priority than Task #2, the kernel will switch back to Task #1. If not, Task #2 continues execution.

15   Message Passing

It is sometimes necessary for a task or an ISR to communicate information to another task. This information transfer is called *intertask communication*. Information can be communicated between tasks in two ways: through global data or by sending messages.

As discussed in the Section 12, when using global variables, each task or ISR must ensure that it has exclusive access to variables. If an ISR is involved, the only way to ensure exclusive access to common variables is to disable interrupts. If two tasks share data, each can

```
OS_SEM   MySem1;
OS_SEM   MySem2;


void Task1 (void)
{
    while (1) {
        :
        OSSemSignal(&MySem2);              (1)
        OSSemWait(&MySem1);                (2)
        :
    }
}


void Task2 (void)
{
    while (1) {
        :
        OSSemSignal(&MySem1);              (3)
        OSSemWait(&MySem2);                (4)
        :
    }
}
```

Listing 13 Implementing a bilateral rendez-vous.

gain exclusive access to variables either by disabling interrupts, using a semaphore, or preferably, using a mutual exclusion semaphore. Note that a task can only communicate information to an ISR by using global variables. A task is not aware when a global variable is changed by an ISR, unless the ISR signals the task or the task polls the contents of a variable periodically.

Messages can be sent to an intermediate object called a *message queue*. Multiple tasks can wait for messages to arrive on a message queue and the kernel generally gives the received message to the highest priority task waiting for a message for that queue. When a task waits for a message to arrive, it does not consume CPU time.

15.1   Messages

A message generally consists of a pointer to data instead of copying the actual data. The pointer can point to a data area or even a function. Obviously, the sender and the receiver must agree as to the contents and the meaning of the message. In other words, the receiver of the message will need to know the meaning of the message received to be able to process it. For example, an Ethernet controller receives a packet and sends a pointer to this packet to a task that knows how to handle the packet.

The message contents must always remain in scope since the data is actually sent by reference (i.e., a pointer to the data) instead of by value. In other words, data sent is not copied. You might consider using dynamically allocated memory for the actual message content but you should avoid using allocated memory from the heap because your heap will eventually become so fragmented that your request for memory might not be satisfied. Alternatively, you can pass a pointer to a global variable, a global data structure, a global array, or a function, etc.

15.2 Message Queues

A message queue is a kernel object allocated by the application. In fact, you can allocate any number of message queues. The only limit is the amount of RAM available. There are a number of operations that the user can perform on message queues but the most typical ones are "create a queue," "send a message through a queue," and "wait for a message to arrive on a queue." An ISR can only send a message to a queue; it cannot wait for a message. A message queue must be created before sending messages through it.

Message queues are drawn as a first-in, first-out (FIFO) pipe. However, some kernels allow messages to be sent in a last-in, first-out (LIFO) order. The LIFO mechanism is useful when a task or an ISR must send an "urgent" message to a task. In this case, the message bypasses all other messages already in the message queue. The size of the message queue (i.e., the number of messages that can be held in a queue waiting for processing) is typically configurable either at runtime or configuration time.

Fig. 20 shows typical operations performed on a message queue (queue creation is not shown). The small hourglass close to the receiving task indicates that the task has an option to specify a timeout. This timeout indicates that the task is willing to wait for a

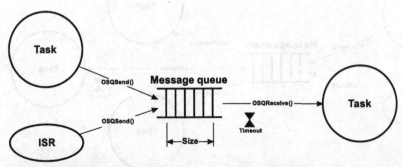

Fig. 20 Sending and receiving data through a message queue.

message to be sent to the message queue within a certain amount of time. If the message is not sent within that time, the kernel resumes the task and returns an error code indicating that the task was made ready-to-run because of a timeout, and not because the message was received. It is possible to specify an infinite timeout and indicate that the task is willing to wait forever for the message to arrive.

The message queue also contains a list of tasks waiting for messages to be sent to the message queue. Multiple tasks can wait on a message queue as shown in Fig. 21. When a message is sent to the message queue, the highest priority task waiting on the message queue receives the message. Optionally, the sender can broadcast a message to all tasks waiting on the message queue. In this case, if any of the tasks receiving the message from the broadcast have a higher priority than the task sending the message (or interrupted task, if the message is sent by an ISR), the kernel will run the highest priority task that is waiting. Notice that not all tasks must specify a timeout; some tasks may want to wait forever.

16 Flow Control

Task-to-task communication often involves data transfer from one task to another. One task produces data while the other consumes it. However, data processing takes time and consumers might not consume data as fast as it is produced. In other words, it is possible for the producer to overflow the message queue if a higher priority task preempts the consumer. One way to solve this problem is to add flow control in the process, as shown in Fig. 22.

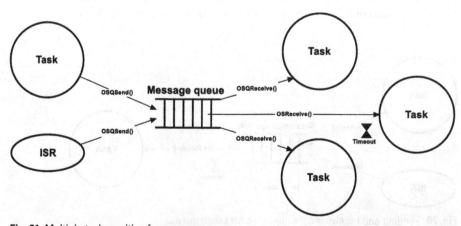

Fig. 21 Multiple tasks waiting for messages.

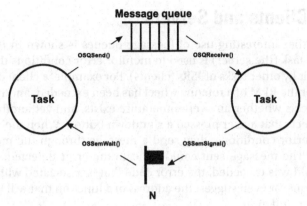

Fig. 22 Multiple tasks waiting for messages.

Here, a counting semaphore is used and initialized with the number of allowable messages that can be received by the consumer. If the consumer cannot queue more than 10 messages, the counting semaphore contains a count of 10.

As shown in the pseudocode of Listing 14, the producer must wait on the semaphore before it is allowed to send a message. The consumer waits for messages and, when processed, signals the semaphore.

```
OS_SEM  MySem;
OS_Q    MyQ;
int     Message;

void  MyProducerTask (void)
{
    while (1) {
        :
        OSSemWait(&MySem);
        OSQSend(&MyQ, (void *)&Message);
        :
    }
}



void  MyConsumerTask (void)
{
    void *p_message;

    while (1) {
        :
        p_message = OSQReceive(&MyQ);
        OSSemSignal(&MySem);
        :
    }
}
```
Listing 14 Message queue flow control.

17 Clients and Servers

Another interesting use of message queues is shown in Fig. 23. Here, a task (the server) is used to monitor error conditions that are sent to it by other tasks or ISRs (clients). For example, a client detects whether the RPM of a rotating wheel has been exceeded, another client detects whether an overtemperature exists, and yet another client detects that a user pressed a shutdown button. When the clients detect error conditions, they send a message through the message queue. The message sent could indicate the error detected, which threshold was exceeded, the error code that is associated with error conditions, or even suggest the address of a function that will handle the error, and more.

17.1 Memory Management

An application can allocate and free dynamic memory using any ANSI C compiler's `malloc()` and `free()` functions, respectively. However, using `malloc()` and `free()` in an embedded real-time system may be dangerous. Eventually, it might not be possible to obtain a single contiguous memory area due to fragmentation. Fragmentation is the development of a large number of separate free areas (i.e., the total free memory is fragmented into small, noncontiguous pieces). Execution time of `malloc()` and `free()` is generally

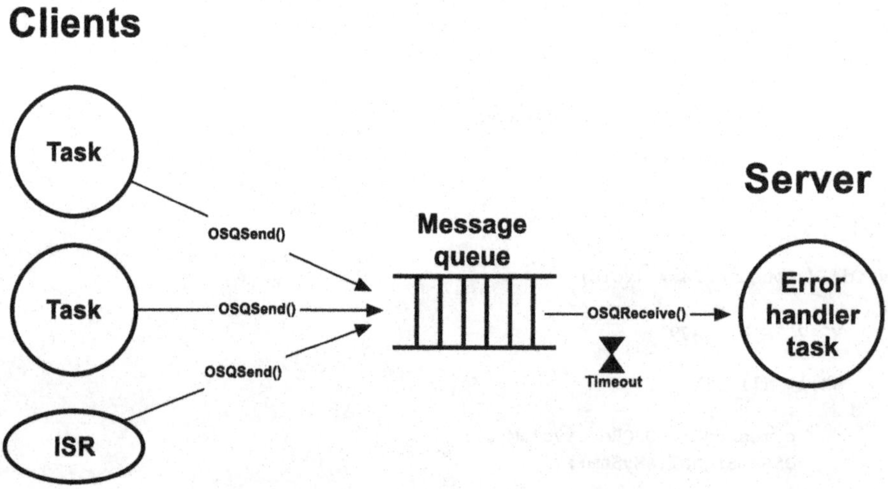

Fig. 23 Client/server using message queues.

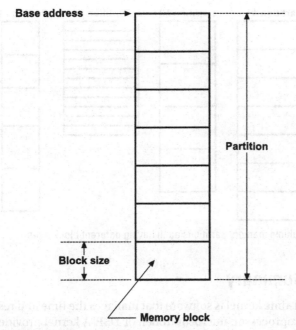

Fig. 24 Fixed-size block memory partition.

nondeterministic given the algorithms used to locate a contiguous block of free memory large enough to satisfy a `malloc()` request.

Kernels provide alternatives to `malloc()` and `free()` by allowing an application to obtain fixed-sized memory blocks from a partition made from a contiguous memory area, as illustrated in Fig. 24. All memory blocks are the same size, and the partition contains an integral number of blocks. Allocation and deallocation of these memory blocks is performed in constant time and is deterministic. The partition itself is typically allocated statically (as an array) but can also be allocated by using `malloc()` as long as it is never freed.

As indicated in Fig. 25, more than one memory partition may exist in an application and each one may have a different number of memory blocks and be a different size. An application can obtain memory blocks of different sizes based upon requirements. However, a specific memory block must always be returned to the partition that it came from. This type of memory management is not subject to fragmentation except that it is possible to run out of memory blocks. It is up to the application to decide how many partitions to have and how large each memory block should be within each partition.

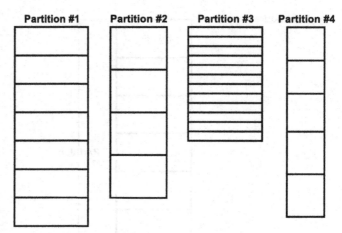

Fig. 25 Multiple memory partitions each having different block sizes.

18 Summary

A real-time kernel is software that manages the time and resources of a microprocessor, microcontroller, or DSP. A kernel provides valuable services to your application (product) through a series of application programming interfaces (APIs). Functions are thus available to manage tasks, manage shared resources, notify tasks that events have occurred, send messages to tasks, suspend execution of tasks for a user specified amount of time, and more.

A kernel allows a CPU to multitask. Multitasking is the process of scheduling (determining which task to run next) and context switching (assigning the CPU to a task) the CPU between several tasks. Multitasking provides the illusion of having multiple CPUs and by doing so, maximizes the use of the CPU and helps in the creation of modular applications.

Most real-time kernels are preemptive meaning that the kernel always runs the highest priority task that is ready-to-run.

One of the world's most popular real-time kernels is called μC/OS-III (pronounced micro-C-OS-three). μC/OS-III is available from Micrium and its source code has been made *source available* by Micrium. Source available means that the source code can be downloaded from the Micrium website and evaluated for free. However, a license is required if μC/OS-III is used commercially (used with the intent to make a profit).

The internals of μC/OS-III are fully described in the book: *μC/OS-III, The Real-Time Kernel* published by MicriumPress (see www.micrium.com). There are in fact many versions of the μC/OS-III book, each of which provides examples of running μC/OS-III on different popular CPU architectures (see the Micrium website for details).

This chapter was excerpted from sections of the μC/OS-III book

7

OPEN-SOURCE SOFTWARE

Jagdish Gediya*, Jaswinder Singh*, Prabhakar Kushwaha*, Rajan Srivastava*, Zening Wang[†]

*NXP Semiconductors, Automotive Division, Noida, India [†]*NXP Semiconductors, Microcontroller Division, Shanghai, China*

Software Engineering for Embedded Systems. https://doi.org/10.1016/B978-0-12-809448-8.00007-2

1   Linux

Linux is one of the most widely used software operating systems across all relevant market segments. Here are a couple of examples that represent the success of Linux: it is used in all 500 of the fastest super-computers across the world [1]; and 85% of mobile phones are now shipped with Android which is based on Linux [2]. In whatever market segments Linux has entered, it has become the most dominant operating system. Fig. 1 shows the domains in which Linux is being used today.

This chapter starts with a brief outline of the Linux journey since its conception. It goes on to explore salient features of Linux to explain why it has become the operating system of choice throughout the world of embedded systems.

1.1   History of Linux

Linux was initially developed as a hobby activity by software enthu-siast Linus Torvalds in the early 1990s. He shared this operating system (which he called "Linux") publicly. Due to its GPL aspect (described

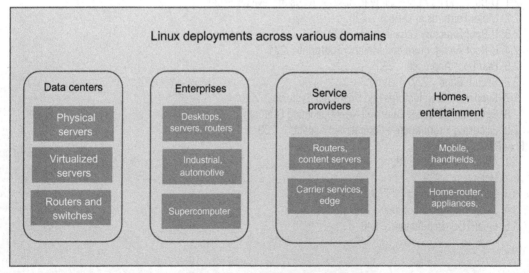

Fig. 1 Major segments where Linux is deployed (embedded and nonembedded).

later in this chapter), it quickly became very popular among the open-source community. Primarily developed for the desktop environment, it was soon adopted by other uses. Toward the late 1990s and early 21st century, most of the biggest commercial desktop suppliers started shipping desktops with preloaded Linux to avoid the need for costly commercial desktop operating systems.

1.1.1 Reason for the Exponential Acceptance of Linux

Linux is one of the most successful products of the open-source software movement driven by the GNU. The availability of the entire source code of the Linux kernel, along with its very visionary licensing approach, has been driving the success of Linux for more than two decades. Since its source code was publicly available immediately after its origination in the early 1990s, many people became Linux experts, creating and adding useful features to it. Under the guidance of strictly disciplined Linux source code maintenance, Linux became an extremely feature-rich and stable operating system within a couple of years of its inception.

Another major reason for the success of Linux was its "free" availability. Because of an existing monopoly by the de facto leader of desktop operating systems, desktop vendors were looking for options to remove the existing omnipresent operating system. In the late 1990s, desktop PCs started to become a commodity. This put pressure on desktop vendors to reduce costs so that the global availability of desktops could be increased. Linux-based desktops were less costly as Linux was freely available for commercial deployment.

Another movement progressed in parallel—in the domain of research. Supercomputer developers wanted an operating system that they could personally tweak to meet their requirements. Linux came to the rescue. All these factors helped make Linux more feature rich, stable, and popular.

1.1.2 Linux and Embedded Systems

Before Linux, commercial operating systems dominated the entire embedded world. In the late 1990s and early 21st century, there were a good number of embedded devices that needed highly feature-rich operating system to support their many features. Prime uses of such products were enterprise and home routers and set-top boxes. We will consider home routers to explain how the transition from commercial operating systems to Linux happened. Routers were supposed to provide networking layers for Ethernet/ATM/Wifi interfaces along with offering various flavors of network security. These embedded devices used commercial operating systems. Commercial operating systems had small footprints and needed very small RAM and flash storage— these reasons were enough for commercial operating systems to capture the embedded market.

Once desktops started arriving in households, the need for home routers became a necessity. The burden of cost reduction started building on home router providers. After the dot-com bubble burst, enterprise routers also started facing cost pressures. One of the major areas that could offer significant cost reduction was the operating system. Vendors were paying heavy royalties to commercial operating system providers. Linux was not feasible for routers because of its big memory footprint. Constant reductions in the cost of flash storage and SDRAM in the early 21st century made it feasible to accommodate Linux within routers [3]. Soon after this, an open-source forum, named OpenWrt, was set up, focusing on Linux deployments in Wi-Fi routers, in 2004.

The remainder of this chapter focuses on the important features of Linux that mean it can be used in various embedded-world use cases.

1.2   How Embedded Linux is Different From Linux?

"Linux" is generally used for desktop computers and servers. These machines are "general-purpose" machines that can do any type of job with reasonable efficiency: spreadsheets, word processing, browsing, games, network access, etc. Since these systems support all kinds of possible use cases, the operating system software also includes such probable features that may be used.

On the other hand, embedded systems are generally customized for one or more specific use cases. In such systems, operating systems also need to be customized based on the requirement to increase throughput performance and decrease memory footprint. During this customization, unwanted software is removed and the configuration of the rest of the software is matched to meet specific use cases. Linux, which is customized for a given use case, is referred to as embedded Linux.

1.3   Major Features of Linux

Most other embedded operating systems are custom solutions for a limited set of use cases. Linux has been designed as a general-purpose operating system and consequently it has most of its features available in desktop, as well as embedded, operating systems.

This section will describe some salient features of Linux.

1.3.1   Portability

Linux is a highly portable operating system, something that is very clear from the list of supported CPU architectures (see subsequent text). Most parts of the core kernel (that includes scheduler, memory management, device driver frameworks, interrupt framework) are written in such a way that can work on any underlying CPU. In addition to the core kernel, most of the drivers for onboard and on-chip devices run without any modifications on any CPU architecture.

1.3.1.1 Supported CPU Architectures

Today Linux is available for all general CPU architectures, for example, for several years it has supported Intel x86, PowerPC, ARM, and MIPS, to name but a few. Linux supports almost all variants of these architectures. For example, it supports all the major cores of Power Architecture: e500, e550, e5500, e6500, etc.

1.3.1.2 POSIX Compliance

There are challenging situations when precompiled applications supported on one operating system fail to execute (or even compile) on other operating systems. This happens because of the incompatibility of system calls (name or prototype of system call) between two operating systems. POSIX standard is meant to establish compatibility between operating systems. This means if an organization has software that is written for any non-Linux POSIX operating system, that application can be reused to run on Linux since most parts of Linux are POSIX compliant.

1.3.2 Support of a Wide Variety of Peripheral Devices

Linux provides excellent support for various communication interfaces, e.g., Ethernet, 802.11 Wi-Fi, Bluetooth, ATM. All popular storage devices are supported: USB storage, Secure Disk (SD), SATA, SSD, SCSI, RAID, etc. For onboard connectivity, it provides support for I2C, SPI (serial peripheral interface), PCIe, etc.

In summary, if you are planning to use any existing on-chip or onboard hardware in your system, it's very likely that its stable driver already exists in Linux. Device drivers contribute 60% of Linux source code—24 million lines of code in Linux kernel 4.19.

1.3.3 Complete Network Stack Solution

For end-to-end connectivity, Linux has a very rich network stack: TCP/UDP, IPv4/v6. Linux has a very feature-rich network stack that is also optimized for throughput performance.

Most of today's high-speed networking devices have network traffic accelerators—these accelerators perform checksum, TCP functions, and even bridging-routing functions. Linux network and device driver frameworks easily support these proprietary accelerators.

1.3.4 Variety of Task Schedulers

The job of task schedulers in operating systems is to execute a task at an appropriate moment. There are scenarios in which a use case demands a specific way of task scheduling. In case of a multiuser system, the CPU should be allocated in a round-robin way to each of the users so that no user feels excessive delays in completion of his or her tasks.

In the case of a desktop with a single user, if the user is running multiple tasks, such as a browser, a word processor, or a print job, they want to see progress in each of these jobs, in parallel; in addition to this the user wants the mouse pointer to immediately change its position on the desktop monitor when the mouse is moved on a mouse mat.

In real-time systems, the task schedule must ensure that real-time jobs finish their work within bounded timelines. For example, in a cellular 4G world, the base station is supposed to transmit the "subframe" exactly at the time defined in the 3GPP specifications. This transmission can only afford delays of less than a microsecond—if the base station incurs higher delays it causes system failure.

For each of the above examples, a custom task scheduler is required. Linux provides task schedulers for each of the above scenarios, and more.

1.3.5 Security

Security is a prime concern when everything is connected—be it the security of the connected device or the security of the data that is being processed by the device. Linux (with the help of underlying CPU architecture) ensures that a device boots with genuine software and also provides enough hooks that prevent the execution of undesired software.

For secure data transfers, Linux provides for the support of security protocols like Ipsec.

Another aspect of security are the firewalls in gateway/router devices. Linux has performance-optimized solutions to ensure that internal/LAN users are not impacted by malicious activities initiated outside of the LAN.

Not only does Linux provide a highly optimized stack to enable the above but it also has a good framework to support proprietary crypto hardware accelerators.

1.3.6 User Space Drivers

Traditionally, device drivers were written to execute in the Linux kernel. Some of the drivers need lots of information to be exchanged with their corresponding user space software. For example, a network interface driver must exchange vast volumes of data each second between the kernel and a webserver running in user space. As we will see later in this chapter, this kind of design incurs a huge penalty on CPU horsepower—Linux provides a user space driver framework where the network driver executes in user space.

1.3.7   Endianness

Some of the product vendors want to use the same onboard and/or on-chip devices on different CPU architectures. For example, a vendor may want to use the same USB device in two separate product lines, one that uses big-endian CPUs and other that uses little-endian CPUs. Maintaining two separate drivers for the same device to handle this situation is a costly proposition. The same scenario arises when one product line uses 32-bit CPU architecture and another uses 64-bit CPU architecture.

The Linux device driver framework and coding practices ensure that when a new driver is added in Linux, endianness and CPU bus sizes of devices do not demand modifications in the drivers.

1.3.8   Debuggability

The success of a design depends on how easily it can be debugged. Debugging doesn't only mean debugging of the hang/crash of software, it also refers to finding places where optimizations can be made for better performance. Linux has several kinds of debugging mechanisms available to developers for various types of issues in networking, file systems, synchronization mechanisms, device drivers, Linux initialization, and schedulers. The debug features allow developers to inspect what is happening in the system in online mode (while Linux is running) and in offline mode (i.e., take the logs and review them separately).

1.4   Benefits of Using Linux

Use of Linux instead of other operating systems (commercial and noncommercial) brings many benefits. The main benefit comes from a reduction in product cost, however, there are several other benefits as well. This section describes the main benefits of using Linux in embedded devices.

1.4.1   Free of Cost

Linux is freely available to everyone—individuals or commercial/noncommercial companies. One can use it as it is; one can modify and sell the modified Linux—no money needs to be paid to Linux or anyone else. All versions of Linux are free for use—long-term stable (LTS) versions or non-LTS versions.

Note that it's not only the Linux kernel that is free, its entire ecosystem is also free, for example, the GNU toolchain that compiles the Linux kernel, the GDB debugger, and root file systems (buildroot, Yocto, OpenWrt, ubuntu, etc.) are all free.

1.4.2 Time to Market

In today's competitive market, return on investment (ROI) also depends on how early you bring your product to market. Software development and testing needs a considerable amount of time in terms of product development therefore it's very important to reduce the software cycle.

Most CPU architectures (old, latest, and upcoming) are generally available in Linux. This means that with Linux you get an operating system that has already been tested for your specific CPU architecture. As mentioned in an earlier section of this chapter, Linux is already deployed in a variety of market segments. This means that required features are likely to be already available in Linux. These aspects reduce the overall product development time.

Other things that reduce time to market include:

1. Quick software team ramp-up. There are many white papers, articles, and videos available on Linux that can be used to quickly ramp-up a project.
2. Linux has a vast open-source community. If software developers face an issue, they can generally do a search on the web and most of the time find some forum or group that has already discussed/solved a particular issue—or can assist through web-posts or chat.
3. Linux's debugging is very feature rich. It can easily help localize problems.

1.4.3 No "Vendor Lock-in"

If a product manufacturer deploys a commercially available operating system in a product, the product manufacturer will have to customize the application software for that particular operating system. Additionally, the software team within the product development organization will have to create expertise in that particular operating system.

Now, if the product manufacturer has to come up with a new product, they will have to use the same commercial operating system since their workforce is already trained in that particular operating system. Even if the product needs a different operating system due to its specific needs, the product manufacturer will prefer to reuse their trained workforce to reduce product development costs. If the manufacturer decides to change the operating system, they will have to retrain the workforce and recustomize the application software for a new operating system.

On the other hand, the vendor of the commercial operating system may take undue advantage of the dependence of the product manufacturer on this operating system.

With Linux, there is no "lock-in" with any operating system vendor. This is because Linux expertise can help in tailoring Linux for different use cases. Therefore retraining of the workforce is not required. Also, there are many software vendors that offer Linux-based software—if a product manufacturer wants to outsource software work, they will have variety of choices for selecting software providers.

1.4.4 Highly Stable Operating System

The Linux open-source community is supposed to be one of the biggest open-source communities. Every new software feature (whether a driver, new framework, new scheduler) undergoes strict reviews by the community. These reviewers and repo maintainers are mostly subject-matter experts and they ensure that new proposed code changes comply with all the rules set by Linux.

Once a Linux version is released, it's tested by several members in the community. Each day, the community keeps adding patches to the mainline Linux and each night the "Linux-next" is tested on a wide range of platforms touching all CPU architectures and almost all features.

The outcome of this effort is that every version of Linux is very stable.

1.4.5 Low Maintenance

There are two aspects of software maintenance that this section describes and in both cases the product vendor achieves low-cost maintenance.

1.4.5.1 Supporting Software Releases After Shipping the Product

Suppose you delivered a production software release and a customer finds a bug. If this issue is a result of core kernel components, you will easily find it discussed on several Internet chat forums. Sometimes, you'll find the fix for such issues in newer Linux kernel versions—it's up to the software developer whether he or she wants to backport the fix to the desired kernel version or migrate the entire software to a new Linux kernel version.

Even if an issue relates to an interface between a component written personally and a Linux kernel framework, you will find matching discussions on the Internet because similar issues are likely to have arisen in similar components written by others.

You can take hints from preexisting discussions (or newer Linux versions) and provide fixes to your customers.

1.4.5.2 Keeping Your Own Drivers up to Date With the Latest Kernel

Supposed a vendor ships a software product and plans to ship the next product a few years after the first. Also suppose a vendor has created a lot of new drivers for the first product. When the vendor starts product development for the second product, they have two options:
1. Use the same software driver and same Linux kernel version as in the first product.
2. Use a new Linux version and port the previously created driver to the new kernel.

Option 2 is the right choice for most as new Linux kernel versions include a lot of new features and bug fixes for previous versions. Porting of your own drivers will require some effort because of the probability that a new kernel framework and features will make your existing drivers incompatible. Linux comes to the rescue of driver developers: one can write the driver, test it, and then get the driver included within the mainline Linux kernel. Once the mainline kernel includes a driver (irrespective of who has written the driver), all new kernel versions always ensure that all the existing drivers are ported to the next kernel version—this forward porting is done by the open-source Linux community.

1.5 Linux Architecture

Linux is designed in a modular way. This approach helps to make it portable across various hardware components and scalable across various use cases. A complete kernel is written in C language (except for very low-level CPU init code that is written in assembly language) making it easily understandable to those who don't understand the complex constructs of object-oriented languages.

A very high-level overview of the Linux kernel architecture is depicted in Fig. 2.

In Linux-based systems, CPU cores execute in two modes:

1. *Privileged mode.* In this mode, software executes with unrestricted privileges. In this mode of execution, the CPU allows software to

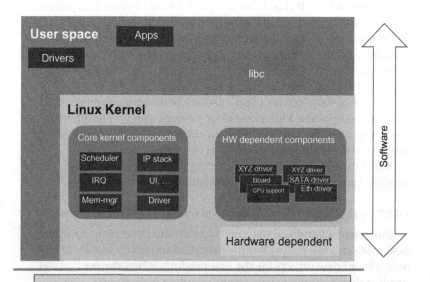

Fig. 2 Linux kernel architecture—a high-level view.

access all hardware resources. The entire Linux kernel executes in this mode.

2. *Unprivileged mode.* The Linux kernel creates a permission setting for all hardware resources and, based on the developer's choice, removes restrictions on some of the resources. Software that runs in unprivileged mode is permitted to access those resources that Linux configures for unrestricted accesses. In Fig. 2, "Apps" running in user space can access kernel resources by directly invoking Linux system calls or by invoking libc (or glibc)-provided API calls. In the latter case, libc (or glibc) invokes a Linux system call. "Drivers" in user space can directly access hardware (more on this later in the chapter), can access a Linux system call, or can access libc (or glibc) APIs. libc (or glibc) is an example library running in user space—there are many more such user space libraries.

1.5.1 Linux Kernel Components

The Linux kernel is composed of several components and in a typical Linux configuration all these components are essential.

1.5.1.1 Device Driver Framework

75% of overall Linux code is for device drivers. This is not a surprise since Linux is ported on a wide range of systems and this deployment has led to the inclusion of drivers of most devices on such systems. These devices include high-complexity hardware like Ethernet and graphics as well as low-complexity devices like EEPROM. To ensure all varieties of devices are plugged into Linux properly, Linux has a device driver framework.

Linux has a generic driver infrastructure that provides a base framework for all kinds of devices. For example, a "struct device" is required by all drivers to help driver modules organize generic resources like interrupts, the bus-type on which such devices exist, and hooks for power management.

The generic driver framework provides further frameworks for each type of device. For example, there is an Ethernet device framework that provides for common jobs associated with a typical Ethernet driver. Any hardware that has Ethernet can leverage from this framework: the actual Ethernet driver becomes smaller in size due to the Ethernet driver framework. Also, the author of an Ethernet device driver does not need to worry about exactly how the driver will exchange Ethernet frames with Linux's network stack.

Fig. 3 shows only a few of the many example device driver frameworks that exist. There are many more in the kernel. Details can be seen in the Linux source tree (visit/drivers at top level directory of the Linux tree).

1.5.1.2 Schedulers

Linux supports several types of task scheduling and one or more can be used at runtime since most of the schedulers are compiled in. Linux assigns a priority to each thread. Here, thread refers to a kernel thread as well as a user thread (please note that even if a user doesn't create an explicit thread via invocation of pthread_create() in his user space program, Linux still considers that process as a thread). This approach of threading allows a very fine control on scheduling priorities across the entire system.

'chrt' is the user program that lets you change the priority of a thread at runtime (the same effect can be made from within the source code of the programs as well). For a high-priority thread, FIFO (first in, first out) should be used for scheduling policy. For a periodic event that needs a fixed amount of processing after the event occurs, DEADLINE scheduling policy should be used. For regular processing (e.g., running a web server or a driver thread) the default scheduling policy OTHER or RR (round-robin) is used.

Note that initially, Linux was not designed to be an RTOS (real-time operating systems) where hard timelines can be met. Gradually, as Linux became popular in embedded domains, some real-time aspects were added from time-to-time to make scheduling and execution more deterministic with respect to timelines. Even today, Linux doesn't guarantee bounded latencies. If your product needs bounded latencies, you can apply a very popular Linux patch "PREEMPT_RT"— more on this patch comes later in the chapter.

1.5.1.3 Interrupt

Devices and timers need CPU attention so that some important actions can be performed at a CPU through an interrupt subroutine of the interrupting device. To simplify interrupt initialization and runtime interrupt handling, Linux implements an interrupt management subsystem. Linux manages the low-level details of interrupt

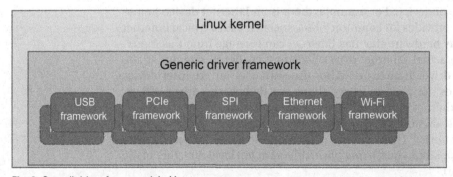

Fig. 3 Overall driver framework in Linux.

management which makes drivers of interrupt controllers and drivers of devices simpler.

This subsystem provides a user interface to check on the statistics of interrupts in the system at runtime and to get/set interrupt affinity to cores.

1.5.1.4 Memory Management

Memory management is comprised of two parts: virtual memory management and memory allocation management. To a good extent, virtual memory management depends on the underlying CPU architecture. Memory allocation management is independent of hardware unless some hardware accelerators are added to the hardware for this purpose. The Linux memory management subsystem is responsible for memory allocations to user space programs as well as kernel space software. For throughput-sensitive modules like Ethernet drivers and network stacks, Linux defines options like slab/slub/slob allocators.

1.5.1.5 Communication Protocol Stack

Linux has native implementations of various types of connectivity protocols like TCP/IP, Wi-Fi, Bluetooth, ATM, MPLS, X25, and so on. This section describes the most commonly used stacks only.

Linux has full networking support—the TCP/IP stack for IPv4 and IPv6. For data protection, Linux has IPsec support using software-based crypto; Linux also has a security framework for systems with hardware-based crypto.

In addition to this, Linux also has the excellent support of a firewall to protect LAN from externally initiated malicious traffic. This framework in Linux is called "NETFILER." It also lets you enable NAT (Network Address Translation) that allows a Linux device to work as a gateway so that local devices can access the Internet even with their local IP addresses.

In addition to networking, it also supports bridging and 802.1Q-based VLANs.

Linux supports traffic classification for a wide variety of rules of bridged and routed traffic.

1.5.1.6 User Interface (UI)

Linux implements a system call interface between the Linux kernel and user space applications, allowing the Linux kernel to be managed. Typically, shells (e.g., bash) use this system call interface to communicate with the Linux kernel for various configurations. There are device-specific, open-source programs that let the user configure the device; e.g., "ethtool" is used to configure an Ethernet interface. The

Linux kernel also provides a well-organized file system "sysfs" that lets the user manage various devices and kernel configurations via the "/sys" directory at the shell prompt.

1.6 Build Environment

This section describes kernel compilation and the root filesystem. The Linux kernel is generally compiled (or cross compiled) on a host x86 machine. Some new embedded systems (e.g., NXP's QorIQ Layerscape series) let you compile the kernel even on the target device.

1.6.1 Kernel Compilation

GNU GCC toolchain, an open-source compilation toolchain, is generally used to compile and link the Linux kernel to all popular CPU architectures. You may also use a commercial toolchain.

Before you compile the kernel, you should configure the kernel to match your CPU architecture by running "make menuconfig" at the top-level directory of the Linux kernel tree on your build machine. This command allows the user to select the CPU architecture of the target, kernel configuration (e.g., virtual memory page size and endianness), protocol stack configuration (e.g., IPv6/v6, firewall), and hardware devices that are present in the system—you may include, exclude, or customize kernel features using "make menuconfig." Subsequent execution of "make" compiles and links the kernel.

1.6.2 Root Filesystem

The Linux kernel alone is not very interesting—it's an operating system without any user-friendly shell interface. Root filesystem provides user-friendly applications, including shells, management applications for devices (e.g., ethtool for Ethernet interfaces), kernel configurations (e.g., "top" to check CPU usage), network stacks (e.g., "ip" for IPv4/IPv6-related configuration), and runtime libraries. Based on user choice, it may include a compiler toolchain and a gdb debugger as well!

There are several ways in which users can generate a root filesystem. This section describes Yocto, one of the most popular frameworks for generating a root filesystem.

1.6.2.1 Yocto

Yocto is an open-source project, sponsored by the Linux Foundation, that provides utilities to create, customize, and build a Linux distribution for an embedded system; the Linux distribution includes boot firmware, the Linux kernel, and a root filesystem. Yocto also allows a user to set up a build environment on a host x86 machine or a target system.

Yocto lets a user fine-tune the root filesystem based on user choice: if the system has very limited interfaces and small memories, the user can generate a tiny root filesystem—in this case the Yocto framework excludes undesired software from the root filesystem. If the target is heavily loaded with memory, the root filesystem can be made extremely feature rich—in this case the root filesystem will be huge!

1.7   Customizing Linux

The default Linux configuration includes many features that are generally required in desktop environments. Embedded devices generally provide a fixed set of capabilities, for example, a router supports routing and IP security–related features, it probably does not need to include graphics, sound, storage features, etc.

Linux provides the means to modulate the kernel so that a designer can attain specific, desired behavior. This section describes how Linux can be configured for desired use cases.

1.7.1   Low Memory Footprint

Low-end embedded devices support a limited set of features. To reduce the cost of these products they have the lowest possible memory resources. Full-blown Linux needs several megabytes of persistent storage (i.e., flash or other media) and several hundred bytes of runtime space (i.e., RAM). It is generally the Linux kernel and root filesystem that take up almost all the memory spaces in a typical embedded system.

1. *Optimizing Linux.* Linux can easily be fine-tuned to meet these limited resources so that it can fit in much smaller persistent and runtime memories. At compiling time, a designer can specify desirable and undesirable features using "make menuconfig." Linux classifies all features in a hierarchical order that allows a designer to either completely remove a feature or remove only part of a feature. For example, you can remove the complete network stack, or just remove IPv6 and retain IPv4. Even within IPv4 you can pick and choose specific features. The compiled Linux image will contain only the selected features.

2. *Optimizing the root filesystem.* This is the biggest memory consumer in most embedded systems. Yocto is the most popular root filesystem builder for Linux-based embedded system products and a full Yocto root filesystem may take up to several hundreds of megabytes of persistent storage. Yocto provides designers with a customization option that allows making a small root filesystem. This is described in more detail later in the chapter.

1.7.2 Boot Performance

When you turn on your home router, you want it to become operational instantly. This is where boot performance comes in, that is, how quickly the software[1] completes all of its initialization. To reduce the boot time, the first thing that a software designer must do is remove undesired software components as described in the previous text.

The next phase of optimization is product specific. Linux starts its multitasking subsystem (i.e., scheduler) very early in its Linux init: boot-critical jobs should be implemented in separate sets of threads and the remainder of jobs should be placed in other threads. If some part of the init system is CPU intensive, and you are operating a multicore system, then you can implement such functions in separate kernel threads and distribute these threads to separate cores. Modules, that are not important for the init system should be moved to separate kernel threads and assigned low priority.

1.7.3 High Throughput Performance

In some categories of products performance is critical. For example, a router is generally expected to route network traffic at Ethernet line rates; similarly, a storage device (USB pen drive) is expected to read/write files as early as possible. There are several options available in Linux that can be enabled at compile and/or runtime to achieve maximum throughput performance.

1.7.3.1 Core Affinity

In SMP systems, the Linux scheduler generally tries to assign a newly ready thread to a CPU core that is currently free. In some cases, designers know that if a job is affined to a specific core, the performance will be better. Linux provides for a runtime user interface (via/proc. files) to allow you to play with the system to discover what affinity configuration works best for your product. Once your experiments yield the right results, you may affine tasks either at compile time or at init time to achieve the discovered affinities of various threads.

1.7.3.2 Interrupt Coalescing

In network-based systems, CPU cores experience interrupts at very high speed when the system is put under heavy load. For example, if a router is subject to small Ethernet frames at maximum supported Ethernet speed, CPU cores become overwhelmed by excessive context switches between their regular task execution and interrupt processing.

---

[1] There are other factors that also contribute to boot performance, for example, in a home router, DSL line training takes several seconds. These kinds of factors are beyond the of scope of this book.

The Linux network device driver subsystem provides for a "NAPI" interface that lets the network device driver process tons of network packets using a single interrupt: while these packets are being processed, interrupts are kept disabled from network interface hardware. This "NAPI" feature increases network throughput manifold.

1.7.3.3 User Space Mapping of Buffers

Generally, user space modules are not permitted to read-write memory resources owned by the kernel. If user space software wants to transfer some memory buffer to a kernel driver, the kernel driver first copies the contents from the user space buffer to the kernel driver-allocated buffer. The same copy operation is required when the kernel driver wants to pass on a buffer to a user space module. This works fine for the user space modules that have a small number of buffers to be transmitted between user and kernel spaces. If the volume of such transfers is big then it consumes many CPU cycles doing a memory copy between the user and kernel buffers.

Linux provides an "mmap" feature that allows user space software to read-write a memory space owned by the kernel. Using "mmap," a user space module can pass on any amount of content to the kernel (and vice versa) without the need for a memory copy.

1.7.3.4 User Space Drivers

Traditionally, device drivers were developed in the Linux kernel. Some of the drivers were complex and when these buggy drivers misbehaved, by accessing memory which a driver was not supposed to access, the result was catastrophic—the entire Linux kernel could hang or crash. This was a problem—no one wanted a crashed/hung system, not even during the software development phase.

There was one further challenge—users didn't want the source code of their driver or software to be publicly available as advised by Linux's GPL license. Hence, such users started looking for alternative ways of using their drivers on Linux-based systems.

User space drivers came to the rescue in these scenarios. In the case of user space drivers, the driver software runs in user space as an application program. User space software doesn't fall under GPL so the user can retain the privacy of their source code in their modules. Also, if the user space driver tries to misbehave by accessing unauthorized regions, Linux detects this and prevents the driver from doing so.

Linux makes this possible by exposing a specific device's configuration space (generally, memory-mapped device configuration registers) and DMA-capable RAM to user space. Since a user space driver can access its device's memory space and DMA-capable regions without involving the Linux kernel, these drivers are very useful in case the device processes a lot of traffic. For example, Ethernet interfaces need huge amounts of network-level processing to cater to network traffic at line rate. In such cases user space network drivers and stacks are becoming popular in the open-source world—DPDK, an open-source project, uses user space drivers to provide maximum throughput. Interrupt handling is a challenge as interrupts force the CPU to enter supervisory mode—hence interrupt routines cannot be implemented completely in user space. The solution to this issue is to have a small Linux kernel driver for the device with an extremely small interrupt subroutine to just notify the user space driver of the interrupt event. Another approach could be to disable the interrupt and let the user space driver do the polling for events. Some solutions use a mix of the two approaches: (1) during high-traffic conditions, use polling mode in user space and keep the interrupt disabled; and (2) during scarce traffic, enable the interrupt and wait for notification from the kernel driver of the device.

The Linux kernel provides "UIO" and "vfio" frameworks to help develop user space drivers.

1.7.4 Latencies

Real-time systems need bounded latencies for handling some of events. Linux, using its default configuration, cannot ensure bounded latencies. The default scheduling scheme in Linux is a sort of nonpreemptive round-robin—if some task is running and a higher priority task becomes runnable due to some event, this new high-priority task will have to wait for the existing low-priority task to complete its scheduling quota. Linux provides some kernel configurations (CONFIG_ PREEMPT...) to make scheduling more deterministic.

If you want Linux to be completely deterministic during scheduling, to ensure bounded latencies, you can use another open-source project "PREEMPT_RT." This is a decade-old project for making Linux an "RTOS"—gradually features of the PREEMPT_RT projects are moving into mainline Linux.

1.8   Linux Development and its Open-Source Ecosystem

According to Greg Kroah-Hartman (one of the maintainers of Linux), 4300 Linux developers from 530 different companies had contributed to Linux by 2017. This represents the biggest collaborative software project. Coordinating among so many developers across the globe needs a well-defined workflow and discipline.

This section describes how the Linux community releases Linux versions and describes other relevant open-source projects. This section also looks at a few forums that are promoting open-source projects.

1.8.1 Linux Versions

Like other open-source communities, Linux developers send, review, and approve features and bug fixes via email. These changes are sent to the community in the form of a "patch" or "patch-set." Every 2 to 3 months, a Linux kernel maintainer adds the approved features and fixes to the existing kernel and comes up with a candidate release for the new Linux kernel version. Once the release candidate is found to be stable enough, a formal Linux kernel release is announced.

Linux kernel version numbers use a template like "a.b.c.," where a, b, and c are natural numbers. For example, the latest Linux version as of October 2018 is 4.18.0. These Linux versions are also called stable kernel releases.

1.8.2 Long-Term Support (LTS) Linux Version

These are the Linux versions that are maintained by Linux maintainers over the long term (approximately 2 years). Every year, one of the stable Linux releases is chosen as an LTS Linux version.

1.8.3 Related Open-Source Communities

The open-source community has created several forums to promote open-source projects, some of these are working around Linux. This section describes a couple of these forums.

1.8.3.1 Linux Foundation

Its primary focus is to build ecosystems around open-source projects to accelerate the commercial adoption of open-source projects. For example, to make Linux suitable for automotive and carrier markets, the Linux Foundation created AGL (Automotive Grade Linux) and CGL (Carrier Grade Linux) working groups. The objectives of such working groups are to identify the gaps in open-source projects like Linux for their deployment in specific market segments. Once the gaps are identified, these groups try to create the requisite groups to fill such gaps.

1.8.3.2 Linaro

Linaro works to promote open-source projects like Linux for various market segments for ARM's generic core–based systems.

1.8.4 Linux-Based Distributions
1.8.4.1 Android

Android is a Linux-based distribution deployed on most smartphones, tablets, and wearables. In its core, Android uses the Linux kernel with some modifications. These Linux modifications are maintained by the Android team and some of these features have been gradually included in Linux.

Android generally uses one of the latest LTS (long-term support) kernel versions of Linux.

1.8.4.2 Ubuntu

Ubuntu is one of the most popular open-source Linux distributions mainly targeted at desktops and servers. Ubuntu is released every 6 months and has LTS (long-term support) for 5 years. Thanks to its popularity, Ubuntu has been ported to several high-end embedded systems based on ARM and Power architectures.

There are many other open-source distros that are based on Linux.

1.9 Coding Guidelines

Linux expects that developers should write the Linux kernel code in such a way that code is as generic as possible (i.e., independent of any specific hardware architecture), is readable, and avoids unnecessary complexity. These guidelines ensure that Linux code is maintainable over the long term and that it increases code reusability across multiple types of hardware. These guidelines are strictly enforced during the patch review process that takes place during patch upstreaming.

Let us see how a good driver code ensures that the Linux driver is usable across two different CPU architectures. "QorIQ" devices from Freescale (now part of NXP) include an Ethernet controller ETSEC, its software driver "gianfar" can be found in the drivers/net/ethernet/freescale/directory of the Linux kernel source. The gianfar driver is written in such a way that whether the CPU core executes in little-endian mode or in big-endian mode, the same driver source code is used. See the code-snippet from gianfar.h below (this is the kernel recommended coding guideline for endian-safe drivers):

```
static inline void gfar_write(unsigned __iomem *addr, u32 val)
{
        iowrite32be(val, addr);
}
```

The ETSEC Ethernet Controller is a big-endian module in several SoC devices of NXP irrespective of whether the CPU cores in those SoC execute in big-endian or little-endian mode. The above driver uses the gfar_write() accessor function (shown in the above code snippet) to write ETSEC configuration registers and this accessor in turn uses an endianness- safe Linux provided accessor iowrite32be()—the Linux accessor is compiled according to the CPU core's endianness defined during the compilation of Linux.

Below is an *incorrect* driver snippet for the same action completed in the previous snippet. Technically, the snippet given below will work but this code includes undesired complexity in terms of its handling of endianness.

```
static inline void bad_example_write(unsigned __iomem *addr, u32 val)
{
#ifdef __BIG_ENDIAN
        *addr = val;
#else
        val = my_swap_bytes(val);
        *addr = val;
#endif
}
```

If a developer sends a patch like the incorrect code-snippet above, reviewers are likely to reject it.

Some important coding style suggestions can be found in a file available in the Linux source: Documentation/process/coding-style.rst.

1.10 Code Review in the Upstream Community

Linux mandates code reviews in the open-source community. The community helps to provide a better coding style, find potential bugs, and identify better architecture frameworks. It eventually makes the code generic enough to be used by the entire Linux community.

Also, if the feature is new or covers several areas of interest then it is easy to find help with testing it within the community.

Let's consider a few examples of how community reviews help with the betterment of a patch.

1. *Example 1* Upstream review of "Upstreaming imx7ulp lpuart support." This feature took four rounds of reviews (which means the author had to send four versions of the patch).

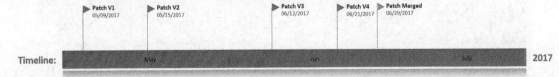

Each revision improved the code in the following manner:

- Round 1: Code clean up and architecture improvement.
- Round 2: Better coding style, eEliminate one unnecessary global variable, make unchangeable variables to "Const."
- Round 3: Better architecture and performance improvement, better driver design to handle different types of SoC devices (e.g., Layerscape lpuart), baud rate calculation algorithm improvement, elimination of another global variable which usually a bad design for per-device routines is.
- Round 4: Fix a small bug caught by 0-day Robot (Community CI). The patch after four rounds of reviews, i.e., the final version [4], had the following improvements compared with the initial version [5]:
- Readability—cleaner code.
- Efficiency—better performance.
- Stability—better stability for different types of SoC devices.
- Scalability—easier to add new types of support for devices.
- Maintainability—better architecture and driver design.

Below is a snapshot of a partial patch showing the differences between initial v1 and final v4: note that the color bar to the right of the snapshot identifies the big differences.

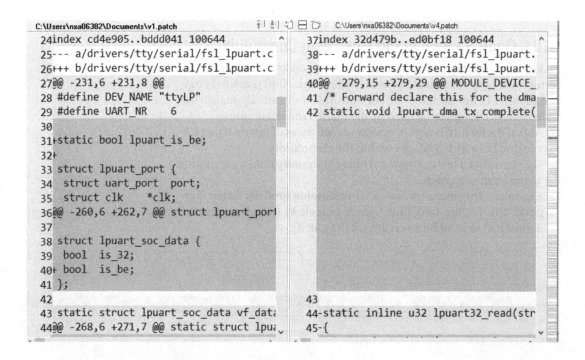

2. *Example 2:* Upstream review of "Upstreaming imx8qxp clock support." This feature took eight rounds of reviewing.

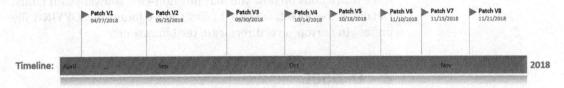

Each revision improved the code in the following manner:

- Round 1: Code clean up and reorganization.
- Round 2: Better coding style, better namespace for exported functions, put device specific SCU service API into device driver.
- Round 3: Proper prefix for exported structure names.
- Round 4: Update header-file path.
- Round 5: Fix potential bugs (memory leak), add more code comments, add missing lock and more code clean up.
- Round 6: Significant architecture improvement, architecture redesign to address a workaround issue, clearer component separation.
- Round 7: Use new kernel API.
- Round 8: Add enough comments to code and clean up the code further.

  The patch after eight rounds of reviews, i.e., the final version [6], had the following improvements compared with the initial version [7]:
- Readability—cleaner code.
- Efficiency—better performance.
- Stability—better stability for different types of SoC devices.
- Scalability—easier to add new types of device support.
- Maintainability—better architecture and driver design.

  To conclude, if the author of the code undertakes all the recommended coding guidelines and various aspects described in this section the code will be accepted by the open-source maintainer in a short period of time. These community reviews help to make code generic and help maintainers gain enough confidence for a change set.

1.11 License

Linux comes under GPL (GNU General Public License) version 2. This license allows anyone to use Linux source code as it is, modify it, and redistribute it (in original and/or modified forms) to others for commercial and noncommercial purposes. GPL asks that if you have modified the Linux source and redistributed it then you are bound to publish the distributed software in source format. It also imposes some restrictions on how you may link non-GPL software with Linux. Details of these GPL version 2 rules can be found in COPYING file present in the top-level directory of the Linux source.

2   U-Boot

2.1   U-Boot and its Applicability to Several Types of Devices

A boot loader is a critical piece of software running on any system. Whenever a computing system is initially powered on, the first piece of code to be loaded and run is the boot loader. It provides an interface for the user to load an operating system and applications.

The open-source ecosystem has lots of boot loaders like GRUB, UEFI, RedBoot, Yaboot, etc. However, U-Boot or Das U-Boot is the most commonly used open-source cross-platform boot loader. It is commonly used in embedded systems with the prime objective of configuring systems and loading next-level operating systems. It supports multiple architectures and has a large following by hardware manufacturers.

U-Boot boot loader typically is loaded by a system's Boot ROM from various boot sources, commonly nonvolatile memory such as NOR flash, SD cards, and SPI flash during power on—taking control of the hardware. Once U-Boot starts execution, it configures the hardware to load next-level images from onboard storage or from a network and then starts loading next-level images. After loading next-level images, U-Boot transfers execution control to next-level executable images. U-Boot also provides users with a "shell"-like interface so that users can play with the hardware configuration before next-level images take over.

2.2   Major Features of U-Boot

Broadly speaking, the major features of U-Boot are multiple boot sources, image upgrades, its shell (user interface), environment variables, scripts, stand-alone applications, and operating system boot commands.

2.2.1   Multiple Boot Source Support

A boot source is a nonvolatile onboard memory from where hardware loads U-Boot into preinitialized memory or transfers control directly for in-place execution (also known as XIP). Later sections in this chapter will share further details on this topic.

U-Boot supports booting from NOR, Serial NOR, SD/MMC, DSPI, NAND, etc.—some systems support multiple boot sources in the same U-Boot executable image while some systems support only one boot source in one U-Boot executable. Once U-Boot executes, it can load next-level images from a desired boot source—U-Boot makes this decision based on the user configuration saved in its "environment variables" (see later sections of this chapter for more on environment variables).

2.2.2 Shell (User Interface)

U-Boot provides a shell (also known as a command-line interface) over its serial interface which lets users manage various U-Boot attributes (e.g., which boot source to use for loading next-level images). This command-line interface provides lots of commands depending upon compile-time configuration. Major supported commands are flash read/write, networking (mdio, dhcp, tftp, ping, etc.), i2c, sdhc, usb, pcie, sata, and memory operations. Users can use these commands to configure the system and access I/O devices. Memory tests can also be initiated using the U-Boot shell.

U-Boot also supports commands for the management of environment variables and the display of runtime system configuration.

2.2.3 Environment Variables

These variables control the hardware configuration and boot behavior. Environment variables are usually stored on nonvolatile memory—they are given a default value at compile time, based on a user's choice for that specific system. Users can also modify these variables at runtime (using U-Boot shell) and save them to nonvolatile memory.

Runtime control and configuration environment variables consist of variables such as the IP address, UART baud rates, Linux bootargs, system MAC address, and bootcmd.

Boot-time control and configurations changes the way devices boot. For example, SDRAM configurations (ECC on/off, type of interleaving) modifies the way SDRAM is initialized during boot-sequence.

2.2.4 Scripts

The scripting feature of U-Boot allows storing multiple command sequences in a plain text file. This plain text file, can be run at the U-Boot user shell by simply invoking the "source" command followed by the script's name. This allows the user to run multiple command sequences in one go.

2.2.5 Stand-Alone Applications

U-Boot supports "stand-alone" applications. These applications can be loaded dynamically during U-Boot execution by bringing in RAM via network or nonvolatile memory. These stand-alone applications can have access to the U-Boot console, I/O functions, and memory allocations.

Stand-alone applications use a jump table, provided by U-Boot, to use U-Boot services.

2.2.6 Operating System Boot Commands

The U-Boot command "bootm" allows booting of next-level executable images (typically an operating system) preloaded in RAM—an operating system image can be obtained via RAM from a network or from onboard nonvolatile memory.

U-Boot also supports file systems. This way, rather than requiring the data that U-Boot will load to be stored at a fixed location on the storage device, U-Boot can read the file system on nonvolatile storage to search for and load specific files (e.g., the kernel and device tree). U-Boot supports all commonly used file systems like btrfs, cramfs, ext2, ext3, ext4, FAT, FDOS, JFFS2, Squashfs, UBIFS, and ZFS.

An operating system boot command is intelligent enough to perform all required prerequisites for operating system boot, such as device tree fix-up, required operating system image and file system uncompressing, and architecture-specific hardware configuration (cache, mmu). Once all prerequisites are completed, it transfers control to the operating system.

2.2.7 Autoboot

This feature allows a system to automatically boot to a next-level image (such as Linux or any user application) without the need for user commands. If any key is pressed before the boot delay time expires, U-Boot stops the autoboot process, provides a U-Boot shell prompt, and waits forever for a user command.

2.2.8 Sandbox U-Boot

The "sandbox" architecture of U-Boot is designed to allow U-Boot to run under Linux on almost any hardware. It is achieved by building U-Boot as a normal C application with a main () and normal C libraries. None of U-Boot's architecture-specific code is compiled as part of the sandbox U-Boot.

The purpose of running sandbox U-Boot under Linux is to test all the generic code—code that is not specific to any one architecture. It helps in creating unit tests which can be run to test upper level code.

The reader is referred to U-Boot documentation for further information on sandbox U-Boot.

2.3   U-Boot Directory Organization

U-Boot code and directory organization is very similar to Linux with customization for boot loaders.

arch/	Architecture-specific files
arc/	Files generic to ARC architecture
arm/	Files generic to ARM architecture
m68k/	Files generic to m68k architecture
microblaze/	Files generic to microblaze architecture
mips/	Files generic to MIPS architecture
nds32/	Files generic to NDS32 architecture
nios2/	Files generic to Altera NIOS2 architecture
openrisc/	Files generic to OpenRISC architecture
powerpc/	Files generic to PowerPC architecture
riscv/	Files generic to RISC-V architecture
sandbox/	Files generic to HW-independent "sandbox"
sh/	Files generic to SH architecture
x86/	Files generic to x86 architecture
api/	Machine/arch-independent API for external apps
board/	Board dependent files
cmd/	U-Boot commands functions
configs/	Board default configuration files
disk/	Code for disk drive partition handling
doc/	Documentation
drivers/	Drivers for on-chip and onboard devices
dts/	Contains Makefile for building internal U-Boot fdt.
examples/	Example code for stand-alone applications, etc.
fs/	Filesystem code (cramfs, ext2, jffs2, etc.)
Include/	Header files
licenses	Various license files
net/	Network-stack code
post/	Power on self-test
scripts/	Various build scripts and Makefiles
tools/	Tools to build S-Record or U-Boot images, etc.

CPU architecture–related code is placed in the arch/folder, while board-related code is placed the board/folder. The folder include/configs contains platform- or system-related header files. It can be used by the user to customize U-Boot for features and commands supported on a platform.

Other folders are self-explanatory, the top-level README file can be referred to for further details.

2.4 U-Boot Architecture and Memory Footprint

U-Boot supports two types of architecture: the single-stage boot loader and two-stage boot loader. The architecture you use depends on the size and nature of the memory available in the hardware. The following sections consider these two types of architecture as well as their applicability.

2.4.1 Single-Stage Boot Loader

If hardware (SoC or board) has XIP (eXecute In Place) nonvolatile (NV) storage, single-stage U-Boot architecture is preferred. In this architecture the entire U-Boot software is compiled into a single U-Boot binary. This single U-Boot binary is stored in XIP NV memory. During a system boot, the hardware transfers control to the abovementioned XIP memory; consequently, U-Boot executes from this memory and configures other desired hardware blocks. In the last stage of execution, U-Boot relocates to a bigger RAM. If the bigger RAM is SDRAM, then U-Boot first configures the SDRAM hardware and then relocates itself from the NV memory to RAM. A detailed flow diagram of this process is given in Fig. 4.

The relocated U-Boot has access to the complete SDRAM, allowing the initialization of the complete system, including drivers, such as USB, PCIe, SATA, and Ethernet. Once all the initialization is completed, U-Boot enters an infinite loop, waiting for user input. Further execution is controlled by the user. Alternatively, relocated U-Boot loads next-level images.

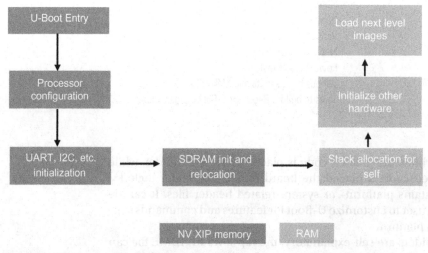

Fig. 4 Single-stage boot loader: U-Boot.

The approximate size of the single-stage boot loader for NXP's QorIQ LS2080ARDB platform is 700 kB.

In terms of design considerations for single-stage boot loader architecture, U-Boot performs some of the tasks before relocation and the remaining tasks after relocation. Execution from XIP NV memory is generally much slower than execution from SRAM or SDRAM. Hence, the user will have to carefully design what should be included in the phase before relocation.

2.4.2 Two-Stage Boot Loader

If the hardware (SoC or board) does not have XIP (eXecute In Place) nonvolatile (NV) storage but has internal SRAM, then two-stage U-Boot architecture is preferred. In this architecture U-Boot is compiled into two sets of binaries known as SPL (Secondary Program Loader) and U-Boot. Here, the SPL and U-Boot binary are stored in nonvolatile memory (such as SD and SPI).

SPL binary is loaded into internal SRAM by the system's BootROM. BootROM further transfers control to SPL. SPL, executing from internal SRAM, configures SDRAM. Once SDRAM is configured it copies U-Boot into SDRAM and transfers control to U-Boot. A detailed flow diagram of this process is given in Fig. 5.

Once U-Boot gets control, it initializes some of the hardware components like USB, PCIe, SATA, Ethernet etc. After complete initialization, U-Boot enters an infinite loop, waiting for user input. Further execution is controlled by the user. A detailed flow diagram of this process is given in Fig. 6.

The approximate sizes of the two-stage boot loaders for NXP's ARMv8 LS2080ardb platforms are ~75 kB for SPL and ~700 kB for U-Boot.

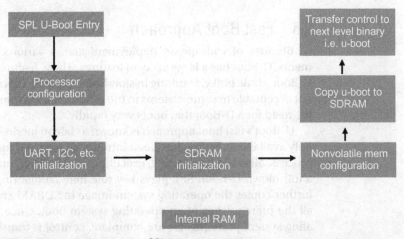

Fig. 5 Two-stage boot loader: SPL.

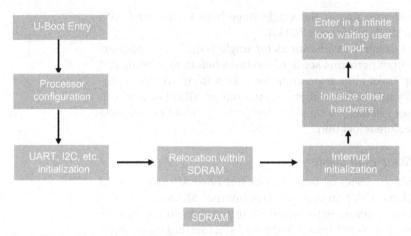

Fig. 6 Two-stage boot loader: U-Boot.

The beauty of U-Boot is that its size is controlled by the user at compile time. Users can customize its size based on system requirements by removing compile-time config options.

Note: selection of type of boot architecture.

The type of boot stage loader used for U-Boot depends upon the system's hardware configurations. If the system's hardware has XIP (eXecute In Place) flash, like NOR flash, then a single-stage boot loader can be used.

However, for cases where a system's hardware does not have XIP (eXecute In Place) memory and has internal RAM of limited size, then the default U-Boot may not be able to fit into the internal RAM. In this case a two-stage boot loader is the best option, i.e., having a small SPL loading normal U-Boot in SDRAM.

2.5 Fast Boot Approach

Because of widespread deployment across various market segments, U-Boot has a large array of features. These features have made U-Boot a little bulky, resulting in slow boot progress. This slow speed is not acceptable to some systems in the production environment, hence the need for a U-Boot that boots very rapidly.

U-Boot's fast boot approach is known as falcon mode. This mode is only available in SPL. It has been introduced to speed up the booting process, allowing the loading/executing of next-level images without a full-blown U-Boot. SPL plays key role here configuring SDRAM. It further copies the operating system image to SDRAM and completes all the prerequisites of an operating system boot. Once all the operating system prerequisites are complete, control is transferred. Fig. 7 represents falcon mode.

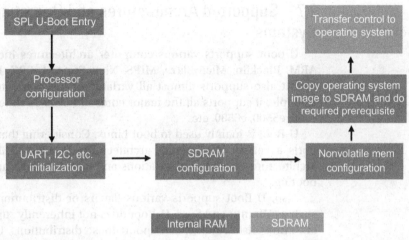

Fig. 7 Falcon mode.

The reader is referred to the U-boot documentation for falcon mode implementation and support.

2.6 Secure Boot

Secure boot or chain-of-trust boot is a mechanism for authenticating and optionally decrypting next-level images while still allowing them to be field upgraded. This feature allows product vendors to ensure that the shipped hardware always executes "genuine" software images ("genuine" software here refers to the software that was distributed/shipped by the specific vendor only).

U-Boot's secure boot depends on two major technologies: cryptographic hashing (e.g., SHA-1) and public key cryptography (e.g., RSA). These two cryptography technologies help product vendors in distributing authentic images, having them verified on target before they are used by the hardware after power-on.

Images can be stored one after another and signed using the cryptographic algorithms mentioned above. For added security the images can be encrypted. After power-on, the hardware authenticates the first image (i.e., U-Boot) and then—if required—decrypts the U-Boot. If hardware finds the U-Boot image to be genuine, it starts U-Boot execution. Next, U-Boot authenticates the next-level image (typically the operating system) and, if desired, decrypts it. U-Boot passes on execution control to the next-level image only if it finds that it is successfully authenticated.

Secure boot is an optional feature—designers need to implement the complete U-Boot flow for the secure boot feature.

2.7   Supported Architectures and Operating Systems

U-boot supports various computer architectures including 68k, ARM, Blackfin, MicroBlaze, MIPS, Nios, SuperH, PPC, RISC-V, and x86. It also supports almost all variants of these architectures. For example, it supports all the major cores of Power Architecture: e500, e550, e5500, e6500, etc.

U-Boot is mainly used to boot Linux. Considering that Linux supports a variety of computer architectures, it does all the required architecture-specific configurations and device tree fix-ups for Linux booting.

Also, U-Boot supports various flavors or distributions of Linux, such as Ubuntu and Suse. U-Boot does not inherently support different Linux distributions. To support these distributions, U-Boot runs a layer of abstraction (EFI). This abstraction (EFI) layer is used by GRUB2 to launch Linux distributions.

2.8   Open-Source Community and New Upcoming Features

U-Boot is maintained by Wolfgang Denx and hosted at www. denx.de/wiki/U-Boot. All discussions, developments, and reviews occur via a U-Boot mailing list [8]. These discussions and developments can be seen on patchwork [9].

The open-source community is continuously evolving U-Boot with the support of new architecture/hardware and features. Considering that the organization of U-Boot is very much the same as Linux, the community used to sync U-Boot's code-base with the Linux code-base, allowing many features to be ported to U-Boot.

U-Boot has lots of new, upcoming features, such as driver model, SPI-NAND framework, EFI layer enhancement for distribution, and device trees for all supported architectures.

2.9   Licensing Information—Commercial Aspects

U-Boot is free software. It has been copyrighted by Wolfgang Denk and many others who have contributed code. It can be redistributed and/or modified under the terms of version 2 of the GNU General Public License as published by the Free Software Foundation. This license does not cover "stand-alone" applications that use U-Boot services by means of the jump table provided by U-Boot exactly for this purpose.

Any organization or individual can freely download this software and customize it for their desired system. Such customized U-Boot software may then be distributed commercially and noncommercially. Organizations and individuals can also send their customized U-Boot back to the mainline U-Boot for long-term maintenance.

3 FreeRTOS

3.1 About FreeRTOS

FreeRTOS is a portable, open-source and tiny footprint real-time kernel developed for small embedded systems commonly used in wearable devices, smart lighting solutions, and IoT solutions.

The FreeRTOS kernel was originally developed by Richard Barry around 2003. Later, the FreeRTOS project was developed and maintained by Real Time Engineers Ltd., a company founded by Richard Barry. In 2017 Real Time Engineers Ltd. passed control of the FreeRTOS project to Amazon Web Services (AWS), however, it is still an open-source project.

FreeRTOS can be built with many open-source compilers, like GCC, as well as many commercial compilers. It supports various architectures, such as ARM, x86, and PowerPC. The FreeRTOS "port" is a combination of one of the supported compilers and architectures. FreeRTOS files that are common to all ports conform to MISRA coding standard guidelines. However, there are a few deviations from MISRA standards.

The FreeRTOS footprint depends on architecture, compiler, and kernel configuration. With full optimizations and the least kernel configurations, its footprint can be as low as ~5 kB [10].

3.2 Licensing

FreeRTOS is provided under an MIT opensource license [11]. Earlier, FreeRTOS kernel versions prior to V10.0.0 were provided under a modified GPLv2 license.

3.3 Commercial Aspects

FreeRTOS has commercial licensing available in form of OpenRTOS. OpenRTOS is the commercial version of FreeRTOS that provides a warranty and dedicated support.

SAFERTOS is a derivative of FreeRTOS designed to meet the requirements of industrial, medical, and automotive standards. It is precertified for standards such as IEC 61508-3 SIL 3 and ISO 26262 ASIL D.

OpenRTOS and SAFERTOS aren't open source.

3.4 Supported Architectures

There are wide range of architectures supported by FreeRTOS, such as ARM7, ARM9, ARM Cortex-M, ARM Cortex R, ARM Cortex-A, AVR, PIC, PowerPC, and x86.

3.5 FreeRTOS Architecture

Fig. 8 describes the FreeRTOS source code directory structure. The source directory contains the common kernel source code and portable layer. The demo directory contains the demo application projects targeted at a specific port.

Fig. 9 describes the typical architecture of a FreeRTOS-based system.

3.6 Portability

Basically, the FreeRTOS kernel has three files called tasks.c, queue.c, and list.c. These three files are present in the source directory. Additional files, i.e., event_groups.c, timers.c, and croutine.c, are only required if software timer, event group, or coroutine functionality are needed. All these files are common between all ports.

Apart from common files, FreeRTOS needs compiler- and architecture-specific code—called port. This code is available in the Source/portable/[compiler]/[architecture] directories, e.g., Source/portable/GCC/ARM_CM4F.

New FreeRTOS ports can also be developed [12].

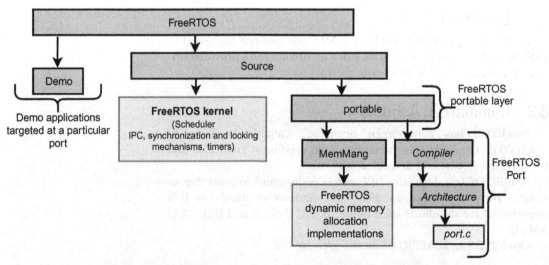

Fig. 8 FreeRTOS source directory structure.

Chapter 7 OPEN-SOURCE SOFTWARE **241**

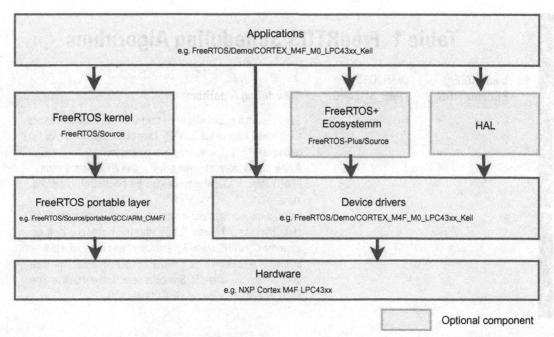

Fig. 9 FreeRTOS-based system architecture.

Each FreeRTOS project requires a file called FreeRTOSConfig.h which contains different configuration macros. This file is used to customize the FreeRTOS kernel.

3.7 Features

3.7.1 Scheduling

FreeRTOS doesn't have any restrictions on the number of real-time tasks that can be created and the number of task priorities that can be used. Multiple tasks can have the same priorities too.

A task can have one of these states: running, ready, blocked, or suspended.

The scheduling algorithm is based on the configUSE_PREEMPTION and configUSE_TIME_SLICING values in FreeRTOSConfig.h (Table 1).

3.7.2 Low Power

FreeRTOS supports tickless idle mode for low-power implementation. A developer can use the idle task hook to enter low-power state. However, power saving using this method is limited because periodically the tick interrupt will be served, hence exit and entry to low-power mode will be frequent. It may introduce an overhead instead of

Table 1 FreeRTOS Scheduling Algorithms

configUSE_PREEMPTION	configUSE_TIME_SLICING	Scheduling Algorithm
0	Any value	Context switch occurs only when the RUNNING state task enters the Blocked state or the RUNNING state task explicitly yields by calling taskYIELD(). Tasks are never preempted.
1	0	A new task is selected to run only if higher priority tasks enter READY state or a running task enters the blocking or suspended state.
1	1	Preempt the running tasks if higher priority tasks enter the READY state. Running tasks enter READY state and higher priority tasks enter the RUNNING state. Equal priority tasks share an equal amount of processing time if they are in READY state. Time slice ends at each tick interrupt. Scheduler selects a new task to enter the RUNNING state during RTOS tick interrupt.

power saving if the tick interrupt frequency is too high. The FreeRTOS tickless idle mode stops the tick interrupt during an idle task to overcome this issue.

FreeRTOS provides for an idle task hook function which is called from the idle task. An application author can use this hook function to enter the device into low-power mode.

3.7.3 Debugging

FreeRTOS provides a mechanism for stack overflow detection. An application needs to provide the stack overflow hook function with a specific prototype and name. The kernel calls the hook function if the stack pointer has a value outside the valid range.

The FreeRTOS kernel contains different types of trace macros that are defined empty by default. An application writer can redefine these macros according to their need to collect application behavioral data.

3.7.4 IPC and Synchronizations

FreeRTOS supports various intertask communication mechanisms like stream and message buffers, task notifications, queues, and event groups.

FreeRTOS supports many synchronization primitives like binary semaphores, counting semaphores, mutexes, and recursive mutexes.

3.7.5   Memory Management

FreeRTOS supports creating different objects, such as tasks, queues, timers, and semaphores, either by using dynamic memory or by an application provided in static memory.

FreeRTOS supports five dynamic memory allocation implementations, i.e., heap_1, heap_2, heap_3, heap_4, and heap_5, which are in the Source/Portable/MemMang directory. An application writer must include only one of these memory allocation implementations in a project.

heap_1 is the simplest and the only implementation which doesn't allow freeing of memory. heap_2 allows freeing memory but doesn't concatenate adjacent free blocks. heap_3 is wrapper around standard "malloc" and "free" interfaces provided by compiler. heap_4 concatenates adjacent blocks to avoid fragmentation. heap_5 allows spanning the heap over multiple nonadjacent memory and concatenates adjacent blocks to avoid fragmentation.

It is also possible to provide your own implementation.

3.8   FreeRTOS+ Ecosystem

There are many add-on software products that are either open source or proprietary, such as filesystems, networking stacks, networking security libraries, command line Interfaces, and I/O frameworks, available to debug and develop FreeRTOS-based embedded systems more rapidly. The source code for these add-on software products is available under the FreeRTOS-Plus/Source directory.

3.9   Debugging

FreeRTOS is widely used in small embedded systems. As a result, FreeRTOS awareness is widely supported in many IDEs, such as DS-5 studio from ARM and Kinetis Design Studio from NXP.

Using these FreeRTOS-aware IDEs and powerful hardware debuggers, like DSTREAM from ARM, application writers can get all the required data for debugging, such as task lists and their status, timer information, queue status, and current data in the queue.

FreeRTOS also provides a mechanism to debug stack overflow problems. However, this mechanism has limitations on certain architectures where the CPU throws exceptions against stack corruption before FreeRTOS checks for an overflow.

Application developers can redefine and use FreeRTOS trace macros to obtain application behavioral data.

3.10   Support

FreeRTOS has a support forum https://sourceforge.net/p/freertos/discussion/ for associated discussions.

Questions

1. Describe how embedded Linux is different from general Linux.
2. What are pros and cons of implementing a device driver in a Linux kernel vs. a user space?
3. What kind of customizations are available in Linux? Explain performance-related customization.
4. What is the significance of the coding guidelines for developers?
5. What are the benefits of upstream code reviewing. Explain providing one suitable example?
6. Why does U-Boot have a two-stage boot load flow for some products and a three-stage boot load flow for others?
7. In what types of device would you use FreeRTOS over Linux?
8. What are different scheduling algorithms supported in FreeRTOS?

References

[1] Linux Runs All of the World's Fastest Supercomputers, Nov 20, 2017.
[2] Worldwide Smartphone OS Market Share, A Report by International Data Corporation (IDC), 2018.
[3] Linux on Wi-Fi Routers, A report by Linux Journal, 2004.
[4] Upstreaming imx7ulp Lpuart Support, Final Patch, https://git.kernel.org/pub/scm/linux/kernel/git/torvalds/linux.git/commit/drivers/tty/serial/fsl_lpuart.c?id=24b1e5f0e83c2aced8096473d20c4cf6c1355f30.
[5] Upstreaming imx7ulp Lpuart Support, Initial Patch, https://source.codeaurora.org/external/imx/linux-imx/commit/drivers/tty/serial/fsl_lpuart.c?h=imx_4.9.88_2.0.0_ga&id=938d46fa07adb548c4bb06ad79024e-de1a363d9d.
[6] Upstreaming imx8qxp Clock Support, Final Patch, https://patchwork.kernel.org/patch/10692625/.
[7] Upstreaming imx8qxp Clock Support, Initial Patch, https://source.codeaurora.org/external/imx/linux-imx/commit/drivers/clk/imx?h=imx_4.9.88_2.0.0_ga&id=be6a7494dc86fcf8eafbed7dadec593bd3b27f99.
[8] U-Boot mailing list: u-boot@lists.denx.de.
[9] U-Boot patchwork link, http://patchwork.ozlabs.org/project/uboot/list/.
[10] https://www.freertos.org/FAQMem.html.
[11] https://www.freertos.org/a00114.html.
[12] https://www.freertos.org/FreeRTOS-porting-guide.html.

SOFTWARE AND COMPILER OPTIMIZATION FOR MICROCONTROLLERS, EMBEDDED PROCESSORS, AND DSPs

Michael C. Brogioli
Polymathic Consulting, Austin, TX, United States

CHAPTER OUTLINE

Software Engineering for Embedded Systems. https://doi.org/10.1016/B978-0-12-809448-8.00008-4

1 Introduction

Optimization for embedded systems can involve several different factors at the software level, many of which directly reflect the underlying hardware. When optimizing embedded applications, the developer must be mindful of algorithmic requirements, supported arithmetic operations and data types, memory system layout, to name a few. In addition, developers must also be mindful of build tools and their optimizing capabilities, and just as important in certain cases, the inability of some tools to optimize. This chapter discusses selected features of modern embedded build tools, the use of data structures, data types, and how to best enable tools to extract the greatest optimization for a given application.

2 Development Tools Overview

It is important to understand the features of development tools as they provide many useful, time-saving opportunities. Modern compilers are increasingly performing better with embedded software, leading to a reduction in required development times. Linkers, debuggers, and other components of the toolchain have useful code-build and debugging features, but in this chapter we will only focus on compilers.

2.1 Compilers, Linkers, Loaders, and Assemblers

From the compiler perspective, there are two basic ways of compiling an application: traditional compilation or global (cross-file) compilation. In traditional compilation, each source file is compiled separately and then the generated objects are linked together. In global optimization, each C file is preprocessed and passed to the optimizer in the same file. This enables greater optimizations (interprocedural optimizations) to be made as the compiler has complete visibility of the program and doesn't have to make conservative assumptions about the external functions and references. Global optimization does have some drawbacks, however. Programs compiled this way will take longer to compile and are harder to debug (as the compiler has taken away function boundaries and moved variables). In the event of a compiler bug, it will be more difficult to isolate and work around when built globally. Global or cross-file optimizations result in full visibility of all the functions, enabling much better optimizations for speed and size. The disadvantage is that since the optimizer can remove function boundaries and eliminate variables, the code becomes difficult to debug. Fig. 1shows the compilation flow for each.

2.1.1 Basic Compiler Configuration

Before building for the first time, some basic configuration will be necessary. Perhaps the development tools come with project stationery and have basic options configured. If not, these items should be checked:

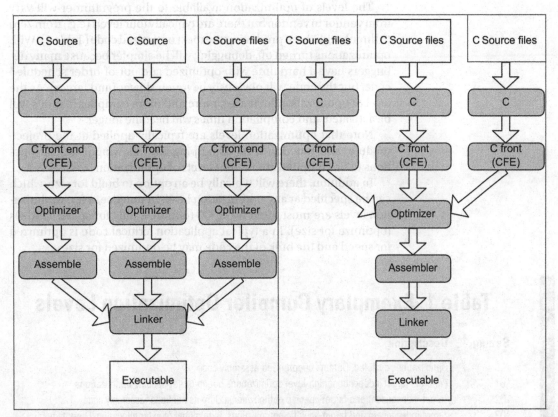

Fig. 1 Compilation tool flow for source code optimization, file level, and global optimization.

- Target architecture: specifying the correct target architecture will allow the best code to be generated.
- Endianness: perhaps the vendor sells silicon with only one endianness, perhaps the silicon can be configured. There will likely be a default option.
- Memory model: different processors may have options for different memory model configurations.
- Initial optimization level: it's best to disable optimizations initially.

2.1.2 Enabling Optimizations

Optimizations may be disabled by default when no optimization level is specified and either new project stationery is created or code is built on the command line. Such code is designed for debugging only. With optimizations disabled, all variables are written and read back from the stack, enabling the programmer to modify the value of any variable via the debugger when stopped. This code is inefficient and should not be used in production code.

The levels of optimization available to the programmer will vary from vendor to vendor, but there are typically four levels (e.g., from zero to three), with three producing the most optimized code (Table 1). With optimizations turned off, debugging will be simpler because many debuggers have a hard time with optimized and out-of-order scheduled code, but the code will obviously be much slower (and larger). As the level of optimization increases, more and more compiler features will be activated, and compilation times will become longer.

Note that optimization levels are typically applied at the project, module, and function level using pragmas, allowing different functions to be compiled at different levels of optimization.

In addition, there will typically be an option to build for size, which can be specified at any optimization level. In practice, a few optimization levels are most often used: O3 (optimize fully for speed) and Os (optimize for size). In a typical application, critical code is optimized for speed and the bulk of the code may be optimized for size.

Table 1 Exemplary Compiler Optimization Levels

Setting	Description
-O0	Optimizations disabled. Outputs unoptimized assembly code
-O1	Performs target independent high-level optimizations but no target-specific optimizations
-O2	Target-independent and target-specific optimizations. Outputs nonlinear assembly code
-O3	Target-independent and target-specific optimizations, with global register allocation. Outputs nonlinear assembly code. Recommended for speed-critical parts of the application

2.2   Peripheral Applications for Performance

Many development environments have a profiler, which enables the programmer to analyze where cycles are spent. These are valuable tools and should be used to find critical areas. The function profiler works in the IDE with the command line simulator.

3   Understanding the Embedded Target Architecture

Before writing code for an embedded processor, it's important to assess the architecture itself and understand the resources and capabilities available. Modern embedded architectures have many features to maximize throughput. Table 2 provides some features that should be understood and some questions the programmer should ask.

The next few sections will cover aspects of embedded architectures and will address how to appropriately optimize an embedded application to take advantage of tools-based optimization as well as select architectural features.

Table 2 Architectural Constructs Candidate for Optimization

Architectural Feature	Description
Instruction set architecture	Native multiply or multiply followed by add?
	Is saturation implicit or explicit?
	Which data types are supported—8, 16, 32, 40?
	Fractional and/or floating-point support?
	SIMD hardware? Does the compiler autovectorize, or are intrinsics required to access SIMD hardware?
	Domain-specific instructions (bit swapping, bit shift, Viterbi, video, etc.). Do these need to be accessed via intrinsics?
Register file	How many registers are there, how many register files comprise the total register set? What are they used for (integer, addressing, floating point)?
	Implication example: How many times can a loop be unrolled before performance decreases due to register pressure? How many live variables can be within a single scope at a time before spill code is generated?
Predication	Does the architecture support predicated execution? How many predicates does the architecture support? More predicates result in better control code performance. How efficient is the compiler at handling code generation for this feature?

Continued

Table 2   Architectural Constructs Candidate for Optimization—cont'd

Architectural Feature	Description
Memory system	What kind of memory is available and what are the speed trade-offs between them? How many busses are there? How many read/write operations can be performed in parallel? Is there a data/instruction cache within the system? Are there small SRAM-based scratch pad buffers? Can bit-reversed addressing be performed? Is there support for circular buffers in the hardware?
Other	Hardware loops?
	Mode bits? If the compiler cannot determine the value of mode bits (saturating arithmetic, nonsaturating arithmetic) it may impact the ability to optimize code.

4   Basic Optimization Goals and Practices

This section contains basic C optimization techniques that will benefit code written for all embedded processors. The central ideas are to ensure the compiler is leveraging all features of the architecture and to communicate to the compiler additional information about the program which is not communicated in C.

4.1   Data Types

It is important to learn about the sizes of the various types on the core before starting to write code. A compiler is required to support all required types but there may be performance implications and reasons to choose one type over another.

For example, a processor may not support a 32-bit multiplication. The use of a 32-bit data type in a multiply operation may cause the compiler to generate a sequence of multiple instructions, versus a single native multiply instruction. If 32-bit precision is not needed, it would be better to use 16-bit. Similarly, using a 64-bit type on a processor which does not natively support it will result in a similar construction of 64-bit arithmetic using 32-bit operations.

4.2   Intrinsics for Leveraging Embedded Processor Features

Intrinsic functions, or intrinsics for short, are a way to express either operations not possible or convenient to express in C or target-specific features (Table 3). Intrinsics in combination with custom data types can allow the use of nonstandard data sizes or types. They can

Table 3 DSP Intrinsic Function Example

Example Intrinsic (C Language)	Generated Assembly Code
int d = L_add(a, b);	iadd d0, d1;

also be used to get application-specific instructions (e.g., Viterbi or video instructions) which cannot be automatically generated from ANSI C by the compiler. They are used like function calls, but the compiler will replace them with the intended instruction or sequence of instructions. There is no calling overhead.

Some examples of features accessible via intrinsics are saturating arithmetic, fractional data types, multiply accumulate, SIMD operations, and so forth. As a general rule, it is advisable to see what architectural features a given embedded processor supports at the ISA level and then review programmer manuals and build tools documentation to see how architectural features are accessed.

For example, an FIR filter can be rewritten to use intrinsics and therefore to specify processor operations natively (Fig. 2; FIR filter example). In this case, simply replacing the multiply and add operations with the intrinsic L_mac (for long multiply-accumulate) replaces two operations with one and adds the saturation function to ensure that digital signal processors (DSP) arithmetic is handled properly.

4.3 Calling Conventions and Application Binary Interfaces

Each processor or platform will have different calling conventions. Some will be stack-based, others register-based or a combination of

```
Short SimpleFir1( short *x, short *y )
{
        int i = 0;
        long acc = 0;
        short ret = 0;


        for(I = 0; i<16; i++)
                // multiply, accumulate, and saturate result
                acc = L_mac(acc, x[i], y[i]) ;
        ret = acc >> 16;

}
```
Fig. 2 FIR filter example.

both. Typically, default calling conventions can be overridden though, which is useful. The calling convention should be changed for functions unsuited to the default, like those with many arguments. In these cases, the calling conventions may be inefficient.

The advantages of changing a calling convention include the ability to pass more arguments in registers rather than on the stack. For example, on some embedded processors, custom calling conventions can be specified for any function through an application configuration file and pragmas. It's a two-step process.

Custom calling conventions are defined by using the application configuration file (a file which is included in the compilation). Once defined a software developer can continue to develop their application as normal, however, if a developer wishes to use a custom-defined calling convention for certain function definitions, they must explicitly do so in the source code, often via #pragmas. For instance, if a custom calling convention named *my_calling_convention* is defined in the application configuration file, and the developer wishes to apply it to the *test_calling_convention()* function, the syntax may appear similar to that shown in Fig. 3:

Developers should always refer to the documentation for their particular IDE and build environment, as well as application configuration files, to see the specific syntax used for a given toolchain and target architecture.

4.4 Memory Alignment

Some embedded processors, like DSPs, support loading of multiple data values across the busses as this is necessary to keep the arithmetic functional units busy. These moves are called multiple data moves (not to be confused with *packed* or *vector* moves). They move adjacent values in memory to different registers. In addition, many compiler optimizations require these multiple register moves because there is so much data to move to keep all the functional units busy.

Typically, however, a compiler aligns variables in memory to their access width. For example, an array of short (16-bit) data is aligned to 16 bits. However, to leverage multiple data moves, the data must have a higher alignment. For example, to load two 16-bit values at once, the data must be aligned to 32 bits.

```
int     test_calling_convention( int a, int b)
{
        Return a + b;
}
#pragma call_conv test_calling_convention my_calling_convention
```
Fig. 3 Calling convention/ABI example.

4.5 Pointers and Aliasing

When pointers are used in the same piece of code, make sure that they cannot point to the same memory location (alias). When the compiler knows the pointers do not alias, it can put accesses to memory pointed to by those pointers in parallel, greatly improving performance. Otherwise, the compiler must assume that the pointers could alias. This can be communicated to the compiler by one of two methods: using the restrict keyword or informing the compiler that no pointers alias anywhere in the program (Fig. 4).

The restrict keyword is a type qualifier that can be applied to pointers, references, and arrays (Figs. 5 and 6). Its use represents a guarantee by the programmer that within the scope of the pointer declaration, the object pointed to can be accessed only by that pointer. A violation of this guarantee can produce undefined results.

```
void foo()
{
    int *a = NULL;
    int *b = NULL;
    int array[4];

    a = &array[1];
    b = &array[1];
    /* rest of func */

}
```

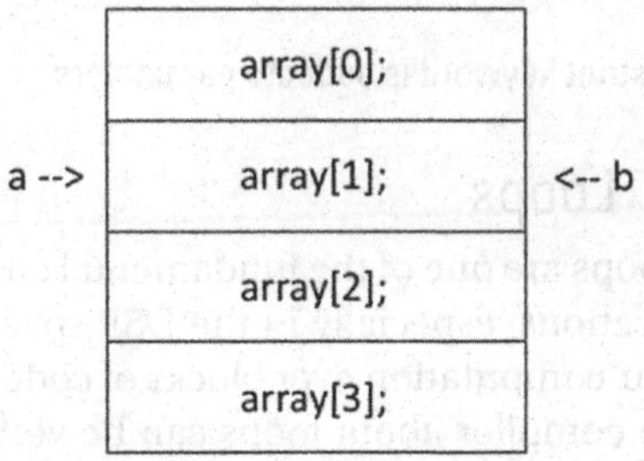

array[0];
array[1];
array[2];
array[3];

a --> array[1]; <-- b

Fig. 4 Illustration of pointer aliasing.

Example loop	Generated assembly code
void foo (short *a, short *b, int N) {	doen3 d4
int i;	FALIGN
	LOOPSTART3
for(i=0; i<N; i++) {	move.w (r0)+, d4
b[i] = shr(a[i], 2);	asrr #<2, d4
}	move.w d4, (r1)+
return;	LOOPEND3
}	

Fig. 5 Example loop before the restrict keyword is added to parameters (DSP code).

Example loop restrict qualifiers added, Note: Pointes a and b must not alias (ensure data is located separately)	Generated assembly code. Note: Now accesses for a and b can be issued in parallel.
| ```
void foo (short * restrict a, short * restrict b, int N) {
  int i;

  for(i=0; i<N; i++) {
    b[i] = shr(a[i], 2);
  }
  return;
}
``` | ```
move.w (r0)+, d4
  asrr #<2, d4
  doensh3 d2
  FALIGN
  LOOPSTART3
  [ move.w d4, (r1)+  ;parallel
      move.w (r0)+, d4  ; accesses
  ]
  asrr #<2, d4
  LOOPEND3
  move.w d4, (r1)
``` |

Fig. 6 Example loop after the restrict keyword is addedto parameters.

4.6 Loops

Loops are one of the fundamental components of many embedded applications, especially in the DSP space where computation is often regular computation over blocks of code. Communicating information to the compiler about loops can be very important in achieving high performance within an application. Pragmas can be used to communicate information to the compiler about loop bounds to help loop optimization. If the loop minimum and maximum are known, for example, the compiler may be able to make more aggressive optimizations.

In the example in Fig. 7, a pragma is used to specify the loop count bounds to the compiler. In this syntax, the parameters are minimum, maximum, and multiple, as shown by the three numerical parameters as part of the loop_count pragma usage. If a nonzero minimum is specified, the compiler can avoid generation of costly zero-iteration checking code. The compiler can use the maximum and multiple parameters to know how many times to unroll the loop if possible.

```
{
    long int L_tmp = 0;
    int i = 0;

    for (i = 0; i < N; i++)
#pragma loop_count (4, 512, 4)
        L_tmp = L_mac (L_tmp, vec1[i], vec2[i]);  *result = round (L_tmp);
}
```

Fig. 7 Use of pragmas and intrinsics.

Reiterating the point that loop structures are a key component of all numerical processing applications and most embedded processing applications, hardware loops are another means of achieving performance within an application.

Hardware loops are mechanisms built into some embedded cores which allow zero- overhead (in most cases) looping by keeping the loop body in a buffer or by prefetching. Hardware loops are faster than normal software loops (decrement counter and branch) because they have less change-of-flow overhead. Hardware loops typically use loop registers that start with a count equal to the number of iterations of the loop, decrease by 1 each iteration (step size of 21), and finish when the loop counter is zero.

Compilers most often automatically generate hardware loops from C even if the loop counter or loop structure is complex. However, there will be certain criteria under which the compiler will be able to generate a hardware loop (which vary depending on compiler/architecture). In some cases the loop structure will prohibit generation, but if the programmer knows about this the source can be modified so the compiler can generate the loop using hardware loop functionality. The compiler may have a feature to tell the programmer if a hardware loop was not generated (compiler feedback). Alternatively, the programmer should check the generated code to ensure hardware loops are being generated for critical code. It is advisable to read the programmers manual for your target architecture and build tools to understand under what conditions the tools can, and cannot, generate hardware loops.

4.7 Advanced Tips and Tricks

The following are some additional tips and tricks that can be used to further increase the performance of compiled code. Note, however, that some of these concepts, like inlining of functions, may have adverse effects if used too aggressively—like impacts on code size.

Memory contention—When data is placed in memory, be aware of how the data is accessed. Depending on the memory type, if two buses issue data transactions in a region/bank/etc., they could conflict and cause a penalty. Data should be separated appropriately to avoid this contention. Scenarios that cause contention are device dependent because memory bank configuration and interleaving differs from device to device.

Unaligned memory accesses—In some embedded processors, devices support unaligned memory access. This is particularly useful for video applications. For example, a programmer might load four byte-values which are offset by one byte from the beginning of an area in memory. Typically, there is a performance penalty for doing this.

Cache accesses—In the caches, place data that is used together side by side in memory so that prefetching the caches is more likely to obtain the data before it is accessed. In addition, ensure that the loading of data for sequential iterations of the loop is in the same dimension as the cache prefetch.

Function Inlining—The compiler normally inlines small functions, but the programmer can force inlining of functions if for some reason it isn't happening (e.g., if size optimization is activated). For small functions the save, restore, and parameter-passing overheads can be significant relative to the number of cycles of the function itself. Therefore inlining is beneficial. Also, inlining functions decreases the chance of an instruction cache miss because the function is sequential to the former caller function and is likely to be prefetched. Note that inlining functions increases the size of the code. On some processors, pragma inline forces every call of the function to be inlined.

5 General Loop Transformations

The optimization techniques described in this section are general in nature. They are critical to taking advantage of modern multi-ALU processors. A modern compiler will perform many of these optimizations, perhaps simultaneously. In addition, they can be applied on all platforms, at the C or assembly level. Therefore throughout this section, examples are presented in general terms, in C and in assembly.

5.1 Loop Unrolling

Loop unrolling is a technique whereby a loop body is duplicated one or more times. The loop count is then reduced by the same factor to compensate. Loop unrolling can enable other optimizations, such as multisampling, partial summation, and software pipelining.

Once a loop is unrolled, flexibility in coding is increased. For example, each copy of the original loop can be slightly changed. Different registers could be used in each copy. Moves can be done earlier and multiple register moves can be used. Fig. 8 shows an example of a for loop that has been unrolled by a factor of four. As can be seen on the right-hand side of the figure, the loop iterations have been reduced by a factor of four, while the amount of instruction level parallelism

Loop prior to unrolling	After unrolling by factor of 4

`for(i=0; i<16; i++)`
`    operation();`
	`for(i=0; i<16; i+=4)`
	`{`
	`        operation();`
	`        operation();`
	`        operation();`
	`        operation();`
	`}`

Fig. 8 Example of loop unrolling.

within the loop has increased, potentially enabling further optimization by the compiler.

5.2   Multisampling

Multisampling is a technique for maximizing the usage of multiple ALU execution units in parallel for the calculation of independent output values that have an overlap in input source data values. In a multisampling implementation, two or more output values are calculated in parallel by leveraging the commonality of input source data values in calculations. Unlike partial summation, multisampling is not susceptible to output value errors from intermediate calculation steps. Multisampling can be applied to any signal-processing calculation of the form:

$$y[n] = \sum_{m=0}^{M} x[n+m]h[n]$$

where:

$$y[0] = x[0+0]h[0] + x[1+0]h[1] + x[2+0]h[2] + \ldots + x[M+0]h[M]$$
$$y[1] = x[0+1]h[0] + x[1+1]h[1] + \ldots + x[M-1+1]h[M-1] + x[M+1]h[M]$$

Using C pseudocode, the inner loop for the output value calculation can be written as:

```
tmp1 = x[n];
for(m = 0;m < M;m+ = 2)
{
        tmp2 = x[n+m+1];
        y[n]+ = tmp1*h[m];
        y[n+1]+ = tmp2*h[m];

        tmp1 = x[k+m+2];
        y[n]+ = tmp2*h[m+1];
        y[n+1]+ = tmp1*h[m+1];
}
tmp2 = x[n+m+1];
y[n+1]+ = tmp2*h[m];
```

As can be seen above, the multisampled version works on *N*output samples at once. Transforming the kernel into a multisample version involves the following changes:

- Changing the outer loop counters to reflect the multisampling by *N*.
- Use of *N*registers for accumulation of the output data.

- Unrolling the inner loop Ntimes to allow for common data elements in the calculation of the Nsamples to be shared.
- Reducing the inner loop counter by a factor of $N$to reflect the unrolling by N.

5.3 Partial Summation

Partial summation is an optimization technique whereby the computation for one output sum is divided into multiple smaller, or partial, sums. The partial sums are added together at the end of the algorithm. Partial summation allows more use of parallelism since some serial dependency is broken, allowing the operation to complete sooner.

Partial summation can be applied to any signal-processing calculation of the form:

$$y[n] = \sum_{m=0}^{M} x[n+m]h[n]$$

where:

$$y[0] = x[0+0]h[0] + x[1+0]h[1] + x[2+0]h[2] + \ldots + x[M+0]h[M]$$

To perform a partial summation, each calculation is simply broken up into multiple sums. For example, for the first output sample, assuming $M = 3$:

$$\text{sum0} = x[0+0]h[0] + x[1+0]h[1]$$
$$\text{sum1} = x[2+0]h[0] + x[3+0]h[1]$$
$$y[0] = \text{sum0} + \text{sum1}$$

Note the partial sums can be chosen as any part of the total calculation. In this example, the two sums are chosen to be the first + the second, and the third + the fourth calculations.

Important note: partial summation can cause saturation arithmetic errors. Saturation is not associative. For example, saturate(a*b) + c may not equal saturate (a*b + c). Care must be taken to ensure such differences do not affect program output.

The partial summed implementation works on Npartial sums at once. Transforming the kernel involves the following changes:

- Use of Nregisters for accumulation of the Npartial sums.
- Unrolling the inner loop will be necessary; the unrolling factor depends on the implementation, how values are reused, and how multiple register moves are used.
- Changing the inner loop counter to reflect the unrolling.

5.4 Software Pipelining

Software pipelining is an optimization whereby a sequence of instructions is transformed into a pipeline of several copies of that sequence. The sequences then work in parallel to leverage more of the available parallelism of the architecture. The sequence of instructions can be duplicated as many times as needed, substituting a different set of registers for each sequence. Those sequences of instructions can then be interwoven.

For a given sequence of dependent operations:

A = operation();
B = operation(A);
C = operation(B);

Software pipelining gives (where operations on the same line can be parallelized):

A0 = operations();
B0 = operation(A); A1 = operation();
C0 = operation(B); B1 = operation(A1);
C1 = operation(B1);

5.5   Advanced Topics

For advanced reading in embedded optimization, and further case study analysis, please refer to Chapter 3. This section performs an in-depth analysis of an architectural breakdown of a wireless application, as well as relevant real-world embedded software optimizations to achieve desired performance results.

6   Code Size Optimization

In compiling a source code project for execution on a target architecture, it is often desirable for the resulting code size to be reduced as much as possible. The reasons for this pertain to both the amount of space in memory the code will occupy at program runtime and the potential reduction in the amount of instruction cache needed by the device. In reducing the code size of a given executable, a number of factors can be tweaked during the compilation process to accommodate this.

6.1   Compiler Flags and Flag Mining

Typically, users will first begin by configuring the compiler to build the program for size optimization, frequently using a compiler command line option like -Os, as available in the GNU GCC compiler version 4.5. When building for code size, it is not uncommon for the compiler to disable other optimizations that frequently result in improvements in the runtime performance of the code. Examples of these might be loop optimizations, such as loop unrolling or software pipelining, which typically are performed in an attempt to increase the runtime performance of the code at the cost of increases in the compiled code size. This is due to the fact that the compiler will insert additional code into the optimized loops, such as prolog and epilog code in the case of software pipelining or additional copies of the loop body in the case of loop unrolling.

In the event that users do not want to disable all optimization or build exclusively at optimization level -O0, with code size optimization enabled, users may also want to disable functionality like function inlining. This can be performed via either a compiler command line option or compiler pragma, depending on the system and functionality supported by the build tools. It is often the case that at higher levels of program optimization, specifically when optimizing for program runtime performance, compilers will attempt to inline copies of a function, whereby the body of the function code is inlined into the calling procedure, rather than the calling procedure being required to make a call into a callee procedure, resulting in a change of program flow and obvious system side effects. By specifying either as a command line option or via a customer compiler pragma, the user can prevent the tools from inadvertently inlining various functions which would result in an increase in the overall code size of the compiled application.

When a development team is building code for a production release, or in a use case scenario when debugging information is no longer needed in the executable, it may also be beneficial to strip out debugging information and symbol table information. In doing this, significant reductions in object file and executable file sizes can be achieved. Furthermore, in stripping out all label information, some level of IP protection may be afforded to the user in that consumers of the executable will have a difficult time reverse engineering the various functions being called within the program.

6.2 Target ISA for Size and Performance Trade-Offs

Various target architectures in the embedded space may afford additional degrees of freedom when trying to reduce the code size of the input application. Quite often it is advantageous for the system developer to take into consideration not only the algorithmic complexity and software architecture of their code but also the types of arithmetic required and how well those types of arithmetic and system requirements map to the underlying target architecture. For example, an application that requires heavy use of 32-bit arithmetic may run functionally on an architecture that is primarily tuned for 16-bit arithmetic; however, an architecture tuned for 32-bit arithmetic can provide a number of improvements in terms of both performance, code size, and perhaps power consumption.

Variable-length instruction encoding is one particular technology that a given target architecture may support, which can be effectively exploited by the build tools to reduce overall code size. In variable-length instruction coding schemes, certain instructions within the target processor's ISA may have what is referred to as "premium encodings," whereby those instructions most commonly used can be represented in a reduced binary footprint. One example of this might be a 32-bit embedded Power Architecture device, whereby frequently used instructions, like integer add, are also represented with a premium 16-bit encoding. When the source application is compiled for size optimization, the build tools will attempt to map as many instructions as possible to their premium encoding counterpart, in an attempt to reduce the overall footprint of the resulting executable.

Freescale Semiconductor supports this feature in the Power Architecture cores for embedded computing, as well as in their StarCore line of DSPs. Other embedded processor designs, such as those by ARM Limited and Texas Instruments' DSP, have also employed variable encoding formats for premium instructions in an effort to curb the size of the resulting executable's code footprint.

It should be mentioned than the reduced-footprint premium encoding of instructions in a variable-length encoding architecture often comes at the cost of reduced functionality. This is due to the reduction in the number of bits that are afforded in encoding the instruction, often reduced from 32 bits to 16 bits. An example of a nonpremium encoding instruction vs. a premium encoding instruction might be an integer arithmetic ADD instruction. On a nonpremium-encoded variant of the instruction, the source and destination operations of the ADD instruction may be any of the 32 general-purpose integer registers within the target architecture's register file. In the case of a premium-encoded instruction, whereby only 16 bits of encoding space are afforded, the premium-encoded ADD instruction may only be permitted to use R0-R7 as source and destination registers, in an effort to reduce the number of bits used in the source and register destination encodings. Although it may not readily be apparent to the application programmer, this can result in subtle, albeit minor, performance degradations. These are often due to additional copy instructions that may be required to move source and destination operations around to adjacent instructions in the assembly schedule because of restrictions placed on the premium-encoded variants.

As evidence of the benefits and potential drawbacks of using variable-length encoding instruction set architectures as a vehicle for code size reduction, benchmarking of typical embedded codes when targeting Power Architecture devices has shown variable-length encoding (VLE)-enabled code to be approximately 30% smaller in code footprint size than standard Power Architecture code, while only exhibiting a 5% reduction in code performance. Resulting minor degradations in code performance are typical, due to limitations in functionality when using a reduced instruction encoding format of an instruction.

6.3 Caveat Emptor: Compiler Optimization Orthogonal to Code Size

When compiling code for a production release, developers often want to exploit as much compile-time optimization of their source code as possible in order to achieve the best performance possible. While building projects with -Os as an option will tune the code for optimal code size, it may also restrict the amount of optimization that is performed by the compiler due to such optimizations resulting in increased code size. As such, a user may want to keep an eye out for errant optimizations performed typically around loop nests and selectively disable them on a one-by-one use case rather than disable them for an entire project build. Most compilers support a list of pragmas that can be inserted to control compile-time behavior. Examples of such pragmas can be found in documentation accompanying the build tools for a processor.

Software pipelining is one optimization that can result in increased code size due to additional instructions that are inserted before and after the loop body of the transformed loop. When the compiler or assembly programmer software pipelines a loop, overlapping iterations of a given loop nest are scheduled concurrently with associated "set up" and "tear down" code inserted before and after the loop body. These additional instructions inserted in the set up and tear down, or prolog and epilog as they are often referred to in the compiler community, can result in increased instruction counts and code sizes. Typically, a compiler will offer a pragma such as "#pragma noswp" to disable software pipelining for a given loop nest, or given loops within a source code file. Users may want to utilize such a pragma on a loop-by-loop basis to reduce increases in code size associated with select loops that may not be performance-critical or on the dominant runtime paths of the application.

Loop unrolling is another fundamental compiler loop optimization that often increases the performance of loop nests at runtime. By unrolling a loop so that multiple iterations of the loop reside in the loop body, additional instruction-level parallelism is exposed for the compiler to schedule on the target processor; in addition, fewer branches with branch delay slots must be executed to cover the entire iteration space of the loop nest, potentially increasing the performance of the loop as well. Because multiple iterations of the loop are cloned and inserted into the loop body by the compiler, however, the body of the loop nest typically grows as a multiple of the unroll factor. Users wishing to maintain a modest code size may wish to selectively disable loop unrolling for certain loops within their code production, at the cost of compiled code runtime performance. By selecting those loop nests that may not be on the performance-critical path of the application, savings in code size can be achieved without impacting performance along the dominant runtime path of the application. Typically compilers will support pragmas to control loop unrolling-related behavior, such as the minimum number of iterations for which a loop will exist or various unroll factors to pass to the compiler. Examples of disabling loop unrolling via a pragma are often in the form "#pragma nonroll." Please refer to your local compiler's documentation for the correct syntax for this and related functionality.

7 Data Structures

Appropriate selection of data structures, before the design of kernels that compute over them, can have significant impact when dealing with high-performance embedded DSP codes. This is often especially true for target processors that support SIMD instruction sets and optimizing compiler technology, as was detailed previously. As an illustrative example, this section details the various trade-offs between using array-of-structure elements vs. structure-of-array elements for commonly used data structures as well as selection of data element sizes.

7.1 Arrays of Data Structures

As a data structure example, we'll consider a set of six dimensional points that are stored within a given data structure as either an array of structures or a structure of arrays, as detailed in Fig. 9.

The array of structures, as depicted on the left-hand side of Fig. 9, details a structure that has six fields of floating-point type, each of

```
/* array of structures*/

struct {

    float x_00;

    float y_00;

    float z_00;

    float x_01;

    float y_01;

    float z_01;

} list[SIZE];
```

```
/* structure of arrays */

struct {

    float x_00[SIZE];  float y_00[SIZE];

    float z_00[SIZE];  float x_01[SIZE];

    float y_01[SIZE];  float z_01[SIZE];

} list;
```

Fig. 9 Array of structures vs. structure of arrays.

which might be the three coordinates of the end points of a line in three-dimensional space. The structures are allocated as an array of SIZE elements. The structure of arrays, which is represented on the right-hand side, creates a single data structure that contains six arrays of floating-point data type, each of which is of SIZE elements. It should be noted that all the data structures above are functionally equivalent but have varying system side effects regarding memory system performance and optimization.

Looking at the array-of-structures example in the previous text, for a given loop nest that is known to access all the elements of a given struct element before moving onto the next element in the list, good locality of data will be exhibited. This will be because as cache lines of data are fetched from memory into the data cache lines, adjacent elements within the data structure will be fetched contiguously from memory and exhibit local reuse.

One downside when using the array-of-structures data structure, however, is that each individual memory reference in a loop that touches all the field elements of the data structure does not exhibit unit memory stride. For example, consider the illustrative loop in Fig. 10.

Each of the field accesses in the loop in Fig. 10 accesses different fields within an instance of the structure and does not exhibit unit stride memory access patterns which would be conducive to compiler-level autovectorization. In addition, any loop that traverses the list of structures and accesses only one or a few fields within a given structure instance will exhibit rather poor spatial locality of data within the cases, due to fetching cache lines from memory that contain data elements which will not be referenced within the loop nest.

In the next section, we see how migrating to an array-of-structures data format may work to the developer's advantage.

```
for(i=0 i<SIZE; ++i)

{

    local_struct[i].x_00 = 0.00;

    local_struct[i].y_00 = 0.00;

    local_struct[i].z_00 = 0.00;

    local_struct[i].x_01 = 0.00;

    local_struct[i].y_01 = 0.00;

    local_struct[i].z_01 = 0.00;

}
```

Fig. 10 Loop iterating over data structure fields.

7.2 Data Structures of Arrays

As seen above, using the array-of-structures format can result in suboptimal stride access patterns at runtime, thereby precluding some compile-time optimization. We can contrast the rather bleak use case depicted earlier by migrating the array-of- structures format to the structure-of-arrays format, as depicted in the loop nest in Fig. 11.

By employing the structure-of-arrays data structure, each field access within the loop nest exhibits unit stride memory references across loop iterations. This is much more conducive to autovectorization by the build tools in most cases. In addition, we still see good locality of data across the multiple array streams within the loop nest. It should also be noted that in contrast to the previous scenario, even if only one field is accessed by a given loop nest, locality within the cache is achieved due to subsequent elements within the array being prefetched for a given cache line load.

While the examples presented previously detail the importance of selecting the data structure that best suits the application developer's needs, it is assumed that the developer or system architect will study the overall application hot spots in driving the selection of appropriate data structures for memory system performance. The result may not be a clear case of black and white, however, and a solution that employs multiple data structure formats may be advised. In these cases, developers may wish to use a hybrid-type approach that mixes and matches between structure-of-array and array-of-structure formats. Furthermore, for legacy code bases which, for various reasons, are tightly coupled to their internal data structures—an explanation of which falls outside the scope of this chapter—it may be worthwhile to runtime convert between the various formats as needed. While the computation required to convert from one format to another is

```
for(i=0 i<SIZE; ++i)
{
        local_struct.x_00[i] = 0.00;

        local_struct.y_00[i] = 0.00;

        local_struct.z_00[i] = 0.00;

        local_struct.x_01[i] = 0.00;

        local_struct.y_01[i] = 0.00;

        local_struct.z_01[i] = 0.00;

}
```

Fig. 11 Loop iterating of data structure of arrays.

nontrivial, there may be use cases where the conversion overhead is dramatically offset by the computational and memory system performance enhancements achieved once the conversion is performed.

7.3 SIMD-Based Optimization and Memory Alignment

In summary, data alignment details the way data is accessed within the computer's memory system. The alignment of data within the memory system of an embedded target can have rippling effects on the performance of the code, as well as on the ability of development tools to optimize certain use cases. On many embedded systems, the underlying memory system does not support unaligned memory accesses, or such accesses are supported with a certain performance penalty. If the user does not take care in aligning data properly within the memory system layout, performance can be lost. Again, data alignment details the way data is accessed within the computer's memory system. When a processor reads or writes to memory, it will often do this at the resolution of the computer's word size, which might be 4 bytes on a 32-bit system. Data alignment is the process of putting data elements at offsets that are some multiple of the computer's word size, so that various fields may be accessed efficiently. As such, it may be necessary for users to put padding into their data structures or for the tools to automatically pad data structures according to the underlying ABI and data type conventions when aligning data for a given processor target.

Alignment can have an impact on compiler and loop optimizations, like vectorization.

For instance, if the compiler is attempting to vectorize computation occurring over multiple arrays within a given loop body, it will need to know whether the data elements are aligned to make efficient use of packed SIMD move instructions, and to know whether certain iterations of the loop nest that execute over nonaligned data elements must be peeled off. If the compiler cannot determine whether the data elements are aligned, it may opt to not vectorize the loop at all, thereby leaving the loop body sequential in schedule. Clearly this is not the desired result for the best performing executable.

Alternatively, the compiler may decide to generate multiple versions of the loop nest with a runtime test to determine at loop execution time whether the data elements are aligned. In this case the benefits of a vectorized loop version are obtained; however, the cost of a dynamic test at runtime is incurred and the size of the executable will increase due to multiple versions of the loop nest being inserted by the compiler.

Users can often do multiple things to ensure that their data is aligned, for instance padding elements within their data structures and ensuring that various data fields lie on the appropriate word boundaries. Many compilers also support sets of pragmas to denote that a given element is aligned. Alternatively, users can put various asserts within their code to compute at runtime whether the data fields are aligned on a given boundary before a version of a loop executes.

7.3.1 Selecting Appropriate Data Types

It is important that application developers also select the appropriate data types for their performance-critical kernels in addition to the strategies of optimization. When the minimal acceptable data type is selected for computation, it may have several secondary effects that can be beneficial to the performance of the kernels. Consider, for example, a performance-critical kernel that can be implemented in either 32-bit integral computation or 16-bit integral computation due to the application programmer's knowledge of the data range. If the application developer selects 16-bit computation using one of the built-in C/C11 language data types, like "short int," then the following benefits may be gained at system runtime.

By selecting 16-bit over 32-bit data elements, more data elements can fit into a single data cache line. This allows fewer cache line fetches per unit of computation and should help alleviate the compute-to-memory bottleneck when fetching data elements. In addition, if the target architecture supports SIMD-style computation, it is highly likely that a given ALU within the processor can support multiple 16-bit computations in parallel vs. their 32-bit counterparts. For example, many commercially available DSP architectures support packed 16-bit SIMD operations per ALU, effectively doubling the computational throughput when using 16-bit data elements vs. 32-bit data elements. Given the packed nature of the data elements, whereby additional data elements are packed per cache line or can be placed in user-managed scratchpad memory, coupled with increased computational efficiency, it may also be possible to improve the power efficiency of the system due to the reduced number of data memory fetches required to fill cache lines.

9

EMBEDDED SOFTWARE QUALITY, INTEGRATION, AND TESTING TECHNIQUES

Mark Pitchford

LDRA, Monks Ferry, United Kingdom

CHAPTER OUTLINE

Software Engineering for Embedded Systems. https://doi.org/10.1016/B978-0-12-809448-8.00009-6

1 What Is Software Test?

There is some inconsistency in how the word "test" is used in the context of software development. For some commentators "software test" implies the execution of software and the resulting confirmation that it performs as was intended by the development team—or not. Such a definition views the inspection or analysis of source code as a different field; that is, one to be contrasted with software test rather than a branch of it.

For the purposes of this chapter the *Oxford English Dictionary*'s definition of the word "test" is applied: "a procedure intended to establish the quality, performance, or reliability of something, especially before it is taken into widespread use."

Any activity which fits that definition can therefore be regarded as a software test, whether it involves code execution or not.

The generic term "static analysis" is used to describe a branch of software test involving the analysis of software without the execution of the code. Conversely, "dynamic analysis" describes a branch of software test in which the code is indeed executed.

2 Why Should We Test Software?

Returning to the definition of "test," software is tested to establish its "quality, performance, or reliability." Testing itself only establishes these characteristics; it does not of itself guarantee that software meets any particular criteria for them.

The aim then is to quantify the standard of the software. Whether that standard is good enough depends very largely on the context in which it will be deployed.

3 How Much Testing Is Enough?

One approach to static analysis focuses on the checking for adherence to coding rules or the achievement of particular quality metrics. Such an approach is usually easy enough to scope. Either code meets the rules or it does not, and if it does not it is either justified or corrected.

Other static analysis tools are designed to predict the dynamic behavior of source code. Such heuristic mechanisms are most commonly used for "bug finding"—looking for evidence of where source code is likely to fail, rather than enforcing standards to be sure that it will not. These are arguably complementary to other static and dynamic techniques and are often easy to apply to get a marked improvement in code quality. However, they sometimes lack the thoroughness sought in the development of mission-, safety-, or security-critical software.

Dynamic testing is less easy to apply in even fairly trivial applications. The possible combinations and permutations of data values and execution paths can be large enough to make it wholly impractical to prove that all possible scenarios are correctly handled.

This means that almost irrespective of how much time is spent performing software tests of whatever nature an element of risk will remain with regard to the potential failure of those scenarios that remain unproven.

Consequently, the decision on what and how much to test becomes a question of cost vs. the impact of the risk outcomes identified. Those risk outcomes include not only the risk of software failure, but also factors such as the risk of delaying the launch of a commercial product and conceding the initiative in the market to a competitor.

Testing is not a cheap activity and there is the cost of both labor and associated test tools to take into account. On the opposite side of the equation lies the consequence of flawed software. What is the likely outcome of failure? Could it kill, maim, or cause temporary discomfort? Could it yield control of the application to bad actors or make personally identifiable information (PII) accessible? Or is a mildly irritating occasional need to restart the application the only feasible problem?

Clearly, the level of acceptable risk to health, safety, and security in each of these scenarios is significantly different, and the analysis is further complicated if there are also commercial risk factors to be added into that equation.

Some standards such as IEC 61508 (see Section 7) define a structured approach to this assessment. In this standard, software integrity level (SIL) 1 is assigned to any parts of a system in continuous use for which a probability of failure on demand of 10^{-5}-10^{-6} is permissible. SILs become more demanding the higher the number assigned, so that SIL2 implies an acceptable probability of failure on demand as 10^{-6}-10^{-7}, SIL 3 as 10^{-7}-10^{-8}, and SIL 4 as 10^{-8}-10^{-9}.

The standard recommends the application of many techniques to varying degrees for each of these SILs on the basis that the proficient application of the specified techniques will provide sufficient evidence to suggest that the maximum acceptable risk level will not be exceeded.

Ultimately, then, the decision is about how the software can be proven to be of adequate quality. In many cases this ethos allows different SIL levels to be applied to different elements of a project depending on the criticality of each such element.

That principle can, of course, be extended outside the realms of high-integrity applications. It always makes sense to apply more rigorous test to the most critical parts of an application.

4   When Should Testing Take Place?

To some extent that depends on the starting point. If there is a suite of legacy code to deal with, then clearly starting with a new test regime at the beginning of development is not an option! However, "the sooner, the better" is a reasonable rule of thumb.

In general terms the later a defect is found in product development, the more costly it is to fix—a concept first established in 1975 with the publication of Brooks' *The* "Mythical Man-Month" and proven many times since through various studies.

The automation of any process changes the dynamic of justification, and that is especially true of test tools given that some are able to make earlier unit test much more feasible (Fig. 1).

5   Who Makes the Decisions?

It is clear that the myriad of interrelated decisions on what, when, why, how, and how much to test is highly dependent on the reasons for doing so.

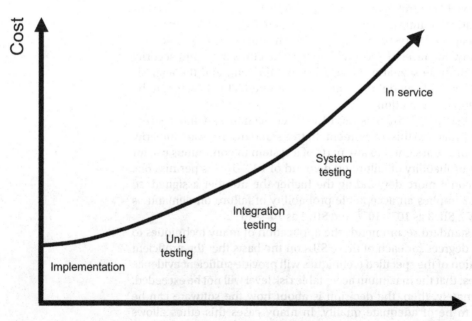

Fig. 1 The later a defect is identified, the higher the cost of rectifying it.

The judgments are perhaps relatively straightforward if an outside agency is involved. For instance, the developers of a control system for use in an aircraft will have to adhere to DO-178C for their product to become commercially available for use in international airspace. It then becomes clear that if the product is to sell, then a level of test that is appropriate to the standard is unavoidable.

This extends further to the qualification or certification of any tools to be used. That can vary quite significantly from one standard to another, but there are usually guidelines or instructions are laid down on what is required.

Conversely, the driver might be an internal one to improve software quality and improve corporate reputation and reduce recall costs. In that case the matter is a decision for management who will need to make judgments as to how much investment in the associated work and tools is appropriate.

6 Available Techniques

Enter the phrase "software test" into any browser and the variation in scope of test techniques and test tools is daunting. Static analysis, coding standards, quality metrics, source code coverage, object code coverage, dynamic analysis, memory leak profiling, abstract interpretation ... the list of buzzwords and techniques is seemingly endless.

The resulting confusion is compounded by the fact that the boundaries between different techniques and approaches are not as clearcut as they might be. "Static analysis" is a prime example of a term that means different things to different observers.

6.1 Static and Dynamic Analysis

The generic term "static analysis" is used only to indicate that analysis of the software is performed without executing the code, whereas "dynamic analysis" indicates that the code is indeed executed. So, simple peer review of source code and functional test fit the definitions of static and dynamic analysis, respectively. The boundaries become blurred when it is understood that static analysis can be used to predict dynamic behavior. As a result, it is a precision tool in some contexts and yet in others it harbors approximations.

To cut through this vague terminology it is useful to consider five key elements of analysis. These are all deployed in one form or another by analysis tools, but many can be and frequently are implemented from first principles usually in combination to provide a "tool kit" of techniques.

The first three are approaches to static analysis. Note that these attributes do not comprehensively describe the categories of static analysis tools. Many tools include more than one of these attributes, and it is possible to approximate each of them without the use of tools at all.

6.1.1 Code Review

Code review traditionally takes the form of a peer review process to enforce coding rules to dictate coding style and naming conventions, and to restrict commands available for developers to a safe subset.

Peer review of software source code was established to achieve effective code review long before any tools automated it, and is still effective today. The key to effective peer reviews is to establish a mutually supportive environment so that the raising of nonconformities is not interpreted as negative criticism.

If manual peer review is to be adopted with such standards as the MISRA ones in mind, then a subset of the rules considered most important to the developing organization is likely to yield the best results.

Many software test tools automate this approach to provide a similar function with benefits in terms of the number and complexity of rules to be checked and in terms of speed and repeatability.

Code review does not predict dynamic behavior. However, code written in accordance with coding standards can be expected to include fewer flaws that might lead to dynamic failure, and assuring a consistent approach from individuals brings its own benefits in terms of readability and maintainability.

Code review can be applied whether the code under development is for a new project, an enhancement, or a new application using existing code. With legacy applications, automated code review is particularly strong for presenting the logic and layout of such code to establish an understanding of how it works with a view to further development. On the other hand, with a new development the analysis can begin as soon as any code is written—no need to wait for a compilable code set, let alone a complete system.

6.1.2 Theorem Proving

Theorem proving defines desired component behavior and individual runtime requirements.

The use of assertions within source code offers some of the benefits of the theorem-proving tools. Assertions placed before and after algorithms can be used to check that the data passing through them meet particular criteria or are within particular bounds.

These assertions can take the form of calls to an "assert" function as provided in languages such as C++, or the form of a user-defined mechanism to perhaps raise an error message or set a system to a safe state.

Automated theorem proof tools often use specially formatted comments (or "annotations") in the native language. These comments can be statically analyzed to confirm the code accurately reflects these definitions, which are ignored by a standard compiler. Because of these annotations verification can concentrate on verification conditions: that is, checking that when one starts under some preconditions and executes such a code fragment the postcondition will be met.

The writing of annotations can be labor intensive and so these tools tend to be limited to highly safety-critical applications where functional integrity is absolutely paramount over any financial consideration (e.g., flight control systems).

Unlike the prediction of dynamic behavior through static analysis the use of "Design by Contract" principles often in the form of specially formatted comments in high-level code can accurately formalize and validate the expected runtime behavior of source code.

Such an approach requires a formal and structured development process, one that is textbook style and has uncompromising precision. Consequently, applying the retrospective application of such an approach to legacy code would involve completely rewriting it.

6.1.3 Prediction of Dynamic Behavior Through Static Analysis

The prediction of dynamic behavior through static analysis mathematically models the high-level code to predict the probable behavior of executable code that would be generated from it. All possible execution paths through that mathematical model are then simulated, mapping the flow of logic on those paths coupled with how and where data objects are created, used, and destroyed.

The net result consists of predictions of anomalous dynamic behavior that could possibly result in vulnerabilities, execution failure, or data corruption at runtime.

Although there is no practical way of exactly performing this technique manually, the use of defensive code and bounds checking within source code offers a different approach to yielding some of the benefits. For example, many of these heuristic tools use a technique called Abstract Interpretation to derive a computable semantic interpretation of the source code. In turn, this is used to analyze possible data ranges to predict any problematic scenarios at runtime. Defensive programming can make no such predictions but assertions, say, placed before and after algorithms can be used to defend against such scenarios by checking that the data passing through them meet particular criteria or are within particular bounds—and that includes checking for the circumstances that may cause runtime errors of the type generally sought out by tools of this nature.

As before, these assertions can take the form of calls to an "assert" function as provided in languages such as C++ or the form of a user-defined mechanism, and it is highly pragmatic to use assertions in the most difficult and complex algorithms where failures are most likely to occur.

When tools are available the static prediction of dynamic behavior works well for existing code or less rigorously developed applications. It does not rely on a formal development approach and can simply be applied to the source code as it stands, even when there is no in-depth knowledge of it. That ability makes this methodology very appealing for a development team in a fix—perhaps when timescales are short, but catastrophic and unpredictable runtime errors keep coming up during system test.

There is, however, a downside. The code itself is not executing, but instead is being used as the basis for a mathematical model. As proven by the works of Church, Gödel, and Turing in the 1930s a precise representation of the code is mathematically insoluble for all but the most trivial examples. In other words the goal of finding every defect in a nontrivial program is unreachable unless approximations are included that by definition will lead to false-positive warnings.

The complexity of the mathematical model also increases dramatically as the size of the code sample under analysis gets bigger. This is often addressed by the application of simpler mathematical modeling for larger code samples, which keeps the processing time within reasonable bounds. But, increases in the number of these "false positives," which has a significant impact on the time required to interpret results, can make this approach unusable for complex applications.

The last two of the "key attributes" concern dynamic analysis. Note that these attributes do not comprehensively describe the categories of dynamic analysis and that many tools include more than one of these attributes.

An overlap between static and dynamic analysis appears when there is a requirement to consider dynamic behavior. At that point the dynamic analysis of code that has been compiled, linked, and executed offers an alternative to the prediction of dynamic behavior through static analysis.

Dynamic analysis involves the compilation and execution of the source code either in its entirety or on a piecemeal basis. Again, while many different approaches can be included, these characteristics complete the list of the five key attributes that form the fundamental "toolbox of techniques."

6.1.4 Structural Coverage Analysis

Structural coverage analysis details which parts of compiled and linked code have been executed, often by means of code instrumentation "probes."

In its simplest form these probes can be implemented with manually inserted print statements as appropriate for the programming language of choice. Although such an approach demands in-depth knowledge of the code under test and carries the potential for human error, it does have a place in smaller projects or when only practiced on a critical subset of an application.

A common approach is for automated test tools to automatically add probes to the high-level source code before compilation.

Adding instrumentation probes obviously changes the code under test, making it both bigger and slower. There are therefore limitations to what it can achieve and to the circumstances under which it can be used, especially when timing errors are a concern. However, within appropriate bounds it has been highly successful and in particular has made a major contribution to the sound safety record of software in commercial aircraft.

Some test tools can perform structural coverage analysis in isolation or in combination with unit, module, and/or integration testing.

6.1.5 Unit, Module, and Integration Testing

Unit, module, and integration testing (referred to collectively as unit testing hereafter) all describe an approach in which snippets of software code are compiled, linked, and built in order that test data (or "vectors") can be specified and checked against expectations.

Traditionally, unit testing involves the development of a "harness" to provide an environment where the subset of code under test can be exposed to the desired parameters for the tester to ensure that it behaves as specified. More often than not in modern development environments the application of such techniques is achieved through the use of automated or semi-automated tools. However, a manual approach can still have a place in smaller projects or when only practiced on a critical subset of an application.

Some of the leading automated unit test tools can be extended to include the automatic definition of test vectors by the unit test tool itself.

Unit testing and structural coverage analysis focus on the behavior of an executing application and so are aspects of dynamic analysis. Unit, integration, and system test use code compiled and executed in a similar environment to that being used by the application under development.

Unit testing traditionally employs a bottom-up testing strategy in which units are tested and then integrated with other test units. In the course of such testing, individual test paths can be examined by means of structural coverage analysis. There is clearly no need to have a complete code set to hand to initiate tests such as these.

Unit testing is complemented by functional testing, a form of top-down testing. Functional testing executes functional test cases, perhaps in a simulator or in a target environment, at the system or subsystem level.

Clearly, these dynamic approaches test not only the source code, but also the compiler, linker, development environment, and potentially even target hardware. Static analysis techniques help to produce high-quality code that is less prone to error, but when it comes to proving correct functionality there is little alternative but to deploy dynamic analysis. Unit test or system test must deploy dynamic analysis to prove that the software actually does what it is meant to do.

Perhaps the most telling point with regard to the testing of dynamic behavior—whether by static or dynamic analysis—is precisely what is being tested. Intuitively, a mathematical model with inherent approximations compared with code being compiled and executed in its native target environment suggests far more room for uncertainty.

If the requirement is for a quick fix solution for some legacy code that will find most problems without involving a deep understanding of the code, then the prediction of dynamic behavior via static analysis has merit. Similarly, this approach offers quick results for completed code that is subject to occasional dynamic failure in the field.

However, if there is a need to prove not only the functionality and robustness of the code, but also provide a logical and coherent development environment and integrated and progressive development process, then it makes more sense to use dynamic unit and system testing. This approach provides proof that the code is robust and that it does what it should do in the environment where it will ultimately operate.

6.2 Requirements Traceability

As a basis for all validation and verification tasks all high-quality software must start with a definition of requirements. This means that each high-level software requirement must map to a lower level requirement, design, and implementation. The objective is to ensure that the complete system has been implemented as defined and that there is no surplus code. Terminology may vary regarding what different requirement tiers are called, but this fundamental element of sound software engineering practice remains.

Simply ensuring that system-level requirements map to something tangible in the requirements decomposition tree, design and implementation is not enough. The complete set of requirements comes from multiple sources, including system-level requirements, high-level requirements, and low-level (or derived) requirements. As illustrated below there is seldom a 1:1 mapping from system-level requirements to source code, so a traceability mechanism is required to map and record the dependency relationships of requirements throughout the requirements decomposition tree (Fig. 2).

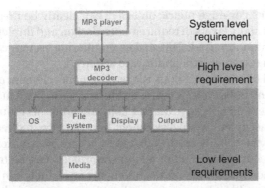

Fig. 2 Example of "1:Many" mapping from system-level requirements through a requirements decomposition tree.

To complicate matters further each level of requirements might be captured using a different mechanism. For instance, a formal requirements capture tool might be used for system-level requirements while high-level requirements are captured in PDF and low-level requirements captured in a spreadsheet.

Modern requirements traceability solutions enable mapping throughout these levels right down to the verification tasks associated with the source code. The screenshot (Fig. 3) shows an example of this. Using this type of requirements traceability tool the 100%

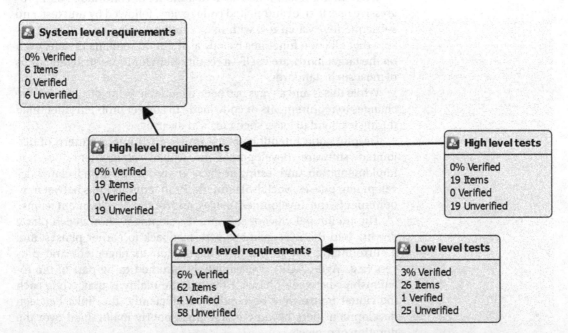

Fig. 3 Traceability from high-level requirements down to source code and verification tasks.

requirements coverage metric objective can clearly be measured, no matter how many layers of requirements, design, and implementation decomposition are used. This makes monitoring system completion progress an extremely straightforward activity.

It would be easy to overlook the requirements element of software development, but the fact is that even the best static and dynamic analysis in tandem do not prove that the software fulfills its requirements.

Widely accepted as a development best practice, bidirectional requirements traceability ensures that not only are all requirements implemented, but also that all development artifacts can be traced back to one or more requirements. Requirements traceability can also cover relationships with other entities such as intermediate and final work products, changes in design documentation, and test plans. Standards such as the automotive ISO 26262 or medical IEC 62304 demand bidirectional traceability, and place constant emphasis on the need for the complete and precise derivation of each development tier from the one above it.

When requirements are managed well, traceability can be established from the source requirement to its lower level requirements and from the lower level requirements back to their source. Such bidirectional traceability helps determine that all source requirements have been completely addressed and that all lower level requirements can be traced to a valid source.

Such an approach lends itself to a model of continuous and progressive use: first, of automated code review, followed by unit test, and subsequently system test with its execution tracing capability to ensure that all code functions exactly as the requirements dictate, even on the target hardware itself—a requirement for more stringent levels of most such standards.

While this is and always has been a laudable principle, last-minute changes to requirements or code made to correct problems identified during test tend to leave such ideals in disarray.

Despite good intentions many projects fall into a pattern of disjointed software development in which requirements, design, implementation, and testing artifacts are produced from isolated development phases. Such isolation results in tenuous links between requirements, the development stages, and/or the development teams.

The traditional view of software development shows each phase flowing into the next, perhaps with feedback to earlier phases, and a surrounding framework of configuration management and process (e.g., Agile, RUP). Traceability is assumed to be part of the relationships between phases. However, the reality is that, while each individual phase may be conducted efficiently, the links between development tiers become increasingly poorly maintained over the duration of projects.

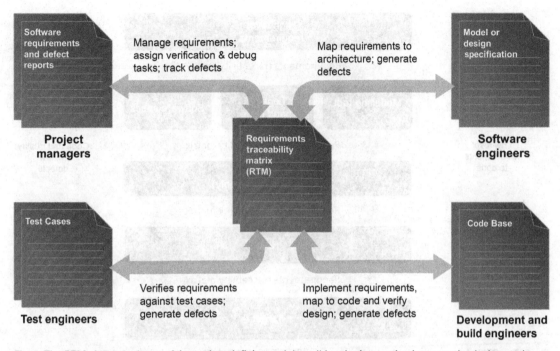

Fig. 4 The RTM sits at the heart of the project defining and describing the interaction between the design, code, test, and verification stages of development.

The answer to this conundrum lies in the requirements traceability matrix (RTM) that sits at the heart of any project even if it is not identified as such (see Fig. 4). Whether or not the links are physically recorded and managed they still exist. For example, a developer creates a link simply by reading a design specification and using that to drive the implementation.

Safety- and security-critical standards dictate that requirements should be traceable down to high-level code and in some cases object code, but elsewhere more pragmatism is usually required. A similar approach can be taken for any project with varying levels of detail depending on criticality both of the project as a whole and within an individual project. The important factor is to provide a level of traceability that is adequate for the circumstance.

This alternative view of the development landscape illustrates the importance that should be attached to the RTM. Due to this fundamental centrality it is vital that project managers place sufficient priority on investing in tooling for RTM construction. The RTM must also be represented explicitly in any life cycle model to emphasize its importance (as Fig. 5 illustrates). With this elevated focus the RTM is constructed and maintained efficiently and accurately.

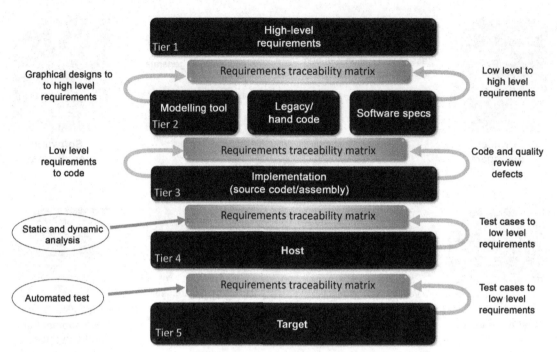

Fig. 5 The RTM plays a central role in a development life cycle model. Artifacts at all stages of development are linked directly to the requirements matrix, and changes within each phase automatically update the RTM so that overall development progress is evident from design through coding and test.

When the RTM becomes the center of the development process it impacts on all stages of design from high-level requirements through to target-based deployment. Where an application is safety critical each tier is likely to be implemented in full, but once again a pragmatic interpretation of the principles can be applied to any project.

Tier 1 high-level requirements might consist of a definitive statement of the system to be developed. This tier may be subdivided depending on the scale and complexity of the system.

Tier 2 describes the design of the system level defined by Tier 1. Above all, this level must establish links or traceability with Level 1 and begin the process of constructing the RTM. It involves the capture of low-level requirements that are specific to the design and implementation and have no impact on the functional criteria of the system.

Tier 3's implementation refers to the source/assembly code developed in accordance with Tier 2. Verification activities include code rule checking and quality analysis. Maintenance of the RTM presents many challenges at this level as tracing requirements to source code files may not be specific enough and developers may need to link to individual functions.

In many cases the system is likely to involve several functions. The traceability of those functions back to Tier 2 requirements includes many-to-few relationships. It is very easy to overlook one or more of these relationships in a manually managed matrix.

In **Tier 4** formal host-based verification begins. Once code has been proven to meet the relevant coding standards using automated code review, unit, then integration and system tests may be included in a test strategy that may be top-down, bottom-up, or a combination of both. Software simulation techniques help create automated test harnesses and test case generators as necessary, and execution histories provide evidence of the degree to which the code has been tested.

Such testing could be supplemented with robustness testing if required, perhaps by means of the automatic definition of unit test vectors or by static prediction of dynamic behavior.

Test cases from Tier 4 should be repeatable at Tier 5, if required.

This is the stage that confirms the software functions as intended within its development environment, even though there is no guarantee it will work when in its target environment. Testing in the host environment first allows the time-consuming target test to merely confirm that the tests remain sound in the target environment.

Tier 5's target-based verification represents the on-target testing element of formal verification. This frequently consists of a simple confirmation that the host-based verification performed previously can be duplicated in the target environment, although some tests may only be applicable in that environment itself.

Where reliability is paramount and budgets permit, the static analysis of dynamic behavior with its "full range" data sets would undoubtedly provide a complementary tool for such an approach. However, dynamic analysis would remain key to the process.

6.3 Static Analysis—Adherence to a Coding Standard

One of the most basic attributes of code that affects quality is readability. The more readable a piece of code is, the more testable it is. The more testable it is, the more likely it will have been tested to a reasonable level of completion. Unfortunately, as The International Obfuscated C Code Contest has demonstrated, there are many ways to create complex and unreadable code for the simplest of applications. This metric is about adopting even a basic coding standard to help enhance code quality by establishing the rules for a minimum level of readability for all the code created within a project.

Modern coding standards go way beyond just addressing readability, however. Encapsulating the wisdom and experience of their creators, coding standards, such as the Motor Industry Software

Reliability Association (MISRA) C and C++ coding standards and the JSF Airborne Vehicle C++ standard or the Barr group (formerly Netrino) Embedded C Coding standard, also identify specific code constructs that can affect overall code quality and reliability, such as areas of C or C++ that the ISO standards state are either undefined or implementation specific.

Coding standards, such as the CERT-C or C++ Secure Coding Standards and the Common Weakness Enumeration list (CWE) also help to identify code constructs that can lead to potentially exploitable vulnerabilities in code.

The optimum coding standard for a project will depend on the project objectives. Fig. 6 provides a simple outline of the objectives for several coding standards.

In practice, most projects will create their own custom standard that uses one or more of these as a baseline, and modify the standard to suit their particular needs. Clearly, software that is safe AND secure is often desirable! Fortunately, the same attributes that make code safe very frequently also make it secure, and it is no coincidence that the MISRA organization, for example, have always focused their guidelines on critical systems, rather than safety-critical systems. The 2016 release of MISRA C:2012 AMD1, "Additional security guidelines for MISRA C:2012," further reinforced that position.

One area, in particular, where these reliability and security-oriented coding standards excel is in identifying code constructs that lead to latent defects, which are defects that are not normally detected during the normal software verification process yet reveal themselves once the product is released. Consider the following simple example:

```
1   #include <stdio.h>
2   #include <stdint.h>
3
4   #define MAX_SIZE 16U
5
6   int32_t main(void)
7   {
8       uint16_t theArray[MAX_SIZE];
9       uint16_t idx;
10      uint16_t *p_var;
```

Coding standard	Language	Objective/application
MISRA	C & C++	High reliability software
JSF Airborne Vehicle C++ Standard	C++	High reliability software
Barr Group Embedded C Standard	C	Defect free Embedded C Code
CERT Secure C Coding Standard	C	Secure Software

Fig. 6 Outline of objectives for several popular coding standards.

```
11   uint16_t UR_var;
12
13   p_var = &UR_var;
14
15   for(idx = 0U; idx < MAX_SIZE; idx += *p_var;)
16   {
17     theArray[idx] = 1U;
18   }
19
20   for(idx = 0U; idx <= MAX_SIZE; idx++)
21   {
22     printf(" %d", theArray[idx]);
23   }
24
25   return(0);
26 }
```

It compiles without warnings using either GCC or Microsoft Visual Studio (the latter requires the user to provide a stdint.h implementation for versions earlier than 2010). On inspection an experienced programmer can find the errors in this fairly simple code which contains both an array-out-of-bounds error (an off-by-one error on line 20 when variable idx == MAX_SIZE) and a reference to an uninitialized variable (on line 11, and line13 for the loop counter increment operator). These violate MISRA C:2012 rules 18.1 (relating to out-of-bounds errors) and 9.1 (which prohibits an object from being read before it has been set), respectively.

At its most basic an array-out-of-bounds error is a buffer overflow, even though it is legal C and/or C++ code. For secure code, buffer overflow is one of the most common vulnerabilities leading to the worst possible type of exploit: the execution of arbitrary code.

Nondeterminism is also a problem with the uninitialized variables in the example. It is impossible to predict the behavior of an algorithm when the value of its variables cannot be guaranteed. What makes this issue even worse is that some compilers will actually assign default values to uninitialized variables! For example, the Microsoft Visual Studio compiler assigns the value 0xCCCC to the variable UR_var by default in debug mode. While this value is meaningless in the context of the algorithm above, it is deterministic so the code will always behave in the same way. Switch to release mode, however, and the value will be undefined resulting in nondeterministic behavior.

Although real code is never as straightforward as this, even in this example there is some isolation from the latter issue as pointer aliasing is used to reference the UR_var variable.

A further MISRA C:2012 violation (rule 14.2: "A *for* loop shall be well-formed") relates to the complexity of the loop counter in this example. It is legitimate C code, but is difficult to understand and hence potentially error prone (Fig. 7).

▲ ◈ main		
◆ Procedure contains UR data flow anomalies. : UR_var	Mandatory	MISRA-C:2012/AMD1/TC1 R.9.1
◆ For loop incrementation is not simple.	Required	MISRA-C:2012/AMD1/TC1 R.14.2
◆ Array bound exceeded. : theArray[*]; accessed=16, range=0-15	Required	MISRA-C:2012/AMD1/TC1 R.18.1

Fig. 7 Static analysis results showing enforcement of the MISRA C:2012 coding standard for the above code, revealing several violations.

In addition to the obvious benefits of identifying and eliminating latent defects the most significant additional benefit of using static analysis tools for coding standards enforcement is that it helps peer review productivity. By ensuring that a piece of code submitted for peer review has contravened no mandatory rules the peer review team can get away from focusing on minutiae and on to what they do best: ensuring that the implementations under inspection are fit for purpose and the best that they can be.

Another area where enforcement of coding standards is beneficial is in identifying unnecessary code complexity.

Complexity is not in and of itself a bad thing; complex problems require complex solutions, but code should not be more complex than necessary. This leads to sections of code that are unnecessarily difficult to read, even more difficult to test, and as a result have higher defect rates than a more straightforward equivalent implementation. It follows that unnecessary complexity is to be avoided.

Several coding standards incorporate maximum code complexity limits as a measure for improving overall code quality. The following case study explains cyclomatic complexity and knots metrics and how they may be used to show how complex a function is. The case study goes on to explain how essential cyclomatic complexity and essential knots metrics can show whether or not a function has been written in a structured manner, in turn giving a measure of complexity.

6.3.1 Essential Knots and Essential Cyclomatic Complexity—Case Study

6.3.1.1 Basic Blocks and Control Flow Branches

It is initially useful to consider how the construction of high-level software code can be described.

A "basic block" is a sequence of one or more consecutive, executable statements in a source program such that the sequence has a start point, an end point, and no internal branches.

In other words, once the first executable statement in a basic block is executed, then all subsequent statements in that basic block can be assumed to be executed in sequence.

Control flow branches and natural succession provide the links between basic blocks.

6.3.1.2 Control, Static, and Dynamic Flow Graphs

The logical flow of source, object, or assembler code can be represented by a control flow graph as first conceived by American computer scientist Frances Elizabeth "Fran" Allen. A control flow graph consists of a number of representations of basic blocks (represented by circular "nodes") interconnected with arrowed lines to represent the decision paths (called "links").

During the static analysis of source code it is possible to detect and understand the structure of the logic associated with it.

Control flow graphs form the basis of static flow graphs to show code structure. Dynamic flow graphs superimpose that representation of the code with execution history information to show which parts have been executed. In both cases the exact color-coding and symbolism used to represent different aspects of these graphs varies between tools—a fact further complicated in this book by the absence of color!

In the following static and dynamic flow graphs diamond-shaped nodes are used where a basic block includes a function call; otherwise circles are used. Relevant use of shading is explained in the context of each illustration.

In languages such as C and C++ it is often necessary to reformat the code to show only one instruction per line. This circumvents the problem of nomenclature for identifying decision points and instructions that occur within a single line of source code (Fig. 8).

6.3.1.3 Calculating a Knots Value

A knot is a point where two control flows intersect.

Knot analysis measures the amount of disjointedness in the code and hence the amount of "jumping about" a code reader will be required to undertake. An excessive number of knots may mean that a program can be reordered to improve readability and reduce complexity. A knot is not in itself a "bad thing," and knots appear in many perfectly acceptable constructs such as for, while, if/else, switch, and exception (Fig. 9).

Because they are a function of the chosen programming style and high-level language the number of knots in a function gives an indication of the complexity added to it as a result of program implementation. An excessive number of knots implies unnecessary complexity.

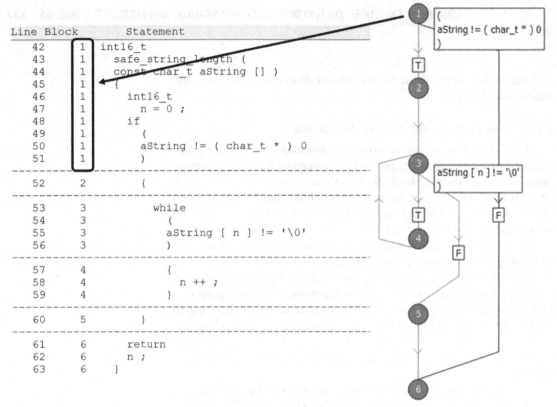

Fig. 8 Reformatted code shows the basic blocks ("nodes") connected by branches ("links").

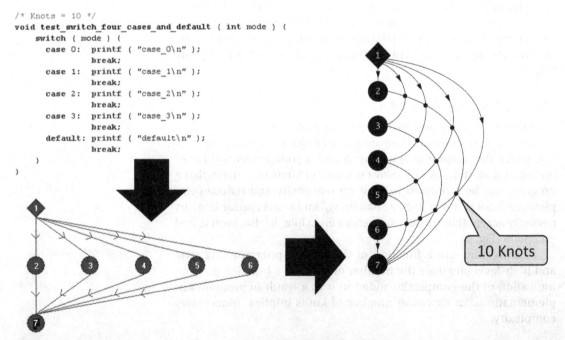

Fig. 9 A switch statement generally generates a high number of knots. There are 10 in this example, as shown on the right-hand side.

6.3.1.4   Calculating a Cyclomatic Complexity Value

Cyclomatic complexity is another measure of how complex a function is. It is a value derived from the geometry of the static flow graph for the function. The absolute value itself is therefore a little abstract and meaningless in isolation, but it provides a comparator to show the relative complexity of the problem addressed by one function vs. another.

Cyclomatic complexity is represented algebraically by the nomenclature $V(G)$ and can be derived in a number of ways, the simplest perhaps being a count of the number of "regions" separated by the links and nodes of the control or static flow graph (Fig. 10).

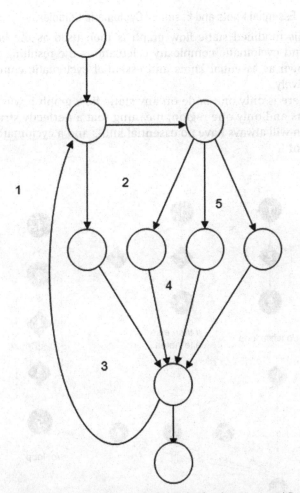

Fig. 10 An example of cyclomatic complexity derivation from a control flow graph. Here the cyclomatic complexity value $V(G) = 5$.

6.3.1.5 Identifying Structured Programming Templates—Structured Analysis

The concept of "structured" programming has been around since the 1960s, derived particularly from work by Böhm and Jacopini, and Edsger Dijkstra. In its modern implementation "structured elements" are defined as those constructs within the code that adhere to one of six "structured programming templates" as illustrated in Fig. 11.

Structured analysis is an iterative process where the static flow graph is repeatedly assessed to see whether it is possible to match one of the structured programming templates to a part of it. If so, that is "collapsed" to a single node as illustrated in Fig. 12.

The process is repeated until no more templates can be matched.

6.3.1.6 Essential Knots and Essential Cyclomatic Complexity

If this modified static flow graph is then used as the basis for knots and cyclomatic complexity calculations the resulting metrics are known as essential knots and essential cyclomatic complexity, respectively.

If there is only one node on any static flow graph it will exhibit no knots and only one region, meaning that a perfectly structured function will always have no essential knots and a cyclomatic complexity of 1.

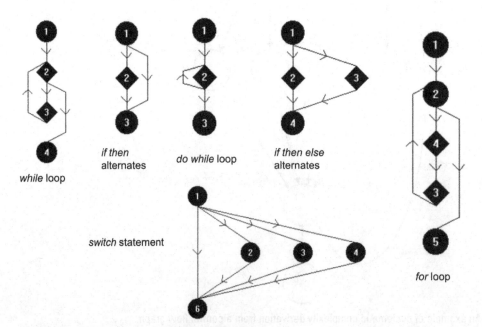

Fig. 11 The six structured programming templates.

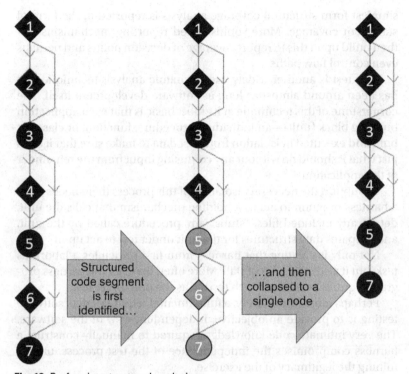

Fig. 12 Performing structured analysis.

The converse of this "perfect" result is that the essential measures will be greater than 0 and 1, respectively, showing that the code is not structured and hence may be unnecessarily complex.

6.4 Understanding Dynamic Analysis

As previously discussed, dynamic analysis involves the execution of some or all of the application source code. It is useful to consider some of the more widely used techniques that fall within this domain.

One such technique is a system-level functional test that defines perhaps the oldest test genre of them all. Simply described, when the code is written and completed the application is exercised using sample data and the tester confirms that everything works as it should.

The problem with applying this approach in isolation is that there is no way of knowing how much of the code has actually been exercised. Structural coverage analysis addresses this problem by reporting which areas of the application source code have been exercised by the test data and, more importantly, which areas have not. In its simplest form structural coverage analysis is reported in the form of statement coverage. More sophisticated reporting mechanisms can then build upon this to report coverage of decision points and perhaps even control flow paths.

Unit test is another widely used dynamic analysis technique that has been around almost as long as software development itself. The cornerstone of this technique at its most basic is that each application building block (unit)—an individual procedure, function, or class—is built and executed in isolation from test data to make sure that it does just what it should do without any confusing input from the remainder of the application.

To support the necessary isolation of this process there needs to be a harness program to act as a holding mechanism that calls the unit, details any included files, "stubs" any procedure called by the unit, and prepares data structures for the unit under test to act upon.

Not only is creating that harness from first principles a laborious task, but it also takes a lot of skill. More often than not the harness program requires at least as much testing as the unit under test.

Perhaps more importantly, a fundamental requirement of software testing is to provide an objective, independent view of the software. The very intimate code knowledge required to manually construct a harness compromises the independence of the test process, undermining the legitimacy of the exercise.

6.5   The Legacy From High-Integrity Systems

In developing applications for the medical, railway, aerospace, and defense industries, unit test is a mandatory part of a software development cycle—a necessary evil. For these high-integrity systems, unit test is compulsory and the only question is how it might be completed in the most efficient manner possible. It is therefore no coincidence that many of the companies developing tools to provide such efficiency have grown from this niche market.

In more mundane environments, perceived wisdom is often that unit testing is a nice idea in principle, but commercially unjustifiable. A significant factor in that stance is the natural optimism that abounds at the beginning of any project. At that stage why would anyone spend money on careful unit testing? There are great engineers in the team, the design is solid, and sound management is in place. What could possibly go wrong?

However, things can and do go wrong, and while unit test cannot guarantee success it can certainly help to minimize failure. It therefore makes sense to consider the principles proven to provide quick and easy unit tests in high-integrity systems in the context of less demanding environments.

6.6 Defining Unit, Module, and Integration Test

For some the terms "unit test" and "module test" are synonymous. For others the term "unit" implies the testing of a single procedure, whereas "module" suggests a collection of related procedures, perhaps designed to perform some particular purpose within the application.

Using the latter definitions manually developed module tests are likely to be easier to construct than unit tests, especially if the module represents a functional aspect of the application itself. In this case most of the calls to procedures are related and the code accesses related data structures, which makes the preparation of the harness code more straightforward.

Test tools render the distinction between unit and module tests redundant. It is perfectly possible to test a single procedure in isolation and equally possible to use the exact same processes to test multiple procedures, a file or multiple files of procedures, a class (where appropriate), or a functional subset of an entire system. As a result the distinction between unit and module test is one that has become increasingly irrelevant to the extent that the term "unit test" has come to include both concepts.

This flexibility facilitates progressive integration testing. Procedures are first unit-tested and then collated as part of the subsystems, which in turn are brought together to perform system tests.

It also provides options when a pragmatic approach is required for less critical applications. A single set of test cases can exercise a specified procedure, all procedures called as a result of exercising the single procedure as illustrated in Fig. 13, or anything in between. The use of test cases that prove the functionality of the whole call chain are easily constructed. Again, it is easy to "mix and match" the processes depending on the criticality of the code under review.

6.7 Defining Structural Coverage Analysis

The structural coverage analysis approach is all about ensuring that enough testing is performed on a system to meet its quality objectives.

For the most complete testing possible it is necessary to ensure that every possible execution path through the code under test is executed at least once. In practice, this is an unachievable aim. An observation made by G.J. Myers in 1976 explains why this is so; Myers described a 100-line program that had 10^{18} unique paths. For comparative purposes he noted that the universe is only about 4×10^{17} s old. With this observation Myers concluded that complete software execution path testing is impossible, so an approximation alternative and another metric are required to assess testing completeness.

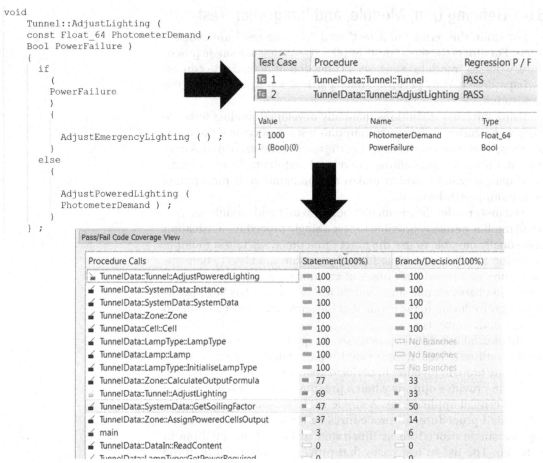

Fig. 13 A single test case (inset) can exercise some or all of the call chain associated with it. In this example "AdjustLighting" is the subject of the test.

Structural coverage analysis has proven to be an excellent technique for that purpose.

The closest structural coverage analysis metric to the 100% execution path ideal is based on the linear code sequence and jump (LCSAJ) software analysis technique, or jump-to-jump path (JJ-path) coverage as it is sometimes described. LCSAJ analysis identifies sections of code that have a single input path and a single output path, referred to as an interval. Within each interval each possible execution path is then identified. A structural coverage analysis metric is then determined by measuring which of these possible execution paths within an interval have been executed.

As with all these metrics the use of tools for measuring structural coverage analysis greatly increases measurement efficiency,

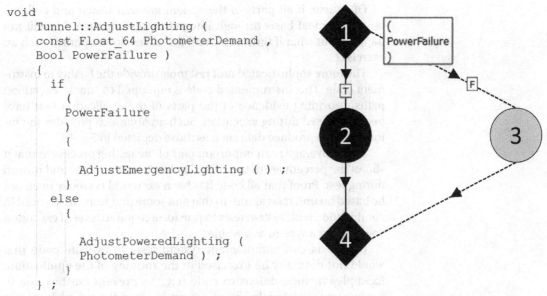

```
void
   Tunnel::AdjustLighting (
   const Float_64 PhotometerDemand ,
   Bool PowerFailure )
   {
     if
        (
        PowerFailure
        )
        {
        AdjustEmergencyLighting ( ) ;
        }
     else
        {
        AdjustPoweredLighting (
        PhotometerDemand ) ;
        }
   } ;
```

Fig. 14 Example coverage analysis results presented against a control flow graph. Black nodes and solid lines represent exercised code to simulate the color-coding used in test tools.

effectiveness, and accuracy. In addition, the visualization of results provides excellent feedback on what additional test cases are required to improve overall coverage measurements. Test tools generally use coloring to represent coverage information. In Fig. 14 and elsewhere in this chapter black nodes and solid branch lines represent exercised code.

From these results it is a straightforward exercise to determine which test data need to be generated to exercise the remaining "cold" paths, making the ability to generate the quality-oriented reports required for certification extremely straightforward.

6.8 Achieving Code Coverage With Unit Test and System Test in Tandem

Traditionally, many applications have been tested by functional means only—and no matter how carefully the test data are chosen the percentage of code actually exercised can be very limited.

That issue is compounded by the fact that the procedures tested in this way are only likely to handle data within the range of the current application and test environment. If anything changes a little—perhaps in the way the application is used or perhaps as a result of slight modifications to the code—the application could be running an entirely untested execution path in the field.

Of course, if all parts of the system are unit-tested and collated on a piecemeal basis through integration testing, then this will not happen. But what if timescales and resources do not permit such an exercise?

The more sophisticated unit test tools provide the facility to instrument code. This instrumented code is equipped to "track" execution paths, providing evidence of the parts of the application that have been exercised during execution. Such an approach provides the information to produce data such as those depicted in Fig. 14.

Code coverage is an important part of the testing process in that it shows the percentage of the code that has been exercised and proven during test. Proof that all code has been exercised correctly need not be based on unit tests alone. To that end some unit tests can be used in combination with system tests to provide a required level of execution coverage for a system as a whole.

Unit tests can complement system tests to execute code that would not normally be exercised in the running of the application. Examples include defensive code (e.g., to prevent crashes due to inadvertent division by zero), exception handlers, and interrupt handlers.

6.8.1 Unit Test and System Test in Tandem—Case Study

Consider the following function, taken from a lighting system written in C++. Line 7 includes defensive code designed to ensure that a divide by zero error cannot occur:

```
1   Sint_32 LampType::GetPowerRequired(const Float_64 LumensRequired) const
2   /* Assume a linear deterioration of efficiency from HighestPercentOutput lm/W output from each lamp at
3   maximum output, down to LowestPercentOutput lm/W at 20% output. Calculate power required based on
4   the resulting interpolation. */
5   {
6      Sint_32 Power=0;
7      if (((mMaximumLumens-mMinimumLumens)>Small) && LumensRequired>=mMinimumLumens))
8         {
9            Power = (Sint_32)(mMinimumPower + (mMaximumPower-mMinimumPower)*
10                  ((LumensRequired-mMinimumLumens)/(mMaximumLumens-mMinimumLumens)));
11        }
12        return Power;
13  }
```

The dynamic flow graph for this function after system test shows that most of the statements and control flow decisions have been exercised as part of system test. However, in a correctly configured system the values of "mMaximumLumens" and "mMinimumLumens" will never be similar enough to force the defensive aspect of the code to be exercised (Fig. 15).

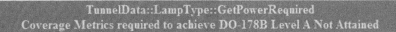

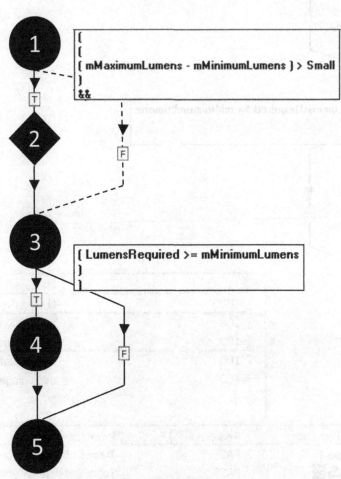

Fig. 15 After system test most of the function has been exercised, but a branch associated with defensive programming remains.

Unit test can be used to complement the code coverage achieved during system test, which forces the defensive branch to be taken (Fig. 16).

The coverage from the unit test and system test can then be combined so that full coverage is demonstrated (Fig. 17).

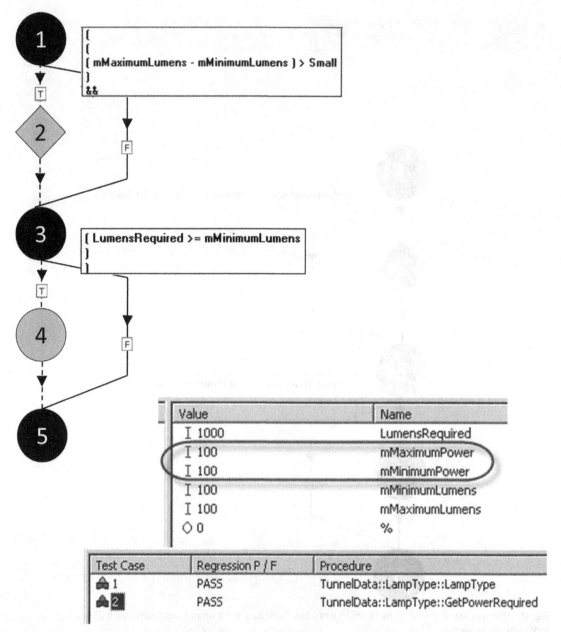

Fig. 16 Unit test exercises the defensive branch left untouched by system test.

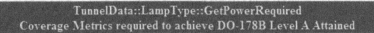

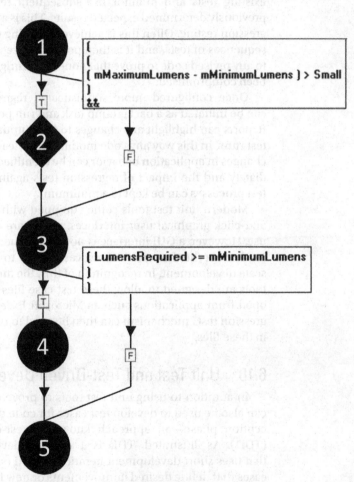

Fig. 17 Full coverage is demonstrated by combining the system test and unit test.

6.9 Using Regression Testing to Ensure Unchanged Functionality

During the course of development, ongoing development can compromise the functionality of software that is considered complete.

As software evolves it is therefore essential to keep reapplying existing tests and monitor the subsequent test outcomes against previously determined expected results. This is a process known as regression testing. Often this is achieved by using test case files to store sequences of tests, and it is then possible to recall and reapply them to any revised code to prove that none of the original functionality has been compromised.

Once configured, more sophisticated regression test processes can be initiated as a background task and run perhaps every evening. Reports can highlight any changes to the output generated by earlier test runs. In this way any code modifications leading to unintentional changes in application behavior can be identified and rectified immediately and the impact of regression tests against other, concurrent, test processes can be kept to a minimum.

Modern unit test tools come equipped with user-friendly, point-and-click graphical user interfaces, which are easy and intuitive to use. However, a GUI interface is not always the most efficient way to implement the thousands of test cases likely to be required in a full-scale development. In recognition of this, the more sophisticated test tools are designed to allow these test case files to be directly developed from applications such as Microsoft Excel. As before, the "regression test" mechanism can then be used to run the test cases held in these files.

6.10 Unit Test and Test-Driven Development

In addition to using unit test tools to prove developed code they can also be used to develop test cases for code that is still in the conception phase—an approach known as test-driven development (TDD). As illustrated, TDD is a software development technique that uses short development iterations based on prewritten unit test cases that define desired improvements or new functions. Each iteration produces code necessary to pass the set of tests that are specific to it. The programmer or team refactors the code to accommodate changes (Fig. 18).

6.11 Automatically Generating Test Cases

Unit tests are usually performed to demonstrate adherence to requirements, to show that elements of the code perform the function they were designed to perform.

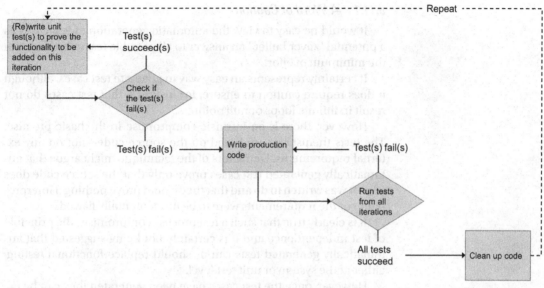

Fig. 18 Unit test tools lend themselves admirably to test-driven development by providing a mechanism to write test cases before any source code is available.

Generally, then, the output data generated through unit tests are important in themselves, but this is not necessarily always the case.

There may be occasions when the fact that the unit tests have successfully been completed is more important than the test data themselves. To address these circumstances as efficiently as possible, the more sophisticated unit test tools can generate test cases automatically, based on information gleaned by means of the initial static analysis of the software under test. For example:

- Source code may be required to pass robustness tests.
- The functionality of source code may already be proven, but the required level of code coverage is unsatisfied.
- A "personality profile" of source code may be required prior to the modification of source code. Sequences of test cases can be generated based on the unchanged code and then exercised again when the source has been modified to prove that there has been no inadvertent detrimental effect on existing functionality.

To tune test cases generated in this way, tools provide a range of options to allow different aspects of the code to be considered. For example, options may include

- generation of test cases to exercise upper and lower boundary values;
- generation of minimum/mean/maximum values; and
- generation of the optimal number of test cases in maximizing code coverage.

6.11.1 A Word of Caution

It would be easy to view the automatic generation of test cases as a potential "silver bullet," an answer to all possible test questions with the minimum of effort.

It certainly represents an easy way to generate test cases, although it does require caution to ensure, for instance, that test cases do not result in infinite loops or null pointers.

However, there is an intrinsic compromise in the basic premise. The tests themselves are based on the source code—not on any external requirements. Detractors of the technique might argue that automatically generated test cases prove only that the source code does what it was written to do and that they would prove nothing if interpretation of the requirements were to be fundamentally flawed.

It is clearly true that such a test process compromises the principle of test independence and it is certainly not being suggested that automatically generated tests can or should replace functional testing, either at the system or unit test level.

However, once the test cases have been generated they can be executed in an identical manner to that provided for conventionally defined unit tests. The input to and output from each of the test cases are available for inspection, so that the correctness of the response from the software to each generated case can be confirmed if required.

7 Setting the Standard

Recent quality concerns are driving many industries to start looking seriously at ways to improve the quality of software development. Not surprisingly, there are marked differences not only in the quality of software in the different sectors, but also in the rate of change of that quality. For example, both the railway and process industries have long had standards governing the entire development cycle of electrical, electronic, and programmable electronic systems, including the need to track all requirements. On the other hand, although a similar standard in the automotive sector was a relatively recent introduction, there has been an explosion in demand for very high-quality embedded software as it moves toward autonomous vehicle production.

The connectivity demanded by autonomous vehicles has also become a significant factor across other safety-critical sectors, too. The advent of the Industrial Internet of Things (IIoT), connected medical devices, and connectivity in aircraft has seen a new emphasis on a need for security in parallel with functional safety.

Clearly, a safety-critical device cannot be safe if it is vulnerable to attack, but the implications go beyond that. For example, the vulnerability of credit card details from an automotive head unit do not represent a safety issue, but demand a similar level of care in software development to ensure that they are protected.

In short, the connected world poses threats to product safety and performance, data integrity and access, privacy, and interoperability.

7.1 The Terminology of Standards

In layman's terms there are documents that define how a process should be managed and standards that dictate the instructions and style to be used by programmers in the process of writing the code itself.

These groups can be further subdivided. For example, there are many collections of these instructions for the use of development teams looking to seek approval for their efforts efficiently. But what are these collections of rules called collectively?

Unfortunately, there is little consensus for this terminology among the learned committees responsible for what these documents are actually called.

The MISRA C:2012 document, for example, is entitled "Guidelines for the use of the C language in critical systems" and hence each individual instruction within the document is a "guideline." These guidelines are further subdivided into "rules" and "directives."

Conversely, the HICC++ document is known as a "coding standards manual" and calls each individual instruction a "rule."

To discuss and compare these documents it is therefore necessary to settle on some umbrella terminology to use for them collectively. For that reason this chapter refers to "Process standards," "Coding standards," and "Coding rules" throughout and distinguishes "internal standards" used within an organization from "recognized standards" such as those established by expert committees.

7.2 The Evolution of a Recognized Process Standard

It is interesting to consider the evolution of the medical software standard IEC 62304 because it mirrors earlier experience in many other sectors.

The US government is well aware of the incongruence of the situation and is considering ways to counter it with the Drug and Device Accountability Act (http://www.govtrack.us/congress/bill.xpd?-bill=s111-882). Recently, the FDA took punitive action against Baxter Healthcare and their infusion pumps, which the FDA has forced the company to recall (https://www.lawyersandsettlements.com/lawsuit/baxter-colleague-infusion-pumps-recall.html).

The net result is that many medical device providers are being driven to improve their software development processes as a result of commercial pressures. In short, they are doing so because it affects the "bottom line."

A common concept in the standards applied in the safety-critical sectors is the use of a tiered, risk-based approach for determining the criticality of each function within the system under development (Fig. 19). Typically known as safety integrity levels there are usually four or five grades used to specify the necessary safety measures to avoid an unreasonable residual risk of the whole system or a system component. The SIL is assigned based on the risk of a hazardous event occurring depending on the frequency of the situation, the impact of possible damage, and the extent to which the situation can be controlled or managed (Fig. 20).

For a company to make the transition to developing certified software they must integrate the standard's technical safety requirements into their design. To ensure that a design follows the standard a company must be able to outline the fulfillment of these safety requirements from design through coding, testing, and verification.

Prominent functional safety standards

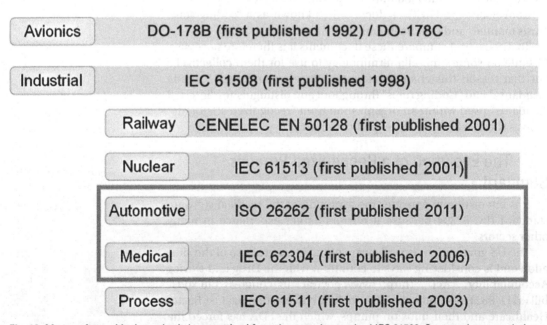

Avionics	DO-178B (first published 1992) / DO-178C
Industrial	IEC 61508 (first published 1998)
Railway	CENELEC EN 50128 (first published 2001)
Nuclear	IEC 61513 (first published 2001)
Automotive	ISO 26262 (first published 2011)
Medical	IEC 62304 (first published 2006)
Process	IEC 61511 (first published 2003)

Fig. 19 Many safety-critical standards have evolved from the generic standard IEC 61508. Sectors that are relative newcomers to this field (highlighted) were advantaged by the availability of established and proven tools to help them achieve their goals.

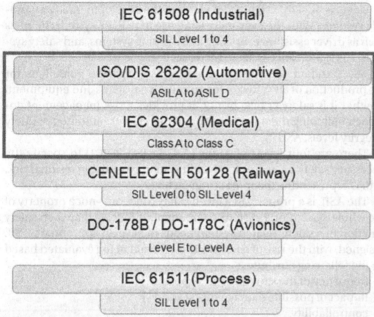

Fig. 20 The example standards all apply some concept of safety integrity levels, although the terminology used to describe them varies. Again, the principles hold true in the highlighted sectors where standards are newest.

To ease the adoption of this standard and manage the shift in requirements many companies use gap analysis. Gap analysis begins by gathering and analyzing data to gauge the difference between where the business is currently and where it wants to be. Gap analysis examines operating processes and artifacts generated, typically employing a third party for the assessment. The outcome will be notes and findings on which the company or individual project may act.

7.2.1 ISO 26262 Recognized Process Standard—Case Study

In parallel with the advances in the medical device sector and in response to the increased application of electronic systems to automotive safety-critical functions the ISO 26262 standard was created to comply with needs specific to the application sector of electrical/ electronic/programmable electronic (E/E/PE) systems within road vehicles.

In addition to its roots in the IEC 61508 generic standard it has much in common with the DO-178B/DO-178C standards seen in aerospace applications. In particular, the requirement for MC/DC (Modified Condition/Decision Coverage—a technique to dictate the tests required to adequately test lines of code with multiple conditions) and the structural coverage analysis process is very similar.

Safety is already a significant factor in the development of automobile systems. With the ever-increasing use of E/E/PE systems in areas such as driver assistance, braking and steering systems, and safety systems this significance is set to increase.

The standard provides detailed industry-specific guidelines for the production of all software for automotive systems and equipment, whether it is safety critical or not. It provides a risk management approach including the determination of risk classes (automotive safety integrity levels, ASILs).

There are four levels of ASILs (A–D in ISO 26262) to specify the necessary safety measures for avoiding an unreasonable residual risk, with D representing the most stringent level.

The ASIL is a property of a given safety function—not a property of the whole system or a system component. It follows that each safety function in a safety-related system needs to have an appropriate ASIL assigned, with the risk of each hazardous event being evaluated based on the following attributes:

- frequency of the situation (or "exposure")
- impact of possible damage (or "severity")
- controllability.

Depending on the values of these three attributes the appropriate ASIL for a given functional defect is evaluated. This determines the overall ASIL for a given safety function.

ISO 26262 translates these safety levels into safety-specific objectives that must be satisfied during the development process. An assigned ASIL therefore determines the level of effort required to show compliance with the standard. This means that the effort and expense of producing a system critical to the continued safe operation of an automobile (e.g., a steer-by-wire system) is necessarily higher than that required to produce a system with only a minor impact in the case of a failure (e.g., the in-car entertainment system).

The standard demands a mature development environment that focuses on requirements that are specified in this standard. To claim compliance to ISO 26262 most requirements need to be formally verified, aside from exceptional cases where the requirement does not apply or where noncompliance is acceptable.

Part 4 of the standard concerns product development at the system level, and Part 6 of the standard concerns product development at the software level. The scope of these documents may be mapped on to any process diagram such as the familiar "V" model (Fig. 21).

Software analysis, requirements management, and requirements traceability tools are usually considered essential for large, international, cost-critical projects.

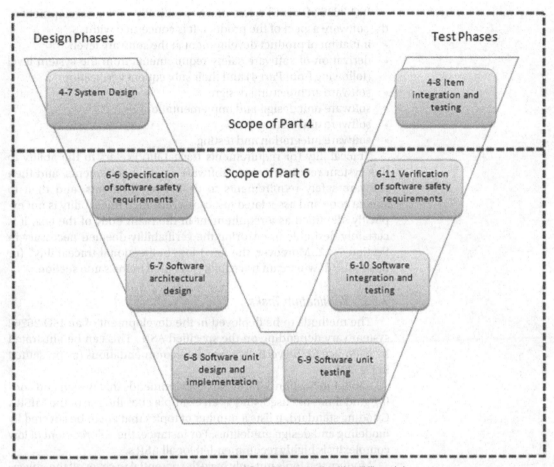

Fig. 21 Mapping the scope of ISO 26262 Part 4 and Part 6 onto the familiar "V" model.

7.2.2 ISO 26262 Process Objectives

ISO 26262 recognizes that software safety and security must be addressed in a systematic way throughout the software development life cycle. This includes the safety requirements traceability, software design, coding, and verification processes used to ensure correctness, control, and confidence both in the software and in the E/E/PE systems to which that software contributes.

A key element of ISO 26262 (Part 4) is the practice of allocating technical safety requirements in the system design and developing that design further to derive an item integration and testing plan and subsequently the tests themselves. It implicitly includes software elements of the system, with the explicit subdivision of hardware and software development practices being dealt with further down the "V" model.

ISO 26262 (Part 6) refers more specifically to the development of the software aspect of the product. It is concerned with:

- initiation of product development at the software level;
- derivation of software safety requirements from the system level (following from Part 4) and their subsequent verification;
- software architectural design;
- software unit design and implementation;
- software unit testing; and
- software integration and testing.

Traceability (or requirements traceability) refers to the ability to link system requirements to software safety requirements, and then software safety requirements to design requirements, and then to source code and associated test cases. Although traceability is not explicitly identified as a requirement in the main body of the text, it is certainly desirable in ensuring the verifiability deemed necessary in Section 7.4.2. Moreover, the need for "bidirectional traceability" (or upstream/downstream traceability) is noted in the same section.

7.2.3 Verification Tasks

The methods to be deployed in the development of an ISO 26262 system vary depending on the specified ASIL. This can be illustrated by reference to the verification tasks recommendations (as presented in Fig. 22).

Table 1 in Section 5.4.7 of Part 6 recommends that design and coding guidelines are used, and as an example cites the use of the MISRA C coding standard. It lists a number of topics that are to be covered by modeling and design guidelines. For instance, the enforcement of low complexity is highly recommended for all ASILs.

Modern test tools not only have the potential to cover all the obligatory elements for each ASIL, but they also have the flexibility in configuration to allow less critical code in the same project to be associated with less demanding standards. That principle extends to mixed C and C++ code, where appropriate standards are assigned to each file in accordance with its extension (Fig. 23).

Table 12 from Section 9 of Part 6 shows, for instance, that measuring statement coverage is highly recommended for all ASILs, and that branch coverage is recommended for ASIL A and highly recommended for other ASILs. For the highest ASIL D, MC/DC is also highly recommended.

Each of these coverage metrics implies different levels of test intensity. For example, 100% statement coverage is achieved when every statement in the code is executed at least once; 100% branch (or decision) coverage is achieved only when every statement is executed at least once, AND each branch (or output) of each decision is tested—that is, both false and true branches are executed.

	Topics	ASIL			
		A	B	C	D
1a	Enforcement of low complexity	++ ✓	++✓	++✓	++✓
1b	Use of language subsets	++✓	++✓	++✓	++✓
1c	Enforcement of strong typing	++✓	++✓	++✓	++✓
1d	Use of defensive implementation techniques	o	+✓	++✓	++✓
1e	Use of established design principles	+✓	+✓	+✓	++✓
1f	Use of unambiguous graphical representation	+✓	++✓	++✓	++✓
1g	Use of style guides	+✓	++✓	++✓	++✓
1h	Use of naming conventions	++✓	++✓	++✓	++✓

"++" The method is highly recommended for this ASIL.

"+" The method is recommended for this ASIL.

"o" The method has no recommendation for or against its usage for this ASIL.

✓ Potential for efficiency gains through the use of test tools

Fig. 22 Mapping the potential for efficiency gains through the use of test tools to "ISO 26262 Part 6 Table 1: Topics to be covered by modeling and coding guidelines."

Statement, branch, and MC/DC coverage can all be automated through the use of test tools. Some packages can also operate in tandem so that, for instance, coverage can be generated for most of the source code through a dynamic system test, and can be complemented using unit tests to exercise defensive code and other aspects that are inaccessible during normal system operation.

Similarly, Table 15 in Section 10.4.6 shows the structural coverage metrics at the software architectural level (Fig. 24).

	Topics	ASIL			
		A	**B**	**C**	**D**
1a	Statement coverage	++✓	++✓	+✓	+✓
1b	Branch coverage	+✓	++✓	++✓	++✓
1c	MC/DC (modified condition/decision Coverage)	+✓	+✓	+✓	++✓

"++" The method is highly recommended for this ASIL.

"+" The method is recommended for this ASIL.

"o" The method has no recommendation for or against its usage for this ASIL.

✓ Potential for efficiency gains through the use of test tools

Fig. 23 Mapping the potential for efficiency gains through the use of test tools to "ISO 26262 Part 6 Table 14: Structural coverage metrics at the software unit level."

	Topics	ASIL			
		A	**B**	**C**	**D**
1a	Function coverage	+✓	+✓	++✓	++✓
1b	Call coverage	+✓	+✓	++✓	++✓

"++" The method is highly recommended for this ASIL.

"+" The method is recommended for this ASIL.

"o" The method has no recommendation for or against its usage for this ASIL.

✓ Potential for efficiency gains through the use of test tools

Fig. 24 Mapping the potential for efficiency gains through the use of test tools to the "ISO 26262 Part 6 Table 17: Structural coverage metrics at the software architectural level."

7.2.4   SAE J3061 and ISO 26262

ISO 26262 requires any threats to functional safety to be adequately addressed, implicitly including those relating to security threats, but it gives no explicit guidance relating to cybersecurity.

At the time of ISO 26262's publication, that omission was perhaps to be expected. Automotive-embedded applications have traditionally been isolated, static, fixed function, device-specific implementations, and practices and processes have relied on that status. But the rate of change in the industry has been such that by the time of its publication in 2016, SAE International's Surface Vehicle Recommended Practice SAE J3061 was much anticipated.

Note the wording here. SAE J3061 is not a standard. It constitutes "recommended practice," and at the time of writing there are plans for it to be replaced by an ISO/SAE 21434 standard—the first result of a partnership between the ISO and SAE standards organizations.

SAE J3061 can be considered complementary to ISO 26262 in that it provides guidance on best development practices from a cybersecurity perspective, just as ISO 26262 provides guidance on practices to address functional safety.

It calls for a similar sound development process to that of ISO 26262. For example, hazard analyses are performed to assess risks associated with safety, whereas threat analyses identify risks associated with security. The considerations for the resulting security requirements for a system can be incorporated in the process described by ISO 26262 Part 8 Section 6: "Specification and management of safety requirements."

Another parallel can be found in the use of static analysis, which is used in safety-critical system development to identify constructs, errors, and faults that could directly affect primary functionality. In cybersecurity-critical system development, static code analysis is used instead to identify potential vulnerabilities in the code.

7.2.4.1   Beyond Functional Safety

Despite the clear synergy between the two standards it is important to note that SAE J3061 does more than simply formalize the need to include security considerations in functional safety requirements. It is easy to focus on an appropriate process once functional, safety, and security requirements are established, but the significance of malicious intent in the definition of those requirements should not be underestimated.

SAE J3061 emphasizes this point throughout. It argues that cybersecurity is likely to be even more challenging than functional safety, stating that "Since potential threats involve intentional, malicious, and planned actions, they are more difficult to address than potential hazards. Addressing potential threats fully, requires the analysts to think like the attackers, but it can be difficult to anticipate the exact moves an attacker may make."

Perhaps less obviously, the introduction of cybersecurity into an ISO 26262-like formal development implies the use of similarly rigorous techniques into applications that are NOT safety critical—and perhaps into organizations with no previous obligation to apply them. SAE J3061 discusses privacy in general and personally identifiable information (PII) in particular, and highlights both as key targets for a bad actor of no less significance than the potential compromise of safety systems.

In practical terms, there is a call for ISO 26262-like rigor in the defense of a plethora of personal details potentially accessed via a connected car, including personal contact details, credit card and other financial information, and browse histories. It could be argued that this is an extreme example of the general case cited by the SAE J3061 standard, which states that "...there is no direct correspondence between an ASIL rating and the potential risk associated with a safety-related threat."

Not only does SAE J3061 bring formal development to less safety-critical domains, it also extends the scope of that development far beyond the traditional project development life cycle. Consideration of incident response processes, over-the-air (OTA) updates, and changes in vehicle ownership are all examples of that.

7.3 Freedom to Choose Adequate Standards

Not every development organization of every application is obliged to follow a set of process or coding standards that has been laid down by a client or a regulatory body. Indeed, it is probably reasonable to suggest that most are not in that position.

However, everyone wants their software to be as sound and as robust as possible. Even if it is a trivial application to be used by the writer as a one-off utility, no developer wants their application to crash. Even if such a utility expands into an application with wider uses, no one wants to have to deal with product recalls. No one wants to deal with irate end users. Even removing all external factors entirely, most people want the satisfaction of a job well done in an efficient manner.

So, it follows that if the use of process and coding standards is appropriate when safety or security issues dictate that software must be robust and reliable, then it is sensible to adopt appropriate standards even if an application is not going to threaten anyone's well-being if it fails.

Once that is established a sound pragmatic approach is required to decide what form those standards should take.

7.4 Establishing an Internal Process Standard

Many recognized standards are most ideally deployed within large organizations. There are many software development teams that consist of two or three people all resident in the same office. It would clearly be overkill to deploy the same tools and techniques here as in a high-integrity team of hundreds of developers spread across the world.

That said, the principles of the requirements traceability matrix (RTM) established earlier in the chapter remain just as valid in either case. The difference lies in the scaling of the mechanisms and techniques used to confirm the traceability of requirements. For that reason an appropriate recognized process standard can prove very useful as a guideline for pragmatic application of similar principles in a less demanding environment.

7.4.1 Establishing a Common Foundation for an Internal Coding Rule Set

The principle of using a recognized standard as the basis for an internal one extends to the realm of coding standards.

Even where there is no legacy code to worry about there is frequently a good deal of inertia within a development team. Something as simple as agreement over the placement of brackets can become a source of great debate among people who prefer one convention over another.

Under these circumstances the establishment of a common foundation rule set that everyone can agree on is a sound place to begin.

For example, practically no one would advocate the use of the "goto" statement in C or C++ source code. It is likely that outlawing the use of the "goto" statement by means of a coding rule will achieve support from all interested parties. Consequently, it is an uncontentious rule to include as part of a common foundation rule set.

Establishing such a set of rules from nothing is not easy, and given that learned organizations are meeting regularly to do just that, neither is it a sensible use of resources.

It therefore makes sense to derive an internal standard from a recognized standard to which the organization might aspire in an ideal world, This does not necessarily imply an intention on the part of the development organization to ever fully comply with that standard, but it does suggest that the rules deployed as part of that subset will be coherent, complementary, and chosen with the relevant industry sector in mind.

7.4.2 Dealing With an Existing Code Base

This principle becomes a little more challenging when there is a legacy code base to deal with.

It is likely that the retrospective enforcement of a recognized coding standard, such as MISRA C:2012, to legacy code is too onerous and so a subset compromise is preferred. In that case it is possible to apply a user-defined set of rules that could simply be less demanding or that could, say, place particular focus on portability issues.

Where legacy code is subject to continuous development a progressive transition to a higher ideal may then be made by periodically adding more rules with each new release, so that the impact on incremental functionality improvements is kept to a minimum.

Test tools enable the correction of code to adhere to such rules as efficiently as possible. Some tools use a "drill down" approach to provide a link between the description of a violation in a report and an editor opened on the relevant line of code.

7.4.3 Deriving an Internal Coding Standard for Custom Software Development—Case Study

In many fields of endeavor for embedded software development—automobiles, aeroplanes, telephones, medical devices, weapons—the software life cycle contributes to a product life cycle of design and development, readying for production, then mass production. In some fields, such as control systems for bespoke plant or machinery, duplicates are the exception—not the rule. That brings a unique set of difficulties.

Imagine a situation where there is a software team of three or four developers within a larger engineering company. The actual value of the software within the context of a typical contract might be very small, but even so the software itself is critical to making sure that the client is satisfied when work is completed.

The original code has been designed to be configurable, but as each sale is made by the sales team the functionality is expanded to encompass new features to clinch the deal. The sales team is motivated by the commission derived from the whole of the contract, meaning that a modification to software functionality is not a primary concern for them.

The developers of the software team are tasked with implementing this expanding functionality for each new contract, often under great pressure from other areas of production.

Commercial milestones loom large for these hapless developers because, despite their small contribution to the overall value of the project, any failure on their part to meet the milestones could result in delay of a stage payment or the triggering of a penalty clause.

To make matters worse, numerous software developers have joined and departed from the team over the years. Each has had their own style and preferences and none has had the time to thoroughly document what they have done.

As a practical example, let us consider that the developers of such a code base, designed to control a widget production machine, have been tasked with establishing the quality of their software and improving it on an ongoing basis.

7.4.3.1 Establishing a Common Foundation

The first step in establishing an appropriate rule set is to choose a relevant recognized standard as a reference point—perhaps MISRA C++:2008 in this case.

Using a test tool the code base can be analyzed to discover the parts of the standard where the code base is already adequate. By opting to include information relating to all rules irrespective of whether the code base complies with them, a subset of the standard to which the code adheres can immediately be derived (Fig. 25).

Typically, the configuration facilities in the test tool can then be used to map the rules that have not been transgressed into a new subset of the reference standard, and the violated rules disabled as illustrated in Fig. 26.

7.4.3.2 Building Upon a Common Foundation

Even if nothing else is achieved beyond checking each future code release against this rule set, it is assured that the standard of the code in terms of adherence to the chosen reference standard will not deteriorate further.

However, if the aim is to improve the standard of the code, then an appropriate strategy is to review the violated rules. It is likely that there are far fewer different rules violated than there are individual violations, and test tools can sometime generate a useful breakdown of this summary information (Fig. 27).

In some cases it may be that a decision is reached that some rules are not appropriate and their exclusion justified.

Prioritizing the introduction of the remaining rules can vary depending on the primary motivation. In the example it is obviously likely to be quicker to address the violations that occur once rather than those that occur 40 or 50 times, which will improve apparent adherence to the standard. However, it makes sense to initially focus on any particular violations that, if they were to be corrected, would address known functional issues in the code base.

Whatever the criteria for prioritization, progressive transition to a higher ideal may then be made by periodically adding more rules with each new release, so that the impact on incremental functionality improvements is kept to a minimum.

8 Dealing With the Unusual

8.1 Working With Autogenerated Code

Many software design tools, such as IBM's Rhapsody and MathWork's Matlab, have the ability to automatically generate high-level source code from a UML or similar design model (Fig. 28).

Number of Violations	LDRA Code	Required Standards	MISRA-C++:2008 Code
0	9 S	Assignment operation in expression.	MISRA-C++:2008 5-0-1,6-2-1
2	11 S	No brackets to loop body.	MISRA-C++:2008 6-3-1
8	12 S	No brackets to then/else.	MISRA-C++:2008 6-4-1
0	32 S	Use of continue statement.	MISRA-C++:2008 6-6-3
0	35 S	Static procedure is not explicitly called in code analysed.	MISRA-C++:2008 0-1-1
0	36 S	Function has no return statement.	MISRA-C++:2008 8-4-3
0	39 S	Unsuitable type for loop variable.	MISRA-C++:2008 6-5-1
0	41 S	Ellipsis used in procedure parameter list.	MISRA-C++:2008 8-4-1
0	43 S	Use of setjmp/longjmp.	MISRA-C++:2008 17-0-5
7	44 S	Use of banned function, type or variable.	MISRA-C++:2008 17-0-1,18-0-2,18-2-1,18-4-1,19-3-1
0	47 S	Array bound exceeded.	MISRA-C++:2008 5-0-16
1	48 S	No default case in switch statement.	MISRA-C++:2008 6-4-6
12	49 S	Logical conjunctions need brackets.	MISRA-C++:2008 5-0-2,5-2-1
0	50 S	Use of shift operator on signed type.	MISRA-C++:2008 5-0-21
0	51 S	Shifting value too far.	MISRA-C++:2008 5-8-1
0	52 S	Unsigned expression negated.	MISRA-C++:2008 5-3-2

Fig. 25 Using source code violation reporting to identify rules that are adhered to. In this report subset only the rules highlighted have been violated.

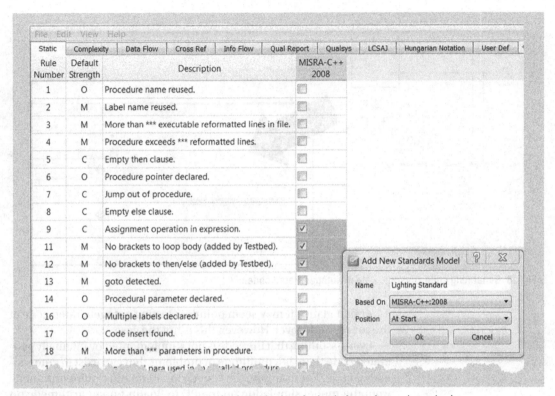

Fig. 26 Using unviolated rules from a recognized standard as the basis for an internal standard.

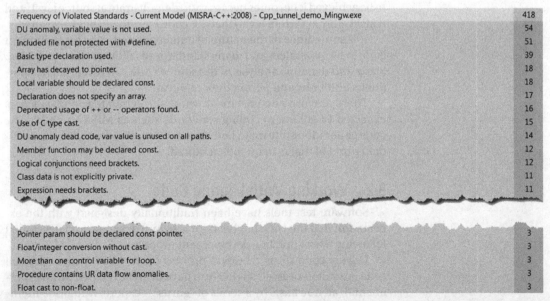

Fig. 27 Summary showing the breakdown of rule violations in a sample code set.

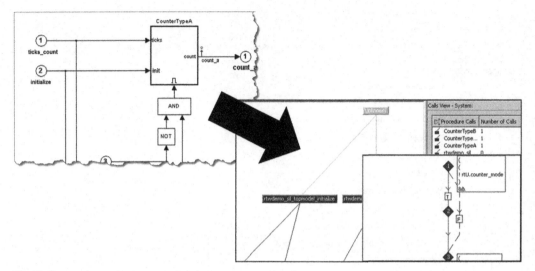

Fig. 28 Generating code coverage data in autogenerated code.

At first glance it may seem pointless testing autogenerated code at the source code level. However, this is not the case.

Even assuming that the code is not supplemented by manually generated code there is a multitude of problems that can exist in autogenerated code. For example, the conversion of floating-point arithmetic from the model simulation on the PC to floating-point arithmetic on the target may be erroneous and so require testing.

When a standard dictates that a particular level of code coverage is to be achieved it becomes necessary to demonstrate at source level (and conceivably at system level) that the code has indeed been executed.

It is sometimes in the nature of autogenerated code for redundant code to be generated, and many standards disallow such an inclusion. Static and dynamic analysis of the source code can reveal such superfluous additions and permit their removal.

There are also circumstances where the generated source code is expected to adhere to coding standards such as MISRA C:2012. The code generation suite may claim to meet such standards, but independent proof of that is frequently required.

8.2 Working With Legacy Code

Software test tools have been traditionally designed with the expectation that the code has been (or is being) designed and developed following a best practice development process.

Legacy code turns the ideal process on its head. Although such code is a valuable asset, it is likely to have been developed on an experimental, ad hoc basis by a series of "gurus"—experts who pride themselves at getting things done and in knowing the application itself, but not necessarily expert at complying with modern development thinking and bored at having to provide complete documentation. That does not sit well with the requirements of such standards as DO-178C.

Frequently, this legacy software—often termed software of unknown pedigree (SOUP)—forms the basis of new developments. The resulting challenges do not just come from extended functionality. Such developments may need to meet modern coding standards and deploy updated target hardware and development tool chains, meaning that even unchanged functionality cannot be assumed to be proven.

The need to leverage the value of SOUP presents its own set of unique challenges.

8.2.1 The Dangers of Software of Unknown Pedigree

Many SOUP projects will initially have been subjected only to functional system testing, leaving many code paths unexercised and leading to costly in-service corrections. Even in the field it is highly likely that the circumstances required to exercise much of the code have never occurred and such applications have therefore sustained little more than an extension of functional system testing by their in-field use.

When there is a requirement for ongoing development of legacy code, previously unexercised code paths are likely to be called into use by combinations of data never previously encountered (Fig. 29).

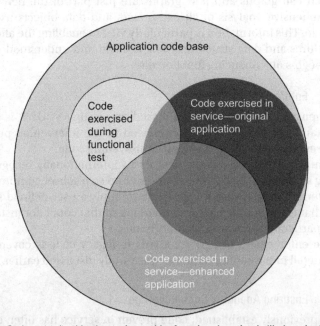

Fig. 29 Code exercised both on-site and by functional testing is likely to include many unproven execution paths. Code enhancements are prone to exercising previously unused paths even in established parts of the system.

The same commercial pressures that rule out a rewrite are likely to rule out the use of all the following options. As ever, they can be used selectively depending on the criticality of an application or its subsections.

8.2.2 Static and Dynamic Analysis of Software of Unknown Pedigree

In the enhancement of SOUP the existing code frequently defines the functionality of the system rather than documentation. In enhancing the code it is therefore vital that the functionality is not unintentionally modified. And, even where all source code remains identical, a new compiler or target hardware can introduce unintentional functionality changes with potentially disastrous results.

The challenge is to identify the building blocks within the test tools that can be used in an appropriate sequence to aid the efficient enhancement of SOUP.

There are five major considerations.

8.2.2.1 Improving the Level of Understanding

The system visualization facilities provided by many modern test tools are extremely powerful. Static call graphs provide a hierarchical illustration of the application and system entities, and static flow graphs show the control flow across program blocks.

Such call graphs and flow graphs are just part of the benefit of comprehensive analysis of all parameters and data objects used in the code. This information is particularly vital to enabling the affected procedures and data structures to be isolated and understood when work begins on enhancing functionality.

8.2.2.2 Enforcing New Standards

When new developments are based on existing SOUP it is likely that standards will have been enhanced in the intervening period. Code review analysis can highlight contravening code.

It may be that the enforcement of an internationally recognized coding standard to SOUP is too onerous and so a subset compromise is preferred. In that case it is possible to apply a user-defined set of rules that could simply be less demanding or that could, for instance, place particular focus on portability issues.

The enforcement of new standards to legacy code is covered in more detail in the coding standards case study discussed earlier.

8.2.2.3 Ensuring Adequate Code Coverage

As previously established, code proven in service has often effectively been subjected only to extensive "functional testing" and may include many previously unexercised and unproven paths.

Structural coverage analysis addresses this issue by testing equally across the sources assuming each path through them has an equal chance of being exercised.

Although not offering a complete solution, system-wide functional testing exercises many paths and so provides a logical place to start.

Commonly a test tool may take a copy of the code under test and implant additional procedure calls ("instrumentation") to identify the paths exercised during execution. Textual code coverage reports are then often complemented with colored graphs to give insight into the code tested and into the nature of data required to ensure additional coverage.

Manually constructed unit tests can be used to ensure that each part of the code functions correctly in isolation. However, the time and skill involved in constructing a harness to allow the code to compile can be considerable.

The more sophisticated unit test tools minimize that overhead by automatically constructing the harness code within a GUI environment and providing details of the input and output data variables to which the user may assign values. The result can then be exercised on either the host or target machine.

To complement system test it is possible to apply code instrumentation to these unit tests and hence exercise those parts of the code that have yet to be proven. This is equally true of code that is inaccessible under normal circumstances such as exception handlers.

Sequences of these test cases can be stored, and they can be automatically exercised regularly to ensure that ongoing development does not adversely affect proven functionality, or to reestablish correct functionality when problems arise in service.

8.2.2.4 Dealing With Compromised Modularity

In some SOUP applications, structure and modularity may have suffered challenging the notion of testing functional or structural subsections of that code.

However, many unit test tools can be very flexible, and the harness code that is constructed to drive test cases can often be configured to include as much of the source code base as necessary. The ability to do that may be sufficient to suit a purpose.

If a longer term goal exists to improve overall software quality, then using instrumented code can help to understand which execution paths are taken when different input parameters are passed into a procedure—either in isolation or in the broader context of its calling tree.

8.2.2.5 Ensuring Correct Functionality

Perhaps the most important aspect of SOUP-based development is ensuring that all aspects of the software functions as expected, despite changes to the code, to the compiler or the target hardware, or to the data handled by the application.

Even with the aid of test tools, generating unit tests for the whole code base may involve more work than the budget will accommodate. However, the primary aim here is not to check that each procedure behaves in a particular way; it is to ensure that there have been no inadvertent changes to functionality.

By statically analyzing the code, test tools provide significant assistance for the generation of test cases, and the more sophisticated tools on the market are able to fully automate this process for some test case types. This assistance, whether partially or fully automated, will help to exercise a high percentage of the control flow paths through the code. Depending on the capabilities of the tool in use, input and output data may also be generated through fully or partially automated means. These data may then be retained for future use.

The most significant future use of these retained data will be in the application of regression tests, the primary function of which is to ensure that when those same tests are run on the code under development there are no unexpected changes. These regression tests provide the cross-reference back to the functionality of the original source code and form one of the primary benefits of the unit test process as a whole. As such the more feature rich of the available unit test tools will often boost the efficiency and throughput of regression tests via the ability to support batch processing.

8.3 Tracing Requirements Through to Object Code Verification (OCV)

With applications whose failure has critical consequences—people's lives could be at risk or there could be significant commercial impact—there is a growing recognition that stopping requirements traceability short of object code raises unanswered questions. There is an implied reliance on the faithful adherence of compiled object code to the intentions expressed by the author of the source code.

Where an industry standard is enforced a development team will usually adhere only to the parts of the standard that are relevant to their application—including OCV. And yet, OCV is designed to ensure that critical parts of an application are not compromised by the object code, which is surely a desirable outcome for any software whatever its purpose.

8.3.1 Industry Standards and Software Certification

Irrespective of the industry and the maturity of its safety standards the case for software that has been proven and certified to be reliable through standards compliance and requirements traceability is becoming ever more compelling.

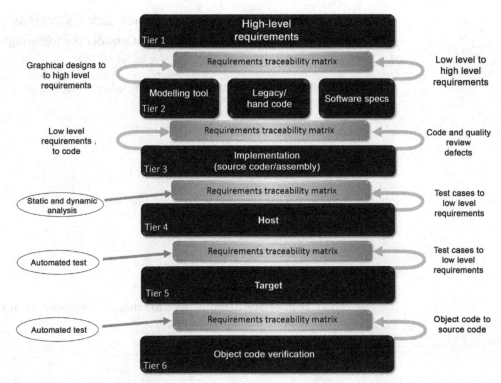

Fig. 30 The requirements traceability matrix (RTM) can be extended through to object code verification at the sixth tier.

When the RTM becomes the center of the development process it impacts all stages of design from high-level requirements through to target-based deployment. The addition of Tier 6 takes the target-based work a stage further, to tie in the comparison of the object and source code as part of the RTM and an extension to it (Fig. 30).

8.3.2 Object Code Verification (OCV)

So what is object code verification? The relevant section of the aerospace DO-178C standard describes the technique as follows:

"Structural coverage analysis may be performed on the Source Code, object code, or Executable Object Code. Independent of the code form on which the structural coverage analysis is performed, if the software level is A and a compiler, linker, or other means generates additional code that is not directly traceable to Source Code statements, then additional verification should be performed to establish the correctness of such generated code sequences."

OCV therefore hinges on how much the control flow structure of the compiler-generated object code differs from that of the application source code from which it was derived.

8.3.3 Object Code Control Flow vs Source Code Control Flow

It is useful to illustrate this variation. Consider the following very simple source code:

```
void f_while4( int f_while4_input1, int f_while4_input2 )
{

    int f_while4_local1, f_while4_local2 ;

    f_while4_local1 = f_while4_input1 ;
    f_while4_local2 = f_while4_input2 ;

    while( f_while4_local1 < 1 || f_while4_local2 > 1 )
        {
            f_while4_local1 ++ ;
            f_while4_local2 -- ;
        }
}
```

This C code can be demonstrated to achieve 100% source code coverage by means of a single call thus:

```
f_while4(0,3);
```

and can be reformatted to a single operation per line like so:

```
1   void

1     f_while4 (
1     int f_while4_input1 ,
1     int f_while4_input2 )
1     {
1       int
1         f_while4_local1 ,
1         f_while4_local2 ;
1       f_while4_local1 = f_while4_input1 ;
1       f_while4_local2 = f_while4_input2 ;
-----------------------------------------------------------------
2     while
2       (
2       f_while4_local1 < 1
2       ||
-----------------------------------------------------------------
3       f_while4_local2 > 1
-----------------------------------------------------------------
4       )
-----------------------------------------------------------------
5       {
5           f_while4_local1 ++ ;
```

```
5              f_while4_local2 -- ;
5         }
---------------------------------------------------------------
6      }
```

The prefix for each of these reformatted lines of code identifies a "basic block"—that is, a sequence of straight line code. The resulting flow graph for the function shows both the structure of the source code and the coverage attained by such a test case with the basic blocks identified on the flowchart nodes (Fig. 31).

In this sequence of illustrations exercised parts of the code are shown using black nodes and solid branch lines.

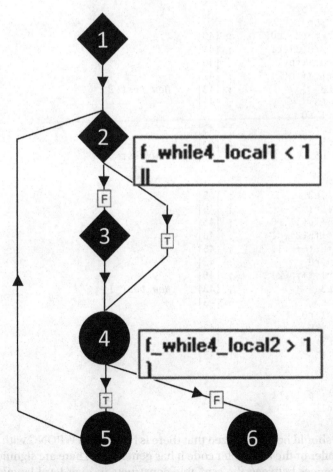

Fig. 31 A dynamic flowgraph showing that all source code has been exercised through a single function call.

The object code generated by a compiler will depend on the optimization setting, the compiler vendor, the target, and a host of other issues. The following shows just one example of resulting (reformatted) assembler code generated by a widely used commercially available compiler with optimization disabled.

It is not necessary to understand all the code to grasp the principle, but it is useful to be aware that the *ble* and *bgt* branch instructions ("branch less than equal" and "branch greater than," respectively) are used to divert control flow to their associated labels (L3 and L5). These branches lie within code that otherwise executes sequentially:

```
39 _f_while4:
40              push     fp
41              ldiu     sp,fp
42              addi     2,sp
43              ldi      *-fp(2),r0        ;  |40|
44              ldiu     *-fp(3),r1        ;  |41|
45              sti      r0,*+fp(1)        ;  |40|
46              sti      r1,*+fp(2)        ;  |41|
47              ble      L3                ;  |43|      New test 2
48 ;*     Branch Occurs to L3             ;  |43|
49              ldiu     r1,r0
50              cmpi     1,r0              ;  |43|
51              ble      L5                ;  |43|
52 ;*     Branch Occurs to L5             ;  |43|      New test 3
53
54 L3:
55              ldiu     1,r0              ;  |45|
56              ldiu     1,r1              ;  |46|
57              addi     *+fp(1),r0        ;  |45|
58              subri    *+fp(2),r1        ;  |46|
59              sti      r0,*+fp(1)        ;  |45|
60              cmpi     0,r0              ;  |43|
61              sti      r1,*+fp(2)        ;  |46|
62              ble      L3                ;  |43|      New test 1
63 ;*     Branch Occurs to L3             ;  |43|
64              ldiu     r1,r0
65              cmpi     1,r0              ;  |43|
66              bgt      L3                ;  |43|
67 ;*     Branch Occurs to L3             ;  |43|
68
69 L5:
70              ldiu     *-fp(1),r1
71              bud      r1
```

It should be emphasized that there is NOTHING WRONG with this compiler or the assembler code it has generated. There are significant differences between the available constructs in high-level languages (such as C and C++) and object code. To interpret the source code a compiler and/or linker therefore have to generate code that is not directly traceable to source code statements.

Consequently, the flowgraph looks different for the assembler code—and, in particular, using the identical test case generates a quite different flowgraph both in terms of appearance and importantly in terms of coverage.

This phenomenon is acknowledged in the DO-178C standard used in the aerospace industry. For the most safety-critical level A code it states that "if ... a compiler, linker or other means generates additional code that is not directly traceable to Source Code statements, then additional verification should be performed to establish the correctness of such generated code sequences." (Fig. 32).

It is clear from the flowchart and the assembler code that more tests are necessary to achieve 100% code coverage:

- New test 1. Line 62. End of block 3. Branch to L3.
  This ble branch always evaluates to false with the existing test data because it only exercises the loop once, and so only one of the two possible outcomes results from the test to see whether to continue. Adding a new test case to ensure a second pass around that loop exercises both true and false cases. A suitable example can be provided thus:

  ```
  f_while4(-1,3);
  ```

- New test 2. Line 47. End of block 1. Branch to L3.
  This code contains an "or" statement in the "while" loop conditions. The existing test cases both result in the code:

  ```
  f_while4_local1 < 1
  ```

  returning a "true" value.
  The addition of a new test case to return a "false" value will address that:

  ```
  f_while4(3,3);
  ```

- New test 3. Line 52. End of block 2. Branch to L5.
  The remaining unexercised branch is the result of the fact that if neither of the initial conditions in the "while" statement is satisfied then the code within the loop is bypassed altogether via the ble branch.
  So, the final test added will provide such a circumstance:

  ```
  f_while4(3,0);
  ```

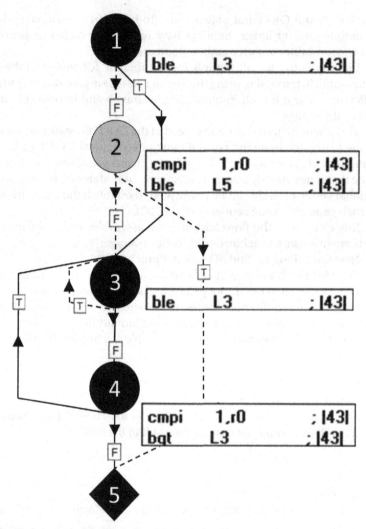

Fig. 32 A dynamic flowgraph showing the assembler code exercised through a single function call.

These three additional tests result in 100% statement and branch coverage of the assembler code (Fig. 33).

- So, to achieve 100% coverage of the assembler code, four tests are required:

```
f_while4(0,3);
f_while4(-1,3);
f_while4(3,3);
f_while4(3,0);
```

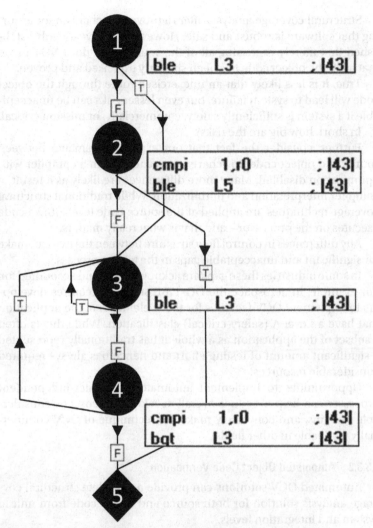

Fig. 33 A dynamic flowgraph showing 100% assembler code exercised through additional function calls.

8.3.3.1 Extending Source Code Coverage to Object Code Verification

If the principle of structural coverage analysis is justified, then it follows that object code verification (OCV) is also worthy of consideration.

In the general case structural coverage analysis provides evidence that ALL of the code base has been exercised. Such an approach has been proven to reduce the risk of failure and consequently is specified in most, if not all, industrial standards concerned with safety.

Structural coverage analysis offers a proven mechanism for ensuring that software is robust and safe. However, we have already established that merely exercising all of the source code does NOT prove that all of the object code has been similarly exercised and proven.

True, it is less likely that an unexercised route through the object code will lead to system failure, but even lesser risks can be unacceptable if a system is sufficiently safety, commercially, or mission critical.

In short, how big are the risks?

Further, consider the fact that our example mismatch between source and object code flowcharts was generated in a compiler with optimization disabled. Many more differences are likely as a result of compiler interpretation and optimization. While traditional structural coverage techniques are applied at the source code level, object code executes on the processor—and that is what really matters.

Any differences in control flow structure between the two can make for significant and unacceptable gaps in the testing process.

In some industries these gaps are acknowledged and accounted for. For example, in aerospace the DO-178B standard requires developers to implement OCV facilities for those elements of the application that have a Level A (safety-critical) classification. While this is often a subset of the application as a whole it has traditionally represented a significant amount of testing effort and hence has always required considerable resources.

Opportunities to implement automated, compiler-independent processes can help to reduce overall development costs by considerable margins, and conversely make the technique of OCV commercially justifiable in other fields.

8.3.3.2 Automated Object Code Verification

Automated OCV solutions can provide a complete structural coverage analysis solution for both source and object code from unit to system and integration levels.

Typical solutions combine both high- and object-level (assembler) code analysis tools, with the object-level tool variant being determined by the target processor that the application is required to run on. A typical example might see C/C++ and PowerPC Assembler analysis tools teamed together to provide the required coverage metrics.

8.3.3.3 Object Code Verification at the Unit Level

Tools are available that enable users to create test cases for structural coverage of high-level source and apply these exact same test cases to the structural coverage of the corresponding object code.

A driver program is generated by such a unit test tool which encapsulates the entire test environment, defining, running, and monitoring the test cases through initial test verification and then subsequent regression analysis. When used for OCV this driver may be linked with either the high-level source unit or the associated object code. In so doing users can apply a uniform test process and compare code to determine any discrepancies or deficiencies.

If any such structural coverage discrepancies or deficiencies are identified at the object level, users are then presented with an opportunity to define additional test cases to close any gaps in the test process. The obvious advantage of identifying and applying corrective action at such an early development stage is that it is much easier and cheaper. It also significantly increases the quality of the code and the overall test process with the latter reaping benefits at the later stages of integration and system testing and then onward in the form of reduced failure rates/maintenance costs when the application is in the field.

While the code is still under development, together with satisfying the necessary OCV requirements in a highly automated and cost-effective manner developers can also benefit from the considerable additional test feedback. The results of these analysis facilities can be fed back to the development team with the possibility that further code and design deficiencies may be identified and rectified, further enhancing the quality of the application as a whole.

8.3.3.4 Justifying the Expense

It is clear that OCV has always involved significant overhead and that even in the aerospace sector it is only enforced as a requirement for the most demanding safety integrity levels. Even then, the elements nominated for object code verification in these applications usually represent a subset of the application as a whole—a specialist niche indeed.

However, there is precedence for this situation. Until quite recently unit test has been considered by many as a textbook nicety for the purposes of the aircraft and nuclear industry. More recently it has found a place in automotive, railway, and medical applications, and now the ever-increasing capabilities and ease of use of automated unit test tools has introduced a commercial justification of such techniques even when risks are lower.

Most applications include key elements in the software: a subset of code that is particularly critical to the success of the application and can be identified in the application requirements. The software requiring OCV can be identified and traced through an extension to the RTM.

The advent of tools to automate the whole of that process from requirements traceability right through to OCV challenges the notion that the overhead involved can only justify the technique in very rare circumstances. Just as for unit test before it perhaps the time has come for OCV to be commercially justifiable in a much broader range of circumstances.

9 Implementing a Test Solution Environment

9.1 Pragmatic Considerations

Like so many other things in business life, ultimately the budget that is to be afforded to the test environment depends on commercial justification. If the project under consideration has to be shown to comply with standards in order to sell, then that justification is straightforward. It is much less clear-cut if it is based entirely on cost savings and enhanced reputation resulting from fewer recalls.

Although vendors make presentations assuming developers are to work on a virgin project where they can pick and choose what they like, that is often not the case. Many development projects enhance legacy code, interface to existing applications, are subject to the development methods of client organizations and their contractual obligations, or are restricted by time and budget.

The underlying direction of the organization for future projects also influences choices:

- Perhaps there is a need for a quick fix for a problem project in the field, or a software test tool that will resolve a mystery and an occasional runtime error crash in final test.
- Maybe there is a development on the order books which involves legacy code requiring a one-off change for an age-old client, but which is unlikely to be used beyond that.
- Perhaps existing legacy code cannot be rewritten, but there is a desire and mandate to raise the quality of software development on an ongoing basis for new developments and/or the existing code base.
- Or perhaps there is a new project to consider, but the lessons of problems in the past suggest that ongoing enhancement of the software development process would be beneficial.

To address a particular situation it is initially useful to consider how each of the five key attributes discussed earlier fit into the development process.

9.2 Considering the Alternatives

Given that vendors are generally not keen to highlight where their own offering falls short some insight into how to reach such a decision would surely be useful.

Fig. 34 superimposes the different analysis techniques on a traditional "V" development model. Obviously, a particular project may use another development model. In truth the analysis is model agnostic and a similar representation could be conceived for any other development process model—waterfall, iterative, agile, etc.

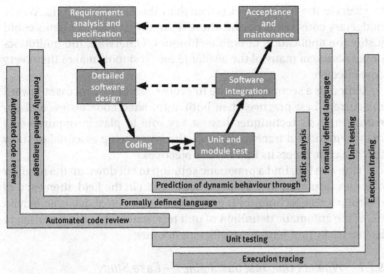

Fig. 34 The five fundamental test tool attributes directly relate to the specific development stages of design, code, test and verification, etc.

The extent to which it is desirable to cover all elements of the development cycle depends very much on the initial state of development and the desired outcome.

Each of the five key test tool attributes has merit.

There is a sound argument supporting traditional formal methods, but the development overheads for such an approach and the difficulty involved in applying it retrospectively to existing code limits its usefulness to the highly safety-critical market.

Automated code review checks for adherence to coding standards and is likely to be useful in almost all development environments.

Of the remaining approaches, dynamic analysis techniques provide a test environment much more representative of the final application than static predictions of dynamic analysis as well as the means to provide functional testing.

Where requirements traceability is key within a managed and controlled development environment the progressive nature of automated code review followed by unit, integration, and system test aligns well within the overall tiered concept of most modern standards. It also fulfills the frequent requirement or recommendation to exercise the code in its target environment.

Where robustness testing is considered desirable and justified it can be provided by means of the automatic definition of unit test vectors, or through the use of the static prediction of dynamic behavior. Each of these techniques has its own merits, with the former exercising code in its target environment and the latter providing a means to exercise the full data set rather than discrete test vectors. Where budgetary constraints permit, these mutually exclusive benefits could justify the application of both techniques. Otherwise, the multifunctional nature of many of the available unit test tools makes them very cost-effective.

If there is a secondary desire to evolve corporate processes toward the current best practice, then both automated code review and dynamic analysis techniques have a key role to play in requirements management and traceability, with the latter being essential to show that the code meets its functional objectives.

If the aim is to find a pragmatic solution to cut down on the number of issues displayed by a problem application in the field, then each of the robustness techniques (i.e., the static analysis of dynamic behavior or the automatic definition of unit test vectors) has the potential to isolate tricky problems in an efficient manner.

9.2.1 When Is Unit Test Justifiable?—Case Study

It is perhaps useful to consider one of the five attributes to illustrate a possible thought process to be applied when deciding where to invest.

Unit testing cannot always be justified. Moreover, sometimes it remains possible to perform unit test from first principles without the aid of any test tool at all.

There are pragmatic judgments to be made.

Sometimes such a judgment is easy. If the software fails, what are the implications? Will anyone be killed, as might be the case in aircraft flight control? Will the commercial implications be disproportionately high, as exemplified by a continuous plastics production plant? Or are the costs of recall extremely high, perhaps in an automobile's engine controller? In these cases extensive unit testing is essential and hence any tools that may aid in that purpose make sense.

On the other hand, if software is developed purely for internal use or is perhaps a prototype, then the overhead in unit testing all but the most vital of procedures would be prohibitive.

As might be expected, there is a gray area. Suppose the application software controls a mechanical measuring machine where the quantity of the devices sold is low and the area served is localized. The question becomes: Would the occasional failure be more acceptable than the overhead of unit test?

In these circumstances it is useful to prioritize the parts of the software that are either critical or complex. If a software error leads to a strangely colored display or a need for an occasional reboot, it may be inconvenient but not in itself justification for unit test. On the other hand, the unit test of code that generates reports showing whether machined components are within tolerance may be vital. Hence, as we have already seen advocated by leading standards such as DO-178B, significant benefit may be achieved through a decision to apply the rigor of unit test to a critical subset or subsets of the application code as a whole.

9.2.2 When Are Unit Test Tools Justifiable?

Again, it comes down to cost. The later a defect is found in the product development, the more costly it is to fix—a concept first established in 1975 with the publication of Brooks' *The Mythical Man-Month* and proven many times since through various studies.

The automation of any process changes the dynamic of commercial justification. That is especially true of test tools given that they make earlier unit test much more feasible. Consequently, modern unit test almost implies the use of such a tool unless only a handful of procedures are involved.

The primary function of such unit test tools is to assist with the generation and maintenance of the harness code that provides the main and associated calling functions or procedures (generically "procedures"), with the more sophisticated tools on the market being able to fully automate this process. The harness itself facilitates compilation and allows unit testing to take place.

The tools not only provide the harness itself, but also statically analyze the source code to provide the details of each input and output parameter or global variable in an easily understood form. Where unit testing is performed on an isolated snippet of code, stubbing of called procedures can be an important aspect of unit testing. This can also often be partially or fully automated to further enhance the efficiency of the approach.

High levels of automation afforded by modern unit test tools makes the assignment of values to the procedure under test a simple process and one that demands little knowledge of the code on the part of the test tool operator. This creates the necessary unit test objectivity because it divorces the test process from that of code development where circumstances require it and from a pragmatic perspective substantially lowers the level of skill required to develop unit tests.

It is this ease of use that means unit test can now be considered a viable arrow in the development quiver, targeting each procedure at the time of writing. When these early unit tests identify weak code it can be corrected while the original intent remains very fresh in the mind of the developer.

10   Summary and Conclusions

There are hundreds of textbooks about software test and many that deal with only a specialist facet of it. It is therefore clear that a chapter such as this cannot begin to cover the whole subject in detail.

Some elements of software test remain unmentioned here. What about testing for stack overflow? Timing considerations? And multi-threaded applications, with their potentially problematic race and lethal embrace conditions?

In preference to covering all such matters superficially the technique demonstrated in this chapter—of "drilling down" into a topic to shed sufficient light on its worth in a particular circumstance—is sound and will remain so even for test techniques as yet unavailable. It is therefore applicable in those matters not covered in any detail here.

In each case these are techniques that can be deployed or not as circumstances dictate; the whole genre of software test techniques and tools constitute a tool kit just as surely as a toolbox holding spanners, hammers, and screwdrivers.

And, just like those handyman's friends, sometimes it is possible to know from a superficial glance whether a technique or test tool is useful, while at other times it needs more detailed investigation.

The key then is to be sure that decisions to follow a particular path are based on sufficient knowledge. Take the time to investigate and be sure that the solution you are considering will prove to be the right one for a particular circumstance.

Consider the option of developing in-house tests, and when commercially marketed test tools are considered be sure to ask for an evaluation copy.

Choosing the wrong technique or the wrong tool can be a very costly and embarrassing mistake indeed.

Questions and Answers

Question 1 Why is it best to identify bugs as early as possible?

Answer 1 The earlier a bug is identified, the less it costs to fix it. Compare the cost of a developer spotting a bug in his automotive code as he is writing it with the cost of a vehicle recall when the software is in service!

Question 2 Identify the industrial sectors to which these functional safety standards apply:
  a. DO-178C
  b. ISO 26262
  c. IEC 62304

Answer 2 **a.** DO-178C, "Software considerations in airborne systems and equipment certification," applies to the aerospace industry
  b. ISO 26262, "Road vehicles—Functional safety," applies to the automotive industry
  c. IEC 62304, "Medical device software—Software life cycle processes," applies to the medical device industry

Question 3   Why is the achievement of 100% branch coverage more demanding than the achievement of 100% statement coverage?

Answer 3   Where 100% branch coverage is achieved both false and true decisions will be executed for each branch (100% statement coverage is achievable without exercising both false and true decisions for each branch)

Question 4   Why does 100% source code coverage not imply 100% object code coverage?

Answer 4   There are significant differences between the available constructs in high-level languages (such as C and C++) and object code. To interpret the source code a compiler and/or linker therefore have to generate code that is not directly traceable to source code statements

Question 5   Are adherents to functional safety standards such as DO-178C, ISO 26262, IEC 61508, and IEC 62304 obliged to use software test tools?

Answer 5   There is no obligation for any developer or development team to use software test tools to comply with any of the functional safety standards. There is a commercial decision to be made in each case whether the efficiency gained from the application of tools justifies expenditure on them

Further Reading

[1]   English Oxford Living Dictionaries, https://en.oxforddictionaries.com/definition/test.

[2]   IEC 61508-1:2010 Functional safety of electrical/electronic/programmable electronic safety-related systems.

[3]   The Mythical Man-Month, Addison-Wesley, ISBN: 0-201-00650-2, 1975.

[4]. DO-178C, Software Considerations in Airborne Systems and Equipment Certification (December 13, 2011), RTCA.

[5]   MISRA C:2012, Guidelines for the Use of the C Language in Critical Systems, March 2013. ISBN 978-1-906400-10-1 (paperback), ISBN 978-1-906400-11-8 (PDF).

[6]   MISRA C++:2008 Guidelines for the Use of the C++ Language in Critical Systems (June 2008), ISBN 978-906400-03-3 (paperback), ISBN 978-906400-04-0 (PDF).

[7]   Computability: Turing, Gödel, Church, and Beyond (January 30, 2015) The MIT Press, Paperback, B. Jack Copeland (Editor, Contributor), Carl J. Posy (Editor, Contributor), Oron Shagrir (Editor, Contributor), Martin Davis (Contributor) et al., ISBN 978-0262527484.

[8]   ISO 26262:2011 Road vehicles—Functional safety.

[9]   IEC 62304 Edition 1.12015–05 Medical device software—Software life cycle processes.

[10] The International Obfuscated C Code Contest, https://www.ioccc.org.

[11] Joint Strike Fighter Air Vehicle C++ Coding Standards for The System Development And Demonstration Program (December 2005), Document Number 2RDU00001 Rev. C, Lockheed Martin Corporation.

[12] Barr group Embedded C Coding Standard, https://barrgroup.com/Embedded-Systems/Books/Embedded-C-Coding-Standard.

[13] SEI CERT C Coding Standard, Rules for Developing Safe, Reliable, and Secure Systems, 2016. Edition.

[14] MISRA C:2012 AMD1, Additional security guidelines for MISRA C:2012, https://www.revolvy.com/page/Frances-E.-Allen.

[15] Just The Facts 101: Discovering Computers, 2010. Complete: First edition. Cram 101.

[16] Classics in Software Engineering, Yourdon Press, ISBN: 0-917072-14-6, 1979. https://cse.buffalo.edu/~rapaport/111F04/greatidea3.html.

[17] Software Reliability, Principles and Practices, G. J. Myers, Wiley, New York, 1976.

[18] Microsoft Excel, https://www.microsoft.com/en-gb/p/excel/cfq7ttc0k7dx?activetab= pivot%3aoverviewtab.

[19] High Integrity C++ Coding Standard Version 4.0, www.codingstandard.com.

[20] S. 882 (111th): Drug and Device Accountability Act of 2009, https://www.gov-track.us/congress/bills/111/s882.

[21] FDA Orders Recall Of Baxter Colleague Infusion Pumps (May 2010), Lucy Campbell, https://www.lawyersandsettlements.com/lawsuit/baxter-colleague-infusion-pumps-recall.html.

[22] DO-178B, Software Considerations in Airborne Systems and Equipment Certification (December 1992), EUROCAE.

[23] CENELEC–EN 50128, Railway applications—Communication, signaling and processing systems—Software for railway control and protection systems, 2011.

[24] IEC 61513, Nuclear power plants—Instrumentation and control important to safety—General requirements for systems, 2011.

[25] IEC 61511, Functional safety—Safety instrumented systems for the process industry sector, 2016.

[26] SAE J3061, Cybersecurity Guidebook for Cyber-Physical Vehicle Systems, 2016.

[27] ISO/SAE 21434, Scope (DRAFT) Road vehicles–Cybersecurity engineering, 2017.

10

EMBEDDED MULTICORE SOFTWARE DEVELOPMENT

Rob Oshana

Vice President Software Engineering R&D, NXP Semiconductors, Austin, TX, United States

CHAPTER OUTLINE

Software Engineering for Embedded Systems. https://doi.org/10.1016/B978-0-12-809448-8.00010-2
© 2019 Elsevier Inc. All rights reserved.

A multicore processor is a computing device that contains two or more independent processing elements (referred to as cores) integrated on to a single device that read and execute program instructions. There are many architectural styles of multicore processors and many application areas such as embedded processing, graphics processing, and networking.

A typical multicore processor will have multiple cores that can be the same (homogeneous) or different (heterogeneous), accelerators (the more generic term is processing element) for dedicated functions such as video or network acceleration, and a number of shared resources (memory, cache, peripherals such as ethernet, display, codecs, and UART) (Fig. 1).

A key algorithm that should be memorized when thinking about multicore systems is the following:

$$\text{High performance} = \text{parallelism} + \text{memory hierarchy} - \text{contention}$$

- Parallelism is all about exposing parallelism in the application.
- Memory hierarchy is all about maximizing data locality in the network, disk, RAM, cache, core, etc.
- Contention is all about minimizing interactions between cores (e.g., locking, synchronization, etc.).

To achieve the best performance we need to achieve the best possible parallelism, use memory efficiently, and reduce the contention. As we move forward we will touch on each of these areas.

1 Symmetric and Asymmetric Multiprocessing

Efficiently allocating resources in multicore systems can be a challenge. Depending on the configuration the multiple software components in these systems may or may not be aware of how other components are using these resources. There are two primary forms of multiprocessing (as shown in Fig. 2):
- Symmetric multiprocessing
- Asymmetric multiprocessing

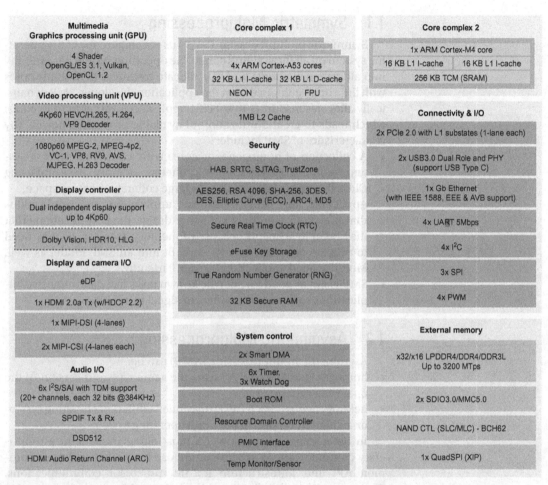

Fig. 1 A heterogeneous multicore system.

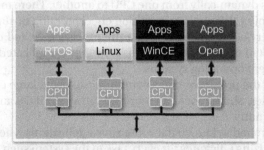
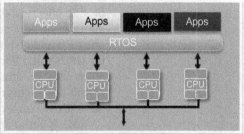

Fig. 2 Asymmetric multiprocessing (left) and symmetric multiprocessing (right).

1.1 Symmetric Multiprocessing

Symmetric multiprocessing (SMP) uses a single copy of the operating system on all the system's cores. The operating system has visibility into all system elements and can allocate resources to multiple cores with little or no guidance from the application developer. SMP dynamically allocates resources to specific applications rather than to cores, which leads to greater utilization of available processing power. Key characteristics of SMP include:

- A collection of homogeneous cores with a common view of system resources such as sharing a coherent memory space and using CPUs that communicate using a large coherent memory space.
- Applicable for general-purpose applications or applications that may not be entirely known at design time. Applications that may need to suspend because of memory accesses, or may need to migrate or restart on any core, fit into an SMP model as well. Multithreaded applications are SMP friendly.
- SMP is not as good for specific known tasks like data-intensive applications such as audio, video, or signal processing.

1.2 Asymmetric Multiprocessing

Asymmetric multiprocessing (AMP) can be:

- homogeneous—each CPU runs the same type and version of the operating system; or
- heterogeneous—each CPU runs either a different operating system or a different version of the same operating system.

In heterogeneous systems you must either implement a proprietary communications scheme or choose two OSs that share a common API and infrastructure for interprocessor communications. There must be well-defined and implemented methods for accessing shared resources.

In an AMP system an application process will always runs on the same CPU, even when other CPUs run idle. This can lead to one CPU being underutilized or overutilized. In some cases it may be possible to migrate a process dynamically from one CPU to another. There may be side effects of doing this such as requiring checkpointing of state information or a service interruption when the process is halted on one CPU and restarted on another CPU. This is further complicated if the CPUs run different operating systems.

In AMP systems the processor cores communicate using large coherent bus memories, shared local memories, hardware FIFOS, and other direct connections.

AMP is better applied to known data-intensive applications where it is better at maximizing efficiency for every task in the system such as audio and video processing. AMP is not as good as a pool of general computing resources.

The key reason there are AMP multicore devices is because they are the most economical way to deliver multiprocessing to specific tasks. The performance, energy, and area envelope is much better than SMP.

2 Parallelism Saves Power

Multicore reduces average power comsumption. It is becoming harder to achieve increased processor performance from traditional techniques such as increasing the clock frequency or developing new architectural approaches to increase instructions per cycle (IPC). Frequency scaling of CPU cores is no longer valid, primarily due to power challenges.

An electronic circuit has a capacitance C associated with it. Capacitance is the ability of a circuit to store energy. This can be defined as:

$$C = \text{charge}(q) / \text{voltage}(V)$$

And the charge on a circuit can therefore be $q = CV$.

Work can be defined as the act of pushing something (charge) across a "distance." In this discussion we can define this in electrostatic terms as pushing the charge from 0 to V volts in a circuit:

$$W = V^* q, \text{or in other terms}, W = V^* CV \text{ or } W = CV^2$$

Power is defined as work over time, or in this discussion it is how many times a second we oscillate the circuit.

$$P = (\text{work}) W / (\text{time}) T \text{ and since } T = 1 / F \text{ then } P = WF \text{ or substituting}, P = CV^2 F$$

We can use an example to reflect this. Let us assume the circuit is as in Fig. 3.

This simple circuit has a capacitance C, a voltage V, a frequency F, and therefore a power defined as $P = CV^2 F$.

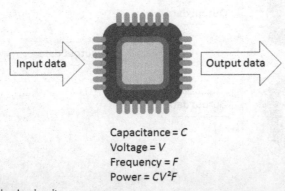

Capacitance = C
Voltage = V
Frequency = F
Power = $CV^2 F$

Fig. 3 A simple circuit.

If we instead use a multicore circuit (as shown in Fig. 4) we can make the following assumptions:

- We will use two cores instead of one.
- We will clock this circuit as half the frequency for each of the two cores.
- We will use more circuitry (capacitance C) with two cores instead of one, plus some additional circuitry to manage these cores, assume 2.1X the capacitance.
- By reducing the frequency we can also reduce the voltage across the circuit. Let's assume we can use a voltage of 0.7 or the single core circuit (it could be half the single core circuit but let's assume a bit more for additional overhead).

Given these assumptions, power can be defined as:

$$P = CV^2F = (2.1)(0.7)^2(0.5) = .5145$$

What this says is by going from one core to multicore we can reduce overall power consumption by over 48%, given the conservative assumptions above.

There are other benefits from going to multicore. When we can use several smaller simpler cores instead of one big complicated core we can achieve more predictable performance and achieve a simpler programming model in some cases.

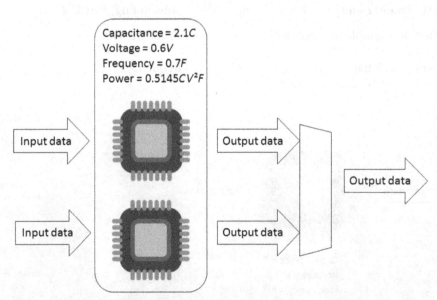

Capacitance = 2.1C
Voltage = 0.6V
Frequency = 0.7F
Power = 0.5145CV^2F

Input data

Output data

Output data

Input data

Output data

Fig. 4 A parallel multicore circuit.

3   Look for Parallelism Opportunities

A computer program always has a sequential part and a parallel part. What does this mean? Let's start with a simple example:

1. A=B+C
2. D=A+2
3. E=D+A
4. For (i=0; i<E; i++)
5. N(i)=0

In this example Steps 1, 2, and 4 are "sequential." There is a data dependence that prevents these three instructions from executing in parallel.

Steps 4 and 5 are parallel. There is no data dependence and multiple iterations of N(i) can execute in parallel.

Even with E a large number—say, 200—the best we can do is to sequentially execute four instructions, no matter how many processors we have available to us.

When algorithms are implemented serially there is a well-defined operation order that can be very inflexible. In the edge detection example for a given data block the Sobel cannot be computed until after the smoothing function completes. For other sets of operations, such as within the correction function, the order in which pixels are corrected may be irrelevant.

Dependencies between data reads and writes determine the partial order of computation. There are three types of data dependencies that limit the ordering: true data dependencies, antidependencies, and output dependencies (Fig. 5).

True data dependencies imply an ordering between operations in which a data value may not be read until after its value has been written. These are fundamental dependencies in an algorithm, although it might be possible to refactor algorithms to minimize the impact of this data dependency.

Antidependencies have the opposite relationship and can possibly be resolved by variable renaming. In an antidependency a data value cannot be written until the previous data value has been read. In Fig. 5 the final assignment to A cannot occur before B is assigned because B needs the previous value of A. In the final assignment variable A is renamed to D, then the B and D assignments may be reordered.

Data dependency	Antidependency	Output dependency
A = 4 * C + 3; B = A + 1; A = 3 * C + 4;	A = 4 * C + 3; B = A + 1; A = 3 * C + 4;	A = 4 * C + 3; B = A + 1; A = 3 * C + 4;
Read after write	Write after read	Write after write

Fig. 5 Key data dependencies that limit parallelism.

Renaming may increase storage requirements when new variables are introduced if the lifetimes of the variables overlap as code is parallelized. Antidependencies are common occurrences in sequential code. For example, intermediate variables defined outside the loop may be used within each loop iteration. This is fine when operations occur sequentially. The same variable storage may be repeatedly reused. However, when using shared memory, if all iterations were run in parallel, they would be competing for the same shared intermediate variable space. One solution would be to have each iteration use its own local intermediate variables. Minimizing variable lifetimes through proper scoping helps to avoid these dependency types.

The third type of dependency is the output dependency. In an output dependency writes to a variable may not be reordered if they change the final value of the variable that remains when the instructions are complete. In the "output dependency" of Fig. 5 the final assignment to A may not be moved above the first assignment because the remaining value will not be correct.

Parallelizing an algorithm requires both honoring dependencies and appropriately matching the parallelism to the available resources. Algorithms with a high number of data dependencies will not parallelize effectively. When all antidependencies are removed and still partitioning does not yield acceptable performance, consider changing algorithms to find an equivalent result using an algorithm that is more amenable to parallelism. This may not be possible when implementing a standard with strictly prescribed algorithms. In other cases there may be effective ways to achieve similar results.

Let's take a look at some examples:

Loopnest1

$$\text{for}(i=0;i<n;i++)\{$$
$$a[i]=a[i-1]+b[i]$$
$$\}$$

```
Loop 1: a [0] =a [-1] +b [0]
Loop 2: a [1] =a [0] +b [1]
......
Loop N: a [N] =a [N-1] +b [N]
```

Here Loop 2 is dependent on the result of Loop 1: to compute a [1] one needs a [0], which can be obtained from Loop 1. Hence, Loop nest 1 cannot be parallelized because there is a loop-carried dependence flow on the other loop.

Loopnest2

$$\text{for}(i=0;i<n;i++)\{$$
$$a[i]=a[i]+b[i]$$
$$\}$$

```
Loop 1: a [0] =a [0] +b [0]
Loop 2: a [1] =a [1] +b [1]
......
Loop N: a [N] =a [N] +b [N]
```

Here Loop nest 2 can be parallelized because the antidependency from the read of an [i] to the write of an [i] has an (=) direction and it's not loop carried.

$$\text{Loop nest 3}$$
$$for(i=0;i<n;i++)\{$$
$$a[4*i]=a[2*i-1]$$
$$\}$$

```
Loop 1: a [0] =a [-1]
Loop 2: a [4] =a [1]
......
Loop N: a [4*N] =a [2*N-1]
```

We can see that there is no dependency between any loops in Loop nest 3. Hence Loop nest 3 can be parallelized.

Multicore architectures have sensitivity to the structure of software. In general, parallel execution incurs overhead that limits the expected execution time benefits that can be achieved. Performance improvements therefore depend on software algorithms and their implementations. In some cases parallel problems can achieve speedup factors close to the number of cores, or potentially more if the problem is split up to fit within each core's cache(s), which avoids the use of the much slower main system memory. However, as we will show, many applications cannot be accelerated adequately unless the application developer spends a significant effort to refactor the portions of the application.

As an example, we can think of an application as having both sequential parts and parallel parts (as shown in Fig. 6).

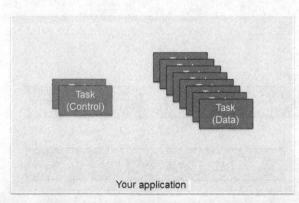

Fig. 6 An application showing sequential (control) parts and data (parallel) parts.

This application, when executed on a single core processor, will execute sequentially and take a total of 12 time units to complete (Fig. 7).

If we run this same application on a dual-core processor (Fig. 8), the application will take a total of 7 time units, limited by the sequential part of the code that cannot execute in parallel due to reasons we showed earlier.

This is a speedup of 12/7 = 1.7X from the single-core processor.

If we take this further to a four-core system (Fig. 9), we can see a total execution time of 5 units for a total speedup of 12/5 = 2.4X from the single-core processor and 7/5 = 1.4X over the two-core system.

If the fraction of the computation that cannot be divided into concurrent tasks is f and no overhead incurs when the computation is divided into concurrent parts, the time to perform the computation with n processors is given by $t_p \geq ft_s + [(1-f)t_s]/n$ (as shown in Fig. 10).

The general solution to this is called Amdahl's Law and is shown in Fig. 11.

Amdahl's Law states that parallel performance is limited by the portion of serial code in the algorithm. Specifically:

$$\text{Speedup} = 1/(S + (1-S)/N)$$

where S is the portion of the algorithm running serialized code, and N is the number of processors running parallelized code.

| Task (Control) | Task (Control) | Task (Data) | Task (Data) | Task (Data) | Task (Data) | Task (Data) | Task (Data) | Task (Data) | Task (Data) | Task (Data) | Task (Data) |

Fig. 7 Execution on a single-core processor, 12 total time units.

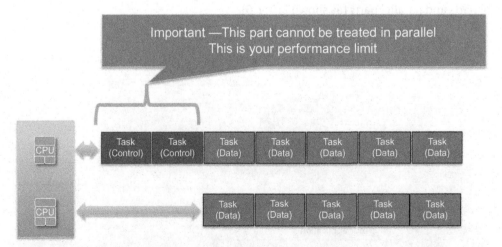

Fig. 8 Execution on a two-core multicore processor, 7 total time units.

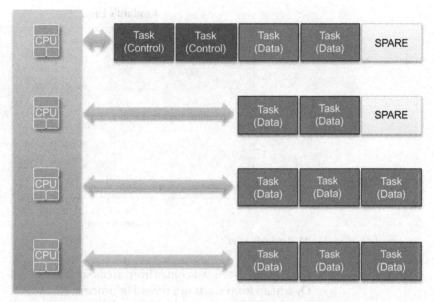

Fig. 9 Execution on a four-core multicore processor, 5 total time units.

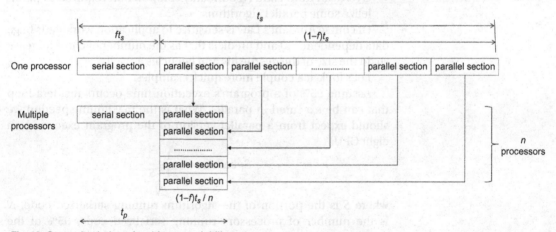

Fig. 10 General solution of multicore scalability.

Amdahl's Law implies that adding additional cores results in additional overheads and latencies. These overheads and latencies serialize execution between communicating and noncommunicating cores by requiring the use of mechanisms such as hardware barriers and resource contention. There are also various interdependent sources of latency and overhead due to processor architecture (e.g., cache coherency), system latencies and overhead (e.g., processor scheduling), and application latencies and overhead (e.g., synchronization).

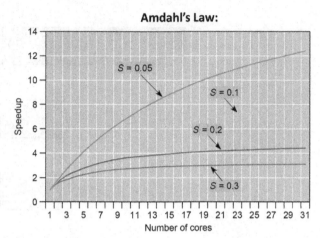

Fig. 11 Amdahl's Law.

Parallelism overhead comes from areas such as:
- Overhead from starting a thread or process.
- Overhead of communicating shared data.
- Overhead of synchronizing.
- Overhead from extra (redundant) computation required to parallelize some parallel algorithms.

Of course Amdahl's Law is sensitive to application workloads (e.g., data dependencies) and predicts that as the number of cores increase so do the size of the overheads and latencies as well.

Let's look at a couple more quick examples.

Assume 95% of a program's execution time occurs inside a loop that can be executed in parallel. What is the maximum speedup we should expect from a parallel version of the program executing on eight CPUs?

$$\text{Speedup} = \frac{1}{S + \frac{1-S}{N}}$$

where S is the portion of the algorithm running serialized code, N is the number of processors running serialized code, 95% of the program's execution time can be executed in parallel, eight CPUs, $S = 1 - 0.95 = 0.05$, and $N = 8$:

$$\text{Speedup} = \frac{1}{0.05 + \frac{1-0.05}{8}}$$

where speedup = 5.9.

Assume 5% of a parallel program's execution time is spent within inherently sequential code. What is the maximum speedup achievable by this program, regardless of how many processing elements are used?

$$\text{Speedup} = \frac{1}{S + \dfrac{1-S}{N}}$$

where 5% parallel program's execution time is spent within inherently sequential code, and $N = \infty$:

$$\text{Speedup} = \frac{1}{0.05 + \dfrac{1-0.05}{N}} = \frac{1}{0.05} = 20$$

3.1 Multicore Processing Granularity

Granularity can be described as the ratio of computation to communication in a parallel program. There are two types of granularity (as shown in Fig. 12).

Fine-grained parallelism implies partitioning the application into small amounts of work leading to a low computation-to-communication ratio. For example, if we partition a "for" loop into independent parallel computions by unrolling the loop, this would be an example of fine-grained parallelism. One of the downsides to fine-grained parallelism is that there may be many synchronization points—for example, the compiler will insert synchronization points after each loop iteration, which may cause additional overhead. Moreover, many loop iterations would have to be parallelized to get decent speedup, but there the developer has more control over load-balancing the application.

Coarse-grained parallelism is where there is a high computation-to-communication ratio. For example, if we partition an application into several high-level tasks that then get allocated to different cores, this would be an example of coarse-grained parallelism. The advantage of this is that there is more parallel code running at any point in time and

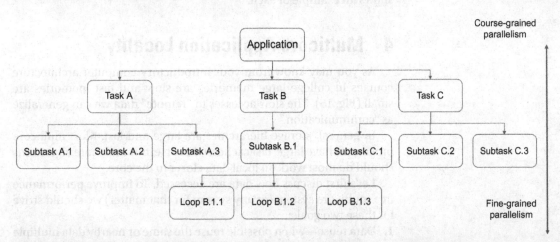

Fig. 12 Course-grained and fine-grained parallelism.

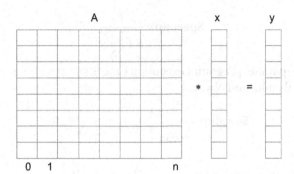

Fig. 13 Matrix multiplication with a vector.

fewer synchronizations are required. However, load-balancing may not be ideal as the higher level tasks are usually not all equivalent as far as execution time is concerned.

Let's take one more example. Let's say we want to multiply each element of an array *A* by a vector X (Fig. 13). Let's think about how to decompose this problem into the right level of granularity. The code for something like this would look like:

```
for (i=0, N-1)
        for (j=0, N-1)
                y[i] = A[i,j] * x[j];
```

From this algorithm we can see that each output element of y depends on one row of A and all of x. All tasks are of the same size in terms of number of operations.

How can we break this into tasks? Course grained with a smaller number of tasks or fine grained with a larger number of tasks. Fig. 14 shows an example of each.

4   Multicore Application Locality

As you may know from your introductory computer architecture courses in college large memories are slow and fast memories are small (Fig. 15). The slow accesses to "remote" data we can generalize as "communication."

In general, storage hierarchies are large and fast. Most multicore processors have large, fast caches. Of course, our multicore algorithms should do most work on local data closer to the core.

Let's first discuss how data are accessed. To improve performance in a multicore system (or any system for that matter) we should strive for these two goals:

1. Data reuse—when possible reuse the same or nearby data multiple times. This approach is mainly intrinsic in computation.

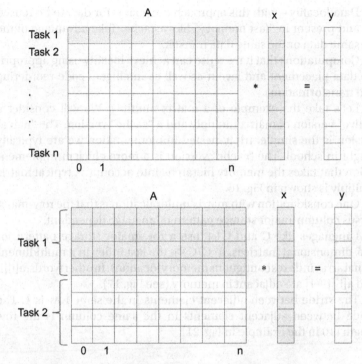

Fig. 14 (A) Fine-grained parallelism. (B) Course-grained parallelism.

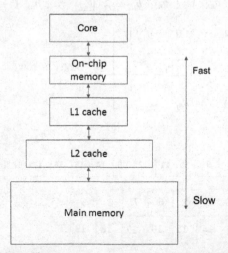

Fig. 15 Memory hierarchies.

2. Data locality—with this approach the goal is for data to be reused and present in "fast memory" like a cache. Take advantage of the same data or the same data transfer.

Computations that have reuse can achieve locality using appropriate data placement and layout as well as intelligent code reordering and transformations.

Let's take the example of a matrix multiply. We will consider a "naive" version of matrix multiply and a "cache" version. The "naive" version is the simple, triply-nested implementation we are typically taught in school. The "cache" version is a more efficient implementation that takes the memory hierarchy into account. A typical matrix multiply is shown in Fig. 16.

One consideration with matrix multiplication is that the row-major versus column-major storage pattern is language dependent.

Languages like C and C++ use a row-major storage pattern for two-dimensional matrices. In C/C++ the last index in a multidimensional array indexes contiguous memory locations. In other words, a[i][j] and a[i][j + 1] are adjacent in memory (see Fig. 17).

The stride between adjacent elements in the same row is 1. The stride between adjacent elements in the same column is the row length (10 in the example in Fig. 21).

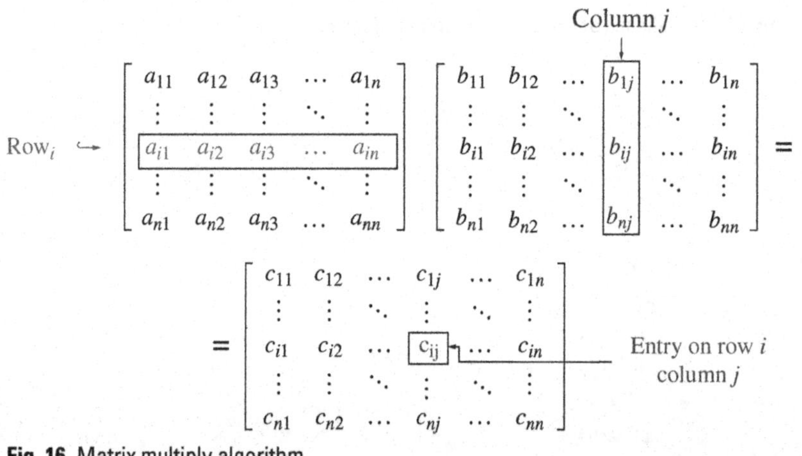

Fig. 16 Matrix multiply algorithm.

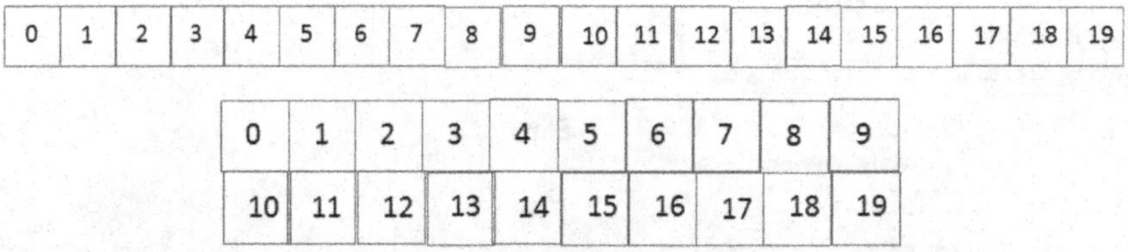

Fig. 17 Row-major storage ordering for C/C++.

Access by rows

```
for (i = 0; i < 5; i++)
        for (j = 0; j < 10; j++)
                a[i][j] = ...
```

Access by columns

```
for (j = 0; j < 10; j++)
        for (i = 0; i < 5; i++)
                a[i][j] = ...
```

Fig. 18 Access by rows and by columns.

This is important because memory access patterns can have a noticeable impact on performance, especially on systems with a complicated multilevel memory hierarchy. The code segments in Fig. 18 access the same elements of an array, but the order of accesses is different.

We can apply additional optimizations including "blocking." "Block" in this discussion does not mean "cache block." Instead, it means a subblock within the matrix we are using in this example.

As an example of a "block" we can break our matrix into blocks ($N = 8$; subblock size = 4):

$$\begin{bmatrix} A_{11} & A_{12} \\ A_{21} & A_{22} \end{bmatrix} \times \begin{bmatrix} B_{11} & B_{12} \\ B_{21} & B_{22} \end{bmatrix} = \begin{bmatrix} C_{11} & C_{12} \\ C_{21} & C_{22} \end{bmatrix}$$

Here is the way it works—instead of the row access model that we just described:

```
/* row access method */
for (i = 0; i < N; i = i+1)
        for (j = 0; j < N; j = j+1)
                {r = 0;
                 for (k = 0; k < N; k = k+1){
                         r = r + y[i][k]*z[k][j];};
                 x[i][j] = r;
                };
```

With the blocking approach we use two inner loops. One loop reads all the $N \times N$ elements of z[]. The other loop will read N elements of one row of y[] repeatedly. The final step is to write N elements of one row of x[].

Subblocks (i.e., A_{xy}) can be treated just like scalars in this example and we can compute:

$C_{11} = A_{11}B_{11} + A_{12}B_{21}$
$C_{12} = A_{11}B_{12} + A_{12}B_{22}$
$C_{21} = A_{21}B_{11} + A_{22}B_{21}$
$C_{22} = A_{21}B_{12} + A_{22}B_{22}$

Now a "blocked" matrix multiply can be implemented as:

```
for (jj=0; jj<n; jj+=bsize) {
 for (i=0; i<n; i++)
  for (j=jj; j < min(jj+bsize,n); j++)
   c[i][j] = 0.0;
 for (kk=0; kk<n; kk+=bsize) {
  for (i=0; i<n; i++) {
   for (j=jj; j < min(jj+bsize,n); j++) {
    sum = 0.0
    for (k=kk; k < min(kk+bsize,n); k++) {
     sum += a[i][k] * b[k][j];
    }
    c[i][j] += sum;
   }
  }
 }
}
```

In this example the loop ordering is bijk. The innermost loop pair multiplies a 1 X b-size sliver of A by a b-size X b-size block of B and sums into a 1 X b-size sliver of C. We then loop over i steps through n row slivers of A and C, using the same B (see Fig. 19).

The results are shown in Fig. 20A. As you can see, row order access is faster than column order access.

Of course, we can also increase the number of threads to achieve higher performance (as shown in Fig. 25 as well). Since this multi-core processor has only four cores, running with more than four threads—*when threads are computer bound*—this only causes the OS to "thrash" as it switches threads across the cores. At some point you can expect the overhead of too many threads to hurt performance and slow an application down. See the discussion on Amdahl's Law a little earlier!

The importance of efficient caching for multicore performance cannot be overstated.

High performance == parallelism + memory hierarchy − contention

You need not only to expose parallelism, but also to take into account the memory hierarchy and work hard to eliminate/minimize

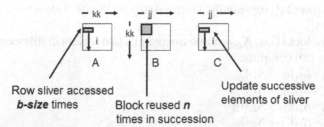

Fig. 19 Blocking optimization for cache.

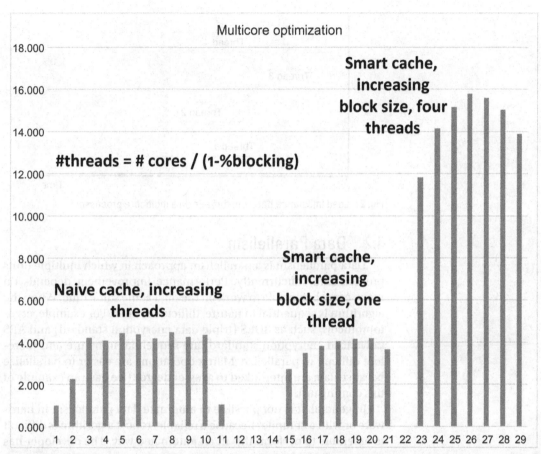

Fig. 20 (A) Performance of naive cache with matrix multiply (column order) and increasing threads, (B) row order and blocking optimizations with just one thread, and (C) row access with blocking caches and four threads of execution.

contention. This becomes increasingly true because as the number of cores grows so does the contention between cores.

4.1 Load Imbalance

Load imbalance is the time that processors in the system are idle due to (Fig. 21):

- Insufficient parallelism (during that phase).
- Unequal size tasks.

Unequal size tasks can include things like tree-structured computations and other fundamentally unstructured problems. The algorithm needs to balance load where possible and the developer should profile the application on the multicore processor to look for load-balancing issues. Resources can sit idle when load-balancing issues are present.

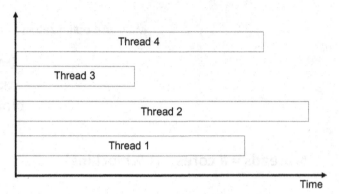

Fig. 21 Load imbalance between threads on a multicore processor.

4.2 Data Parallelism

Data parallelism is a parallelism approach in which multiple units process data concurrently. Performance improvement depends on many cores being able to work on the data at the same time. When the algorithm is sequential in nature, difficulties arise. For example, cryptoprotocols, such as 3DES (triple data encryption standard) and AES (advanced encryption standard) are sequential in nature and therefore difficult to parallelize. Matrix operations are easier to parallelize because data are interlinked to a lesser degree (we have an example of this coming up).

In general, it is not possible to automate data parallelism in hardware or with a compiler because a reliable, robust algorithm is difficult to assemble to perform this in an automated way. The developer has to own part of this process.

Data parallelism represents any kind of parallelism that grows with the data set size. In this model the more data you give to the algorithm, the more tasks you can have and operations on data may be the same or different. But the key to this approach is its scalability.

Fig. 22 shows the scalable nature of data parallelism.

In the example given in Fig. 23 an image is decomposed into sections or "chunks" and partitioned to multiple cores to process in parallel. The "image in" and "image out" management tasks are usually performed by one of the cores (an upcoming case study will go into this in more detail).

4.3 Task Parallelism

Task parallelism distributes different applications, processes, or *threads to* different units. This can be done either manually or with the help of the operating system. The challenge with task parallelism is how to divide the application into multiple threads. For systems with

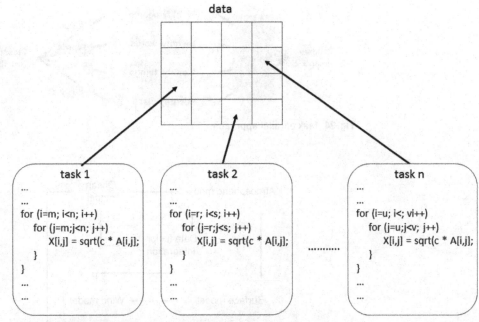

Fig. 22 Data parallelism is scalable with the data size.

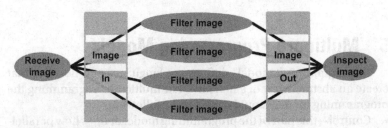

Fig. 23 Data parallel approach.

many small units, such as a computer game, this can be straightforward. However, when there is only one heavy and well-integrated task the partitioning process can be more difficult and often faces the same problems associated with data parallelism.

Fig. 24 is an example of task parallelism. Instead of partitioning data to different cores the same data are processed by each core (task), but each task is doing something different on the data.

Task parallelism is about functional decomposition. The goal is to assign tasks to distinct functions in the program. This can only scale to a constant factor. Each functional task, however, can also be data parallel. Fig. 25 shows this. Each of these functions (atmospheric, ocean, data fusion, surface, wind) can be allocated to a dedicated core, but only the scalability is constant.

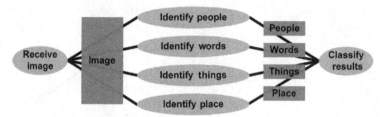

Fig. 24 Task parallel approach.

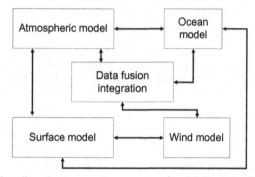

Fig. 25 Function allocation in a multicore system (scalability limited).

5   Multicore Programming Models

A "programming model" defines the languages and libraries that create an abstract view of a machine. For multicore programming the programming model should consider the following:

- Control—this part of the programming model defines how parallelism is created and how dependencies (orderings) are enforced. An example of this would be to define the explicit number of threads of execution.
- Data—this part of the programming model defines how and whether data can be shared or kept private. For shared data this also defines whether the data are shared data accessed or private data communicated. For example, what is the access to global data from multiple threads? What is the control of data distribution to execution threads?
- Synchronization—this part of the programming model defines which operations can be used to coordinate parallelism and which are atomic (indivisible) operations. An example of this is communication (e.g., which data transfer parts of the language will be used or which libraries are used). Another example would be to define explicitly the mechanisms to regulate access to data.

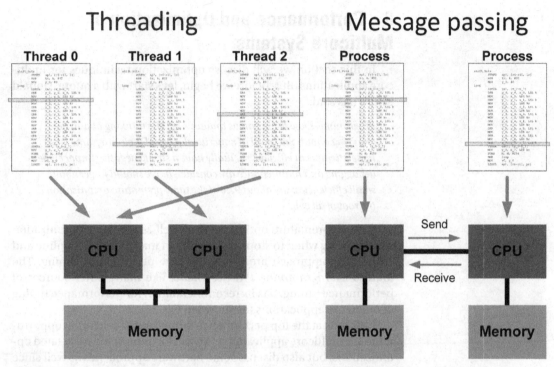

Fig. 26 Multicore programming model decision (threading use of shared memory or message passing).

Fig. 26 is an example of a programming model decision between threading (shared memory) and message passing. As we will soon see this drives decisions on which technology to use in multicore development.

The shared memory paradigm in some ways is similar to sequence programming, but the developer must explicitly specify parallelism and use some mechanism (locks/semaphors) to control access to the shared memory. Sometimes this is called directive based and we can use technologies like OpenMP to help us with this.

The choice is explicit from a parallel-programming perspective. We can use pthreads that are common to a shared memory system and focus on synchronization, or we can use a technology like Message Passing Interface (MPI), which is based on message-passing systems where the focus is on communication—not so much synchronization.

Determining the right programming model for a multicore system is dependent on several factors:

- The type of multicore processor—different multicore processors support different types of parallelism and programming.
- The level of abstraction required—from "do it yourself" (DIY) multicore to using abstraction layers.

6 Performance and Optimization of Multicore Systems

In this section we will discuss optimization techniques for multicore applications. But, before we begin, let's start with a quote. Donald Knuth has said:

Programmers waste enormous amounts of time thinking about, or worrying about, the speed of noncritical parts of their programs, and these attempts at efficiency actually have a strong negative impact when debugging and maintenance are considered. We should forget about small efficiencies, say about 97% of the time: premature optimization is the root of all evil.

Indeed, premature optimization as well as excessive optimization (not knowing when to stop) are harmful in many ways. Discipline and an iterative approach are key to effective performance tuning. The *Multicore Programming Practice Guide*, like many other sources of performance tuning, has its recommendation for performance tuning of multicore applications (as shown in Fig. 27).

Let's look at the top performance tuning and acceleration opportunities for multicore applications. We will focus on software-related optimizations, but also discuss some hardware approaches as well since they ultimately are related to software optimizations. The list we will discuss is:

1. Select the right "core" for your multicore
2. Improve serial performance before migrating to multicore
3. Achieve proper load balancing (SMP Linux)

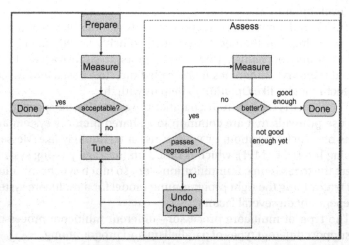

Fig. 27 Performance-tuning process from the *Multicore Programming Practice Guide*.

4. Improve data locality
5. Reduce or eliminate false sharing
6. Use affinity scheduling when necessary
7. Apply the proper lock granularity and frequency
8. Remove sync barriers where possible
9. Minimize communication latencies
10. Use thread pools
11. Manage thread count
12. Stay out of the kernel if at all possible
13. Use parallel libraries (pthreads, openMP, etc.)
   Let's explore these one by one.

6.1   Select the Right "Core" for Your Multicore

Do you need a latency-oriented core or a throughput-oriented core? Do you need hardware acceleration or not? Is a heterogeneous architecture needed or a homogeneous one? It makes a big difference from a performance perspective. I put this in this section as well because it makes software programming more efficient without having to resort to fancy tips and tricks to get decent performance. But it does require benchmarking and analysis of the core and software development tools such as the compiler and the operating system.

6.2   Improve Serial Performance Before Migrating to Multicore (Especially Instruction-Level Parallelism)

Early in the development process, before looking at optimizations specific to multicore, it's necessary to spend time improving the serial (single-core) application performance. Sequential execution must first be efficient before moving to parallelism to achieve higher performance. Early sequential optimization is much easier and less time consuming and less likely to introduce bugs.

Many performance improvements obtained in serial implementation will close the gap on the parallelism required to achieve your goals when moving to multicore. It's much easier to focus on parallel execution alone during this migration, instead of having to worry about both sequential and parallel optimization at the same time.

Just be careful not to introduce serial optimizations that degrade or limit parallelism (such as unnecessary data dependencies) or over-exploiting details of the single-core hardware architecture (such as cache capacity).

Focus on instruction-level parallelism (ILP). The compiler can help. The main goal of a compiler is to maintain the functionality of the application and support special functionality provided by the target and the application such as pragmas, instrinsics, and other capability like OpenMP, which we will discuss further. For example, the "restrict" keyword (C99 standard of the C programming language) is used in pointer declarations, basically telling the compiler that for the lifetime of the pointer only it or a value derived from it (such as pointer + 1) can be used to access an object it points to. This limits the effects of memory disambiguation or pointer aliasing, which enables more aggressive optimizations. An example of this is given below. In this example stores may alias loads. This forces operations to be executed sequentially:

```
void VectorAddition(int *a, int *b, int *c)
{
        for (int i = 0; i < 100; i++)
                a[i] = b[i] + c[i];
}
```

In this example the "restrict" keyword allows independent loads and stores. Operations can now be performed in parallel:

```
void VectorAddition(int restrict a, int *b, int *c)

{
        for (int i = 0; i < 100; i++)
                a[i] = b[i] + c[i];
}
```

The "restrict" keyword can enable more aggressive optimizations such as software pipelining. Software pipelining is a powerful loop optimization usually performed by the compiler back end. It consists of scheduling instructions across several iterations of a loop. This optimization (Fig. 28) enables instruction-level parallelism, reduces pipeline stalls, and fills delay slots. Loop iterations are scheduled so that an iteration starts before the previous iteration has completed. This approach can be combined with loop unrolling (see below) to achieve significant efficiency improvements.

Loop transformations also enable ILP. Loops are typically the hotspots of many applications. Loop transformations are used to organize the sequence of computations and memory accesses to better fit the processor internal structure and enable ILP.

One example of a loop transformation to enable ILP is called loop unrolling. This transformation can decrease the number of memory accesses and improve ILP. It unrolls the outer loop inside the inner loop and increases the number of independent operations inside the loop body.

```
//Software pipelining is enabled again by using the restrict
//keyword that informs the compiler that no a lias exist (object
//with restrict can only be accessed by that pointer).
    .....

    doen3 #99

    dosetup3 L7
    ....   //some address computation code not shown here.
[   move.w  (r2)+,d2   // Loop prolog
    move.w  (r4)+,d0  ]
    add    d2,d0,d3
LOOPSTART3        //Loop is composed of 2 VLES
L7
[   move.w  (r2)+,d0
    move.w  (r4)+,d4  ]
[   add    d0,d4,d3
    move.w  d3,(r1)+ ] // This store is for the previous iteration
LOOPEND3
    move.w  d3,(r1)    // Loop epilog
    ......
```

```
Short* restrict a;
Short* restrict b;
Short* restrict c;
.....
int i;
for (i = 0; i <100; i ++)
    a++ = b++ + c++;
.....
```

Fig. 28 Software pipelining enables instruction-level parallelism, reduces pipeline stalls, and fills delay slots.

Below is an example of a doubly nested loop:

```
for (i=0; i<N; i+++)
{
    for (j=0; j<N; j++)
    {
        a[i][j] = b[j][i];
    }
}
```

Here is the same loop with the loop unrolled, enabling more ILP:

```
for (i=0; i<N; i+++)
{
    for (j=0; j<N; j++)
    {
        a[i][j] = b[j][i];
        a[i+1][j] = b[j][i+1];
    }
}
```

This approach improves the spatial locality of b and increases the size of loop body and therefore the available ILP. Loop unrolling also

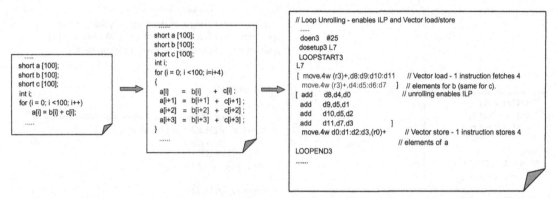

Fig. 29 Loop unrolling can enable vectorization, which improves performance even more.

enables more aggressive optimizations such as vectorizations (SIMD) by allowing the compiler to use these special instructions to improve performance even more (as shown in Fig. 29 where the "move.4w" instructions are essentially SIMD instructions operating on four words of data in parallel).

As with all optimizations you need to use the right ones for the application. Take the example of video versus audio algorithms. For example, audio applications are based on a continuous feed of data samples. There are lots of long loops to process. This is a good application structure to use software pipelining, which works well in this situation.

Video applications, on the other hand, are designed to break up video frames into smaller blocks (like the example earlier, we called this minimal coded units, MCUs). This type of structure uses small loop counts and many blocks of processing. Software pipelining in this case is not as efficient due to the long prologs required for pipelining. There is too much overhead for each loop. In this case it's best to use loop unrolling instead, which works better and more efficiently for these smaller loops.

There are many other examples of sequential optimization. The key message is to make sure you apply these first before worrying about other parallel optimizations.

6.3 Achieve Proper Load Balancing (SMP Linux) and Scheduling

Multicore-aware operating systems like Linux have infrastructure to support multicore performance at the system level such as SMP schedulers, different forms of synchronization, load-balancers for interrupts, affinity-scheduling techniques and CPU isolation. If used properly, overall system performance can be optimized. However, they all have some inherent overhead so they must be used properly.

Linux is a multitasking kernel that allows more than one process to exist at any given time. The Linux process scheduler manages which process runs at any given time. The basic responsibilities are:

- Share the cores equally among all currently running processes.
- Select the appropriate process to run next (if required), using scheduling and process priorities.
- Rebalance processes between multiple cores in SMP systems if necessary.

Multicore applications are generally categorized to be either CPU bound or I/O bound. CPU-bound applications spend a lot of time using the CPU to do computations (like server applications). I/O-bound applications spend a lot of time waiting for relatively slow I/O operations to complete (e.g., like a smartphone waiting for user input, network accesses, etc.). There is obviously a trade-off here. If we let our task run for longer periods of time, it can accomplish more work but responsiveness (to I/O) suffers. If the time period for the task gets shorter, our system can react faster to I/O events. But now more time is spent running the scheduling algorithm between task switches. This leads to more overhead and efficiency suffers.

6.4   Improve Data Locality

Although this was discussed earlier, there are some additional comments to be made concerning software optimizations related to data locality. This is a key focus area for multicore optimization. In many applications this requires some careful analysis of the application. For example, let's consider a networking application that is using the Linux operating system.

In the Linux operating system all network-related queues and buffers use a common data structure called sk_buff. This is a large data structure that holds all the control information needed for a network packet. sk_buff structure elements are organized as a doubly linked list. This allows efficient movement of sk_buff elements from the beginning/end of a list to the beginning/end of another list.

The standard sk_buff has information spread over three or more cache lines. Data-plane applications require only one cache line's worth of information.

We can take advantage of this by creating a new structure that packs/aligns to a single cache line. If we are smart and make this part of the packet buffer headroom, then we don't have to worry about cache misses and flushes each time we access the large sk_buff structure. Instead we use a small portion of this that fits neatly into the cache, improving data locality and efficiency. This is shown in Fig. 30.

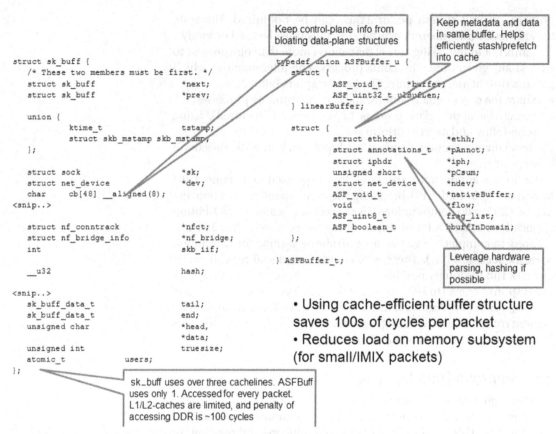

Fig. 30 Creating an efficient data structure to improve locality and performance.

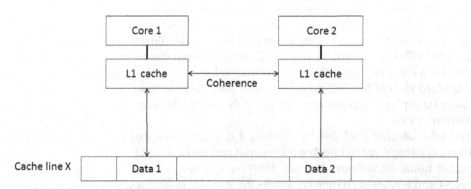

Fig. 31 False sharing in SMP systems.

6.5 Reduce or Eliminate False Sharing

False sharing occurs when two software threads manipulate data that are on the same cache line. As we discussed earlier the memory system of a multicore processor must ensure cache coherency in SMP systems. Any modifications made to shared cache must be flagged to the memory system so each processor is aware the cache has been modified. The affected cache line is "invalidated" when one thread has changed data on that line of cache. When this happens the second thread must wait for the cache line to be reloaded from memory (Fig. 31).

The code below shows this condition. In this example sum_temp1 may need to continually reread "a" from main memory (instead of from cache) even though inc_b's modification of "b" should be irrelevant.

In this situation if the extra prefetched words are not needed and another processor in this cache-coherent, shared memory system must immediately change these words, this extra transfer has a negative impact on system performance and energy consumption:

```
struct data
{
    volatile int a;
    volatile int b;
};

data f;

int sum_temp1()
{
    int s = 0;
    for (int i = 0; i < 1000; i++)
            s += f.a;
    return s;
}

void inc_b()
{
    for (int i = 0; i < 1000; i++)
            ++f.b;
}
```

One solution to this condition is to pad the data in the data structure so that the elements causing false-sharing performance degradation will be allocated to different cache lines:

```
struct data
{
        volatile int a;
        volatile int b;
};
```

```
struct data
{
        volatile int a;
        unsigned char
        padding[CACHE_LINE__SIZE – sizeof(int)];
        volatile int b;
};
```

6.6   Use Affinity Scheduling When Necessary

In some applications it might make sense to force a thread to continue executing on a particular core instead of letting the operating system make the decision whether or not to move the thread to another core. This is referred to as "processor affinity." Operating systems like Linux have APIs to allow a developer to control this, allowing them the ability to map certain threads to cores in a multicore processor (Fig. 32).

On many processor architectures any migration of threads across cache, memory, or processor boundaries can be expensive (flushing the cache, etc.). The developer can use APIs to set the affinities for certain threads to take advantage of shared caches, interrupt handing, and to match computation with data (locality). A snippet of code that shows how to do this is:

```
#define _GNU_SOURCE
#include <sched.h>
long
sched_setaffinity(pid_t pid, unsigned int len,
                  unsigned long *user_mask_ptr);
long
sched_getaffinity(pid_t pid, unsigned int len,
                  unsigned long *user_mask_ptr);
```

In this code snippet the first system call will set the affinity of a process. The second system call retrieves the affinity of a process.

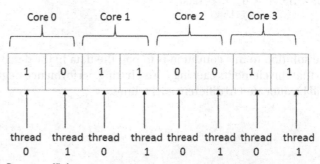

Fig. 32 Process affinity.

The PID argument is the process ID of the process you want to set or retrieve. The second argument is the length of the affinity bitmask. The third argument is a pointer to the bitmask. Masks can be used to set the affinity of the cores of interest.

If you set the affinity to a single CPU this will exclude other threads from using that CPU. This also takes that processor out of the pool of resources that the OS can allocate. Be careful every time you are manually controlling affinity scheduling. There may be side effects that are not obvious. Design the affinity-scheduling scheme carefully to ensure efficiency.

6.7 Apply the Proper Lock Granularity and Frequency

There are two basic laws of concurrent execution:
1. The program should not malfunction.
2. Concurrent execution should not be slower than serial execution.

We use locks as a mutual exclusion mechanism to prevent multiple threads getting simultaneous access to shared data or code sections. These locks are usually implemented using semaphores or mutexes, which are essentially expensive API calls into the operating system.

Locks can be fine grained or coarse grained. Too many locks increases the amount of time spent in the operating system and increases the risk of deadlock. Coarse-grained locking reduces the chance of deadlock, but can cause additional performance degradations due to locking large areas of the application (mainly system latency).

Avoiding heavy use of locks and semaphores due to performance penalties. Here are some guidelines:

- Organize global data structures into buckets and use a separate lock for each bucket.
- Design the system to allow threads to compute private copies of a value and then synchronize only to produce the global result. This will require less locking.
- Avoid spinning on shared variables waiting for events.
- Use atomic memory read/writes to replace locks if the architecture supports this.
- Avoiding atomic sections when possible (an atomic section is a set of consecutive statements that can only be run by one thread at a time).
- Place locks only around commonly used fields and not entire structures if this is possible.
- Compute all possible precalculations and postcalculations outside the critical section, as this will minimize the time spent in a critical section.
- Make sure that locks are taken in the same order to prevent deadlock situations.

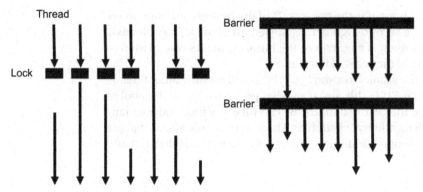

Fig. 33 Locks versus barriers.

- Use mechanisms that are designed to reduce overhead in the operating system. For example, in POSIX-based systems use the try-lock() function that allows the program to continue execution to handle an unsuccessful lock. In Linux use the "futex" (fast user space mutex) to do resource checks in user space instead of the kernel (see Fig. 33).

6.8   Remove Sync Barriers Where Possible

A synchronization barrier causes a thread to wait until the other threads have reached the barrier and is used to ensure that variables needed at a given execution point are ready to be used. It differs from a lock for this very reason (Fig. 33).

Barriers can have the same performance problem as locks if not used properly. Oversynchronizing can negatively impact performance. This is the reason barriers should only be used to ensure that data dependencies are respected and/or where the execution frequency is the lowest.

A barrier can be used to replace creating and destroying threads multiple times when dealing with a sequence of tasks (i.e., replacing join and create threads). Hence when used in this way certain performance improvements can be realized.

6.9   Minimize Communication Latencies

When possible limit communication in multicore systems. Even extra computing is often more efficient than communication in many cases. For example, one approach to minimize communication latencies is to distribute data by giving each CPU has its own local data set that it can work on. This is called per-cpu data and is a technique used in the Linux kernel for several critical subsystems. For example, the kernel slab allocator uses per-cpu data for fast CPU local memory allocation. The disadvantage is higher memory overhead and increased complexity dealing with CPU hotplug.

In general, the communication time can be estimated using the following:

$$T_{com}(n) = \alpha + \beta^* n$$

where n is the size of the message, α is the startup time due to the latency, and β is the time for sending one data unit limited by the available bandwidth.

Here are a few other tips and tricks to reduce communication latency:

- Gather small messages into larger ones when possible to increase the effective communications bandwidth (reduce β, reduce α, increase n).
- Sending noncontiguous data is usually less efficient than sending contiguous data (increases α, decreases n).
- Do not use messages that are too large. Some communication protocols change when messages get too large (increases α, increases n, increases β).
- The layout of processes/threads on cores may affect performance due to communication network latency and the routing strategy (increases α).
- Use asynchronous communication techniques to overlap communication and computation. Asynchronous communication (nonblocking) primitives do not require the sender and receiver to "rendezvous" (decrease α).
- Avoid memory copies for large messages by using zero-copy protocols (decrease β).

6.10 Use Thread Pools

When using a peer or master/worker design, like the one we discussed earlier, users should not create new threads on the fly. This causes overhead. Instead have them stopped when they are not being used. Creating and freeing processes and threads is expensive. The penalty caused by the associated overhead may be larger than the benefit of running the work in parallel.

In this approach a number of threads are allocated to a thread pool. N threads are created to perform a number of operations M, where N $\ll$ M. When a threads completes the task it's working on, it will then request the next task from the thread pool (usually organized into a queue) until all the tasks have been completed. The thread can then terminate (or sleep) until new tasks become available. Fig. 34 is a conceptual diagram of this.

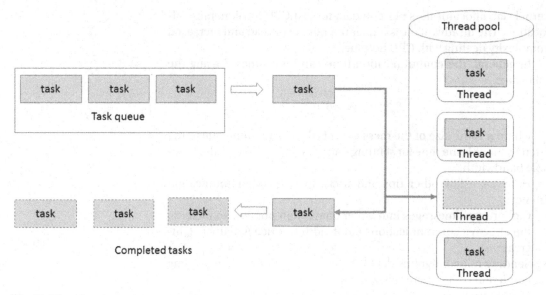

Fig. 34 Thread pool concept.

A thread pool is implemented as a data structure as:

```
typedef struct _threadpool{
        int num_threads;               //number of active threads
        int qsize;                     //queue size
        pthread_t *threads;            //pointer to threads
        work_t* qhead;                 //queue head pointer
        work_t* qtail;                 //queue tail pointer
        pthread_mutex_t qlock;         //lock on the queue list
        pthread_cond_t q_not_empty;    //nonempty and empty condition vairiables
        pthread_cond_t q_empty;
        int shutdown;
        int dont_accept;
} thread_pool;
```

The associated C code can be implemented to control access to the thread pool's data structure. This is not shown here, but there are plenty of examples you can find that show reference implementations.

6.11 Manage Thread Count

As discussed earlier, parallel execution always incurs some overhead resulting from functions such as task startup time, intertask synchronization, data communications, hardware bookkeeping (e.g., memory consistency), software overhead (libraries, tools, runtime system, etc.), and task termination time.

As a general rule, small tasks (fine grain) are usually inefficient due to the overhead to manage many small tasks. Large tasks could lead to load imbalance. In many applications there is a trade-off between having enough tasks to keep all cores busy and having enough computation in each task to amortize the overhead.

The optimal thread count can also be determined by estimating the average blocking time of the threads running on each core:

Thread count = number of cores / (1 − perentage average blocking time of threads)

6.12 Stay Out of the Kernel If at All Possible

We gain a lot of support from operating systems, but when optimizing for performance it's sometimes better to not go in and out of the operating system often, as this incurs overhead. There are many tips and tricks for doing this, so study the manual for the operating system and learn about the techniques available to optimize performance. For example, Linux supports SMP multicore using mechanisms like "futex" (fast user space mutex).

A futex is comprised of two components:

- A kernelspace wait queue.
- An integer in userspace.

In multicore applications the multiple processes or threads operate on the integer in userspace and only use expensive system calls when there is a need to request operations on the wait queue (Fig. 35). This would occur if there was a need to wake up waiting processes, or put the current process in the wait queue. Futex operations do not use

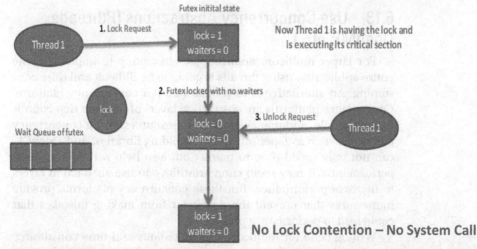

Fig. 35 A futex prevents calls into the kernel.

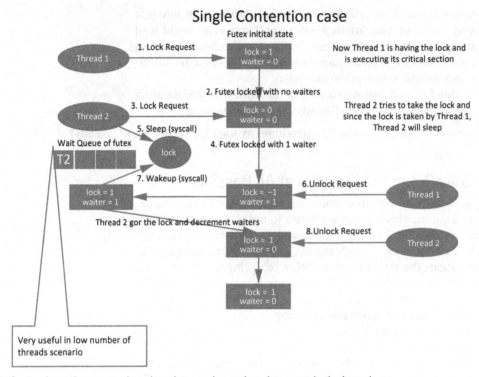

Fig. 36 Contention only occurs when there is a need to update the queue in the kernel.

system calls except when the lock is contended. However, since most operations do not require arbitration between processes this will not happen in most cases (Fig. 36).

6.13 Use Concurrency Abstractions (Pthreads, OpenMP, etc.)

For larger multicore applications, attempting to implement the entire application using threads is going to be difficult and time consuming. An alternative is to program atop a concurrency platform. Concurrency platforms are abstraction layers of software that coordinate, schedule, and manage multicore resources. Using concurrency platforms, such as OpenMP, thread-building libraries, and OpenCL, can not only speed time to market but also help with system-level performance. It may seem counterintuitive to use abstraction layers to increase performance. But these concurrency platforms provide frameworks that prevent the developer from making mistakes that could lead to performance problems.

Writing code for multicore can be tedious and time consuming. Here is an example showing the sequential code for a dot product:

```
#define SIZE 1000
Main() {
        double a[SIZE], b[SIZE];
        // Compute a and b
        double sum = 0.0;
        for(int i=0, i < SIZE; i++)
                sum += a[i] * b[i];
        // use sum....
}
```

Now let's implement this same dot product using pthreads for a four-core multicore processor. Here is the code:

```
#include <iostream>
#include <pthread.h>
#define THREADS 4
#define SIZE 1000
using namespace std;
double a[SIZE], b[SIZE], sum;
pthread_mutex mutex_sum;
void *dotprod(void *arg) {
        int my_id = (int)arg;
        int my_first = my_id * SIZE/THREADS;
        int my_last = (my_id + 1) * SIZE/THREADS;
        double partial_sum = 0;
        for(int i = my_first; i < my_last && i < SIZE; i++)
                partial_sum += a[i] * b[i];
        pthread_nmutex_lock(&mutex_sum);
        sum += partial_sum;
pthread_mutex_unlock(&mutex_sum);
pthread_exit((void*)0);
}
int main(int argc, char *argv[]) {
// compute a and b...
pthread_attr_t attr;
pthread_t threads[THREADS];
pthread_mutex_init(&mutex_sum, NULL);
pthread_attr_init(&attr);
pthread_attr_setdetachstate(&attr, PTHREAD_CREATE_JOINABLE);
sum = 0;
for(int i = 0; i < THREADS; i++)
pthread_create(&threads[i], &attr, dotprod, (void*)i);
        pthread_attr_destroy(&attr);
        int status;
        for(int i=0, i < THREADS; i++)
                pthread_join(threads[i], (void**)&status);
        // use sum....
        pthread_mutex_destroy(&mutex_sum);
        pthread_exit(NULL);
}
```

As you can see, this implementation, although probably much faster on a multicore processor, is rather tedious and difficult to implement correctly. This do-it-yourself approach can definitely work, but there are abstractions that can be used to make this process easier. That is what this section is all about.

Multicore programming can be made easier using concurrency abstractions. These include but are not limited to:

- Language extensions
- Frameworks
- Libraries

There exists a wide variety of frameworks, language extensions, and libraries. Many of these are built upon pthreads technology. Pthreads is the API of POSIX-compliant operating systems, like Linux, used in many multicore applications.

Let's take a look at some of the different approaches to provide levels of multicore programming abstraction.

7 Language Extensions Example—OpenMP

OpenMP is an example of language extensions. It's actually an API that must be supported by the compiler. OpenMP uses multithreading as the method of parallelizing, in which a master thread forks a number of slave threads and the task is divided among these threads. The forked threads run concurrently on a runtime environment that allocates threads to different processor cores.

This requires application developer support. Each section of code that is a candidate to run in parallel must be marked with a preprocessor directive. This is an indicator to the compiler to insert instructions into the code to form threads before the indicated section is executed. Upon completion of the execution of the parallelized code the threads join back into the master thread, and the application then continues in sequential mode again. This is shown in Fig. 37.

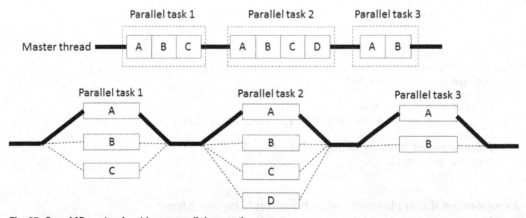

Fig. 37 OpenMP mode of multicore parallel execution.

```
int a, b;
main ()
{
  // serial code

  # pragma omp parallel num_threads (8)  (a) shared (b)
  {
      // parallel segment
  }
  // remaining serial code
}
```

```
# include <omp.h>
int a, b;
main ()
{
  // serial code

  for (i=0; i<8; i++)
      pthread_create (....., thread_function_name, ..);
  for (i=0; i<8; i++)
      pthread_join (....);

  // remaining serial code
}

void *thread_function_name (void *packaged_argument)
{
    int a;
    // parallel segment
}
```

Fig. 38 Instrumenting code to achieve parallelism in OpenMP. Original code is shown on the left, and the modified code is shown on the right.

Each thread executes the parallelized section of code independently. There exist various work-sharing constructs that can be used to divide a task among the threads so that each thread executes its allocated part of the code. Both task parallelism and data parallelism can be achieved this way. The example code below shows an example of how code can be instrumented to achieve this parallelism. The execution graph is shown in Fig. 38.

Using basic OpenMP programming constructs can lead you toward realizing Amdahl's Law by parallelizing those parts of a program that the application developer has identified to be concurrent and leaving the serial portions unaffected (Fig. 39).

OpenMP programs start with an initial master thread operating in a sequential region. When a parallel region is encountered (indicated by the compiler directive "#pragma omp parallel" in Fig. 40) new threads called worker threads are created by the runtime scheduler. These threads execute simultaneously on the block of parallel code. When the parallel region ends the program waits for all threads to terminate (called a join), and then resumes its single-threaded execution for the next sequential region.

The OpenMP specification supports several important programming constructs:

1. Support of parallel regions
2. Worksharing across processing elements
3. Support of different data environments (shared, private, ...)
4. Support of synchronization concepts (barrier, flush, ...)
5. Runtime functions/environment variables

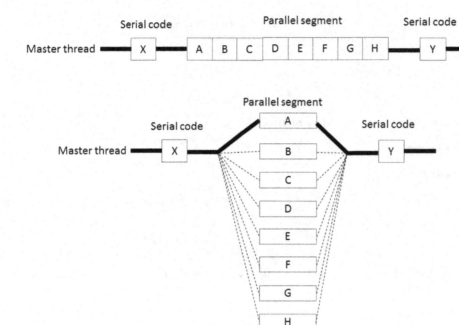

Fig. 39 Execution flow for the example in Fig. 38.

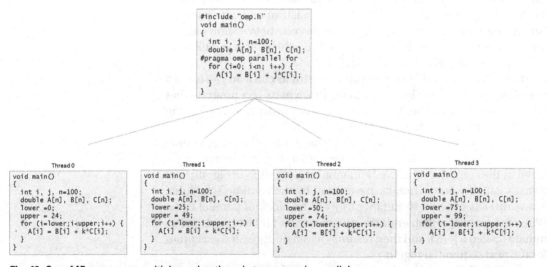

Fig. 40 OpenMP can spawn multiple worker threads to process in parallel.

We will not go into the details of all the APIs and programming constructs for OpenMP, but there are a few items to keep in mind when using this approach:

Loops must have a canonical shape in order for OpenMP to parallelize it. Be careful using loops like for (i=0; i<max; i++) zero[i] = 0;

It is necessary for the OpenMP runtime system to determine loop iterations. Moreover, no premature exits from loops are allowed (i.e., *break, return, exit, goto*, etc.).

The number of threads that OpenMP can create is defined by the OMP_NUM_THREADS environment variable. The developer should set this variable to the maximum number of threads you want OpenMP to use, which should be at least one per core/processor.

8 Pulling It All Together

In this section we will take a look at the process and steps to convert a sequential software application to a multicore application. There are many legacy sequential applications that may be converted to multicore. This section shows the steps to take to do that.

Fig. 41 is the process used to convert a sequential application to a multicore application.

Step 1: Understand requirements

Of course, the first step is to understand the key functional as well as nonfunctional requirements (performance, power, memory footprint, etc.) for the application. When migrating to multicore should the results be bit exact or is the goal equivalent functionality? Scalability and maintainability requirements should also be considered. When migrating to multicore keep in mind that good enough is good enough. We are not trying to overachieve just because we have multiple cores to work with.

Step 2: Sequential analysis

In this step we want to capture design decisions and learnings. Start from a working sequential code base. Then iteratively refine implementation. Explore the natural parallelism in the application. In this phase it also makes sense to tune implementation to target platform. Move from stable state to stable state to ensure you do not break anything. Follow these stages:

- Start with optimized code.
- Ensure libraries are thread safe.

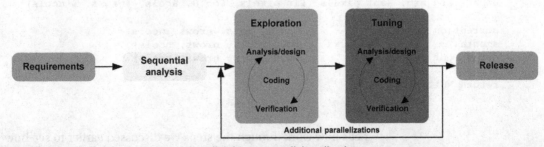

Fig. 41 Process flow from a sequential application to a parallel application.

- Profile to understand the structure, flow, and performance of the application. To achieve this attack hotspots first and then elect optimal cut points (a set of locations comprising a cut-set of a program) such that each cycle in the control flow graph of the program passes through some program location in the cut-set).

Step 3: Exploration

In this step we explore different parallelization strategies. The best approach is to use quick iterations of analysis/design, coding, and verification. We aim to favor simplicity over performance. Remember to have a plan for verification of each iteration so you do not regress. The key focus in this step is on dependencies and decomposition.

Step 4: Code optimization and tuning

The tuning step involves the identification and optimization of performance issues. These performance issues can include:

- Thread stalls
- Excessive synchronization
- Cache thrashing

Iteration and experimentation are key to this step in the process. Examples of experiments to try are:

- Vary threads and the number of cores
- Minimize locking
- Separate threads from tasks

8.1 Image-Processing Example

Let's explore this process in more detail by looking at an example. We will use an image-processing example shown in Fig. 42. We will use an edge detection pipeline for this example.

A basic sequential control structure for the edge detection example is:

```
static void *edge_detect(void *argv) {
  ed_arg_t *arg = (ed_arg_t *)argv;;
  char unsigned *out_pixels, *in_pixels;
  int col0, cols, nrows, ncols;

  unpack_arg(arg, &out_pixels, &in_pixels, &col0, &cols, &nrows, & ncols);

  correct(out_pixels, in_pixels, col0, cols, nrows, ncols);
  smooth(in_pixels, out_pixels, col0, cols, nrows, ncols);
  detect(out_pixels, in_pixels, col0, cols, nrows, ncols);

  return NULL;
}
```

Let's now work through the steps we discussed earlier to see how we can apply this approach to this example.

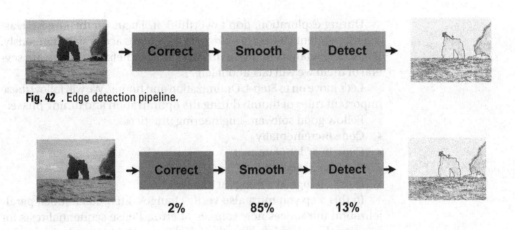

Fig. 42 . Edge detection pipeline.

Fig. 43 Exploration phase of edge detection algorithm showing the profile of processing steps.

We will start with Step 2: Sequential analysis. Fig. 43 shows that the three processing steps are unbalanced when it comes to total processing load. The "Smooth" function dominates the processing, followed by the "Detect" function and then the "Correct" function. By looking at the code above, we can also come to the conclusion that each function is embarrassingly data parallel. There is constant work per pixel in this algorithm, but also a very different amount of work per function (as can be seen in Fig. 43). This is also the time to make sure that you are only using thread-safe C libraries in this function, as we will be migrating this to multicore software using threads.

Let's move on to Step 3: Exploration. As mentioned previously this is the step where we explore different parallelization strategies such as:
- Quick analysis/design, coding, and verification iterations
- Favor simplicity over performance
- Verify each iteration
- Focus on dependencies and decomposition granularity
  We decompose to expose parallelism. Keep in mind these rules:
- Prioritize hotspots
- Honor dependencies
- Favor data decomposition
  Remember we may also need to "recompose" (we sometimes call this agglomeration) considering:
- Workload granularity and balance
- Platform and memory characteristics

During exploration, don't overthink it. Focus on the largest, least interdependent chunks of work to keep N cores active simultaneously. Use strong analysis and visualization tools to help you (there are several of them we will talk about later).

Let's move on to Step 4: Optimization and tuning. We will follow these important rules of thumb during the optimization and tuning phase:

- Follow good software-engineering practices
- Code incrementally
- Code in order of impact
- Preserve scalability
- Don't change two things at once

In this step you must also verify changes often. Remember, parallelization introduces new sources of error. Reuse sequential tests for functional verification even though the results may not be exact due to computation order differences. Check for common parallelization errors such as data races and deadlock. Perform stress-testing where you change both the data as well as the scheduling at the same time. Perform performance bottleneck evaluation.

8.2 Data Parallel; First Attempt

The strategy for attempt 1 is to partition the application into N threads one per core. Each thread handles $width/N$ columns. The different image regions are interleaved. We create a thread for each slice as shown in line 10 of the code below and then we join them back together in line 13. How this is partitioned is shown in Fig. 44.

```
1  void *edge_detect(char unsigned *out_pixels, char unsigned *in_pixels,
2   int nrows, int ncols) {
3  ed_arg_t arg[NUM_THREADS];
4  pthread_t thread[NUM_THREADS];
5  int status, nslices, col0, i;
6  nslices = NUM_THREADS;
7  for (i = 0; i < nslices; ++i) {
8   col0 = i;
9   pack_arg(&arg[i], out_pixels, in_pixels, col0, nslices, nrows, ncols);
10   pthread_create(&thread[i], NULL, edge_detect_thread, &arg[i]);
11  }
12  for (i = 0; i < nslices; ++i) {
13   pthread_join(thread[i], (void *)&status);
14  }
15  return NULL;
16     }
```

Fig. 44 Parallel execution of the edge detection pipeline algorithm on two cores.

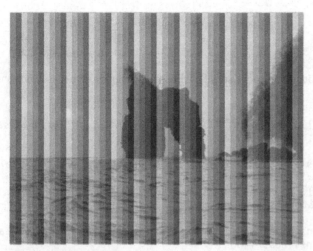

Fig. 45 RAW data dependence error.

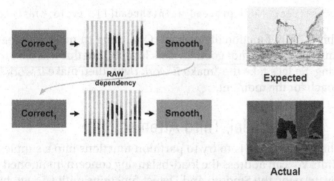

Fig. 46 Visual of the RAW data dependency error introduced during attempt 1.

How does this work? Well, not so well. I have introduced a functional error into the code. Diagnosis (Fig. 45) shows that I have introduced a RAW data dependency. Basically there is a race condition in which $Smooth_0$ can read the image data before being written by the Correct function. You can see this error in Fig. 46.

8.3 Data Parallel; Second Attempt

Let's try this again. The fix to this specific problem is to finish each function before starting the next to prevent the race condition. What we will do is "join" the threads after each function before starting the subsequent function to alleviate this problem. The code below shows this. We create threads for the Correct, Smooth, and Detect functions in lines 3, 8, and 13, respectively. We add "join" instructions at lines 5, 10, and 15. We can see how this works graphically in Fig. 47.

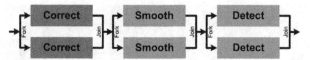

Fig. 47 Join instructions added after each function fixes the race condition.

```
1  for (i = 0; i < nslices; ++i) {
2    pack_arg(&arg[i], out_pixels, in_pixels, i, nslices, nrows, ncols);
3    pthread_create(&thread[i], NULL, correct, &arg[i]);
4  }
5  for (i = 0; i < nslices; ++i) pthread_join(thread[i], (void *)&status);
6  for (i = 0; i < nslices; ++i) {
7    pack_arg(&arg[i], in_pixels, out_pixels, i, nslices, nrows, ncols);
8    pthread_create(&thread[i], NULL, smooth, &arg[i]);
9  }
10 for (i = 0; i < nslices; ++i) pthread_join(thread[i], (void *)&status);
11 for (i = 0; i < nslices; ++i) {
12   pack_arg(&arg[i], out_pixels, in_pixels, i, nslices, nrows, ncols);
13   pthread_create(&thread[i], NULL, detect, &arg[i]);
14 }
15 for (i = 0; i < nslices; ++i) pthread_join(thread[i], (void *)&status);
```

This is now functionally correct. One thing to note. Interleaving columns probably is not good for data locality, but this was easy to get working. So let's take the "make it work right, then make it work fast" approach for the moment.

8.4 Task Parallel; Third Attempt

The next strategy is to try to partition functions into a simple task pipeline. We can address the load-balancing concern mentioned earlier by delaying the Smooth and Detect functions until enough pixels are ready. Fig. 48 shows this pipelining approach.

The code for this is shown below. We create our queues in lines 3 and 5. Three threads are created for each function in the algorithm in line 7. We fill our queues in lines 10 and 11, and then join the threads back together in line 13.

```
1  stage_f stage[3] = { correct, smooth, sobel };
2  queue_t *queue[4];
3  queue[0] = queue_create(capacity);
4  for (i = 0; i < 3; ++i) {
5    queue[i + 1] = queue_create(capacity);
6    pack_arg(&arg[i], queue[i + 1], queue[i], nrows, ncols);
7    pthread_create(&thread[i], NULL, stage[i], &arg[i]);
```

Fig. 48 Pipelining architecture for the edge detect algorithm.

```
8  }
9  while (*in_pixels) {
10  queue_add(queue[0], *in_pixels++);
11 } queue_add(queue[0], NULL);
12 for (i = 0; i < 3; ++i) {
13      pthread_join(thread[i], &status);
14      }
```

The results are improved throughput but not latency. The throughput is limited by the longest stage (1/85% = 1.12x). This approach would be difficult to scale with a number of cores. The latency is still the same but, like a pipeline, throughput improves. Remember that the stages are very unbalanced, so performance is not very good (Fig. 49). We could consider doing data parallelism within the longest stage to better balance the stages. A comparison of sequential and pipelined schedules is shown in Fig. 50.

8.5 Exploration Results

So far we have some interesting exploration results. We should go with the data decomposition approach. We should match the number of threads to the number of cores, since the threads are compute

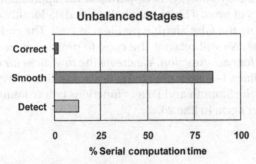

Fig. 49 Unbalanced stages in the pipelining approach.

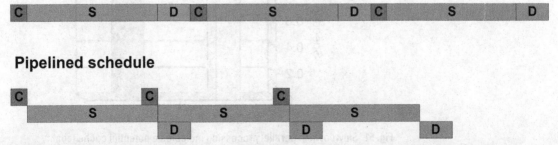

Fig. 50 Comparison of sequential and pipelined schedules.

intensive with few data dependencies to block them. There are still some tuning opportunities that we can take advantage of. We have seen some interleaving concerns in one of our approaches and experienced some delay required for the pipelining design.

8.6 Tuning

We eventually want to identify and optimize performance issues. These could include thread stalls, excessive synchronization, and cache thrashing.

This is where we iterate and experiment. Vary the number of threads and number of cores. Look for ways to minimize locking. Try to separate threads from tasks in different ways to see the impact.

Analysis shows the results of this recent approach made performance worse than the sequential code (Fig. 51)! Diagnosis shows false sharing between caches and significant performance degradation.

8.7 Data Parallel; Fourth Attempt

Let's try a new strategy. Let's partition the application into contiguous slices of rows. This should help with data locality and potentially eliminate the false-sharing problem as well. The code for this is shown below. We will refactor the code to process slices of rows for each thread for each function. We create the row slices for the Correct function in lines 1–4, then create the threads in line 5. We do a similar thing for the Smooth and Detect functions before joining them all back together again in line 26.

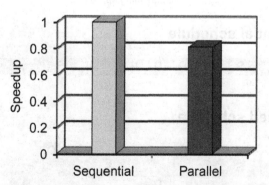

Fig. 51 Slowdown in parallel processing indicates a potential cache locality problem.

```
1  for (i = 0; i < nslices; ++i) {
2    row0 = i * srows;
3    row1 = (row0 + srows < nrows)? row0 + srows : nrows;
4    pack_arg(&arg[i], out_pixels, in_pixels, row0, row1, nrows, ncols);
5    pthread_create(&thread[i], NULL, correct_rows_thread, &arg[i]);
6  }
7  for (i = 0; i < nslices; ++i) {
8    pthread_join(thread[i], (void *)&status);
9  }
10 for (i = 0; i < nslices; ++i) {
11   row0 = i * srows;
12   row1 = (row0 + srows < nrows)? row0 + srows : nrows;
13   pack_arg(&arg[i], in_pixels, out_pixels, row0, row1, nrows, ncols);
14   pthread_create(&thread[i], NULL, smooth_rows_thread, &arg[i]);
15 }
16 for (i = 0; i < nslices; ++i) {
17   pthread_join(thread[i], (void *)&status);
18 }
19 for (i = 0; i < nslices; ++i) {
20   row0 = i * srows;
21   row1 = (row0 + srows < nrows)? row0 + srows : nrows;
22   pack_arg(&arg[i], out_pixels, in_pixels, row0, row1, nrows, ncols);
23   pthread_create(&thread[i], NULL, detect_rows_thread, &arg[i]);
24 }
25 for (i = 0; i < nslices; ++i) {
26   pthread_join(thread[i], (void *)&status);
27 }
```

8.8 Data Parallel; Fourth Attempt Results

The results of this iteration are better. We have good localization of data and there is good scalability as well. Data locality is important in multicore application. Traditional data layout optimization and loop transformation techniques apply to multicore since they

- minimize cache misses
- maximize cache line usage
- minimize the number of cores that touch a data item

Use the guidelines discussed earlier to achieve the best data locality for the application and significant performance results will be achieved. Use profiling and cache measurement tools to help you.

8.9 Data Parallel; Fifth Attempt

We are now at a point where we can continue to make modifications to the application to improve cache performance. We can split rows into slices matching the L1 cache width of the device (an example of this is blocking where the image can be decomposed into the appropriate block sizes as shown in Fig. 52). We can process rows by slices for better data locality. In the end, additional data layout improvements are possible. Some of these could have been sequential optimizations,

Fig. 52 Blocking and other cache optimizations can improve performance.

so make sure you look for those opportunities first. When migrating to multicore it makes it easier having already incorporated key optimizations into the sequential application.

8.10 Data Parallel; Work Queues

As an additional tuning strategy we can try using work queues (Fig. 53). In this approach we attempt to separate the number of tasks from the number of threads. This enables finer grained tasks without extra thread overhead. Locking and condition variables that can cause extra complexity are hidden in work queue abstractions.

When we use this approach the results get better as slices get smaller (Fig. 54). The key is to tune empirically. In this example we sacrificed some data locality—the work items are interleaved, but empirically this is minor. The results are improved load balancing, some loss of data locality, easy-to-tune thread, and work item granularity.

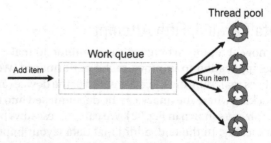

Fig. 53 Work queues can help with performance and abstraction.

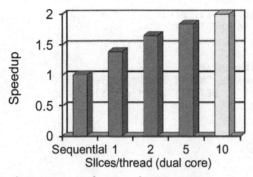

Fig. 54 Track results to assess performance improvement.

```
1  #define SLICES_PER_THREAD 5
2  nslices = NUM_THREADS * SLICES_PER_THREAD;
3  srows = nrows / nslices; if (nrows % nslices) ++srows;
4  crew = work_queue_create("Work Crew", nslices, NUM_THREADS);
5  for (i = 0; i < nslices; ++i) {
6    r0 = i * srows;
7    r1 = (r0 + srows < nrows)? r0 + srows : nrows;
8    pack_arg(&arg[i], out_pixels, in_pixels, r0, r1, nrows, ncols);
9    work_queue_add_item(crew, edge_detect_rows, &arg[i]);
10     } work_queue_empty(&crew);
```

8.11 Going too Far?

Can we do even more? Probably. We can use task parallel learning to compose three functions into one. We can do some more complex refactoring like lagging row processing to honor dependencies (Fig. 55). This could lead to a threefold reduction in thread creates and joins. This is possible with only minor duplicated computations at the region edge. This modest performance improvement comes with increased coding complexity. See the code below to implement lag processing in this application.

This code is certainly more efficient, but it's noticeably more complex. Is it necessary? Maybe or maybe not. Is the additional performance worth the complexity? This is a decision that must be made at some point. "Excessive optimization" can be harmful if the performance is not needed. Be careful with "premature optimization" and "excessive optimization."

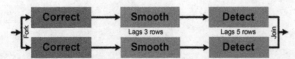

Fig. 55 Lag processing in the Correct/Smooth/Detect functions.

```
1  void *edge_detect_rows(pixel_t out_pixels[], pixel_t in_pixels[],
2      int row0, int row1, int nrows, int ncols) {
3    int rc, rs, rd, rs_edge, rd_edge;
4    int nlines, scols, c0, c1, i;
5    pixel_t *corrected_pixels, *smoothed_pixels;
6    corrected_pixels = MEM_MALLOC(pixel_t, SMOOTH_SIDE * ncols);
7    smoothed_pixels = MEM_MALLOC(pixel_t, DETECT_SIDE * ncols);
8    // determine number of line splits
9    nlines = ncols / CACHE_LINESIZE;
10   scols = ncols / nlines;
11   if (ncols % nlines) ++scols;
12   for (i = 0; i < nlines; ++i) {
13     c0 = i * scols;
14     c1 = (c0 + scols < ncols)? c0 + scols : ncols;
15     log_debug1("row %d:%d col %d:%d\n", row0, row1, c0, c1);
16     for (rc = row0; rc < row1 + SMOOTH_EDGE + DETECT_EDGE; ++rc) {
17       if (rc < nrows) {
18         correct_row(corrected_pixels, in_pixels, rc % SMOOTH_SIDE, rc,
20           c0, c1, ncols);
21       }
22       rs = rc - SMOOTH_EDGE;
23       if (0 <= rs && rs < nrows) {
24         rs_edge = (rs < SMOOTH_EDGE || nrows - SMOOTH_EDGE <= rs) ? 1 : 0;
25         smooth_row(smoothed_pixels, corrected_pixels, rs % DETECT_SIDE,
26             rs_edge, rs % SMOOTH_SIDE, SMOOTH_SIDE, c0, c1, ncols);
27       }
28       rd = rs - DETECT_EDGE;
29       if (0 <= rd && rd < nrows) {
30         rd_edge = ((rd < DETECT_EDGE) ? -1 : 0) +
31             ((nrows - DETECT_EDGE <= rd) ? 1 : 0);
32         detect_row(out_pixels, smoothed_pixels, rd, rd_edge,
33             rd % DETECT_SIDE, DETECT_SIDE, c0, c1, ncols);
34   } } }
35   MEM_FREE(corrected_pixels);
36   MEM_FREE(smoothed_pixels);
37   return NULL;
38 }
```

11

SAFETY-CRITICAL DEVELOPMENT

Mark Kraeling

CTO Office, GE Transportation, Melbourne, FL, United States

CHAPTER OUTLINE

Software Engineering for Embedded Systems. https://doi.org/10.1016/B978-0-12-809448-8.00011-4

1 Introduction

Embedded systems that are sold or upgraded may need to comply with a variety of safety standards based on the market and intended use. These standards can also outline requirements that need to be met based on international standards. Standards, such as ones based on IEC, attempt to develop a common set of guidelines, so that each individual country and/or market doesn't have separate requirements.

This chapter is devoted to looking at various safety-critical software development strategies that could be used with a variety of safety requirements. Some of the strategies may not make sense for your product or market segment.

The first part of the chapter goes over some basic strategies that can be used for the up-front project planning for a safety-critical project.

The second part discusses fault, hazard, and risk analyses. For safety-critical projects the early and continuous focus should be on what fault scenarios exist, the hazard that could occur if failures occur, and what risk it poses to the product and its environment.

The third part of the chapter goes over the basics of safety-critical architectures that are used and the pros/cons of each.

The last part concentrates on strategies in software development and implementation.

Getting a clear picture of the various standards that your project will need to meet up-front, following the appropriate implementation strategies listed, and watching out for the certification "killers" will help to make safety-critical product launch more successful.

1.1 Which Safety Requirements?

One of the most important aspects of developing safety-critical software is determining which requirements and standards are going to be followed.

Depending on the understanding of the safety requirements of your product or the intended market, you may need to get outside help to determine what needs to be met. Consider following the steps below to aid in your safety certification effort:

(1) Customer interaction—If you are entering a new market, the intended customer for that market probably knows the starting point of which safety requirements need to be met. They may be able to provide information on the safety standards that a similar product already meets. If the end customer is just using the product without a lot of technical background, then this step should be skipped.

(2) Similar product in same intended market—It may be more straightforward to see what safety requirements and standards a similar product meets. For instance, if your company or a partner already sells a medical device to the same market and your product is similar, this may be a good place to start.

(3) Competitive intelligence—Doing basic research on the Internet or from freely available open information from marketing materials may help determine a good starting point as well. Often, paperwork needs to be filed with agencies on which specific standards should be met.

(4) Professional assistance—Each market or market segment normally has agencies or contract facilities that can aid in determining which standards need to be met. Paying a little up-front, especially after gathering necessary information from steps 1–3, will help make this step pay off in the long run.

After gathering this information you should have a good idea about which standards need to be met. During this investigation also determine whether it is a self-certification activity, a standardized assessment activity, or a full-fledged independent assessment certification.

For the sets of requirements that need to be met the team should develop a strategy and initial analysis of how they will comply with the requirements. This means of compliance could be by design, by analysis, or by testing.

As an example, for design there may be requirements for redundancy. The design could include a dual-processor design or redundant communications paths. As an example, for analysis, if a certain bit error rate needs to be detected, then an appropriate length cyclic redundancy check (CRC) could be calculated. Through a mathematical equation the bit error rate could be determined. Using testing as a means of compliance is self-explanatory, as each of the requirements listed would have a corresponding test plan and procedure.

Finally, the team should determine how the evidence should be organized and presented.

All standards and delivery dates need to be listed and agreed on regardless of whether a self-certification, utilizing an auditor, or full-fledged independent assessment is performed. If this specific list can be put into a contract and signed, then it should be done to protect the team. If there is no customer contract that would list this type of documentation, then even an agreement between the project team and management could be written.

If using an independent assessor (which is normally paid for by the product team), then agree to the set of documentation, the means of compliance, and the evidence that needs to be provided up-front. Also agree on which party will determine if a newer standard is released while in the early project stages. Also agree in principle (and in writing) on when the project is far enough along so that the standards list and specification dates can be frozen. If this is all discussed and agreed upon up-front, safety certification becomes much easier.

1.2   Certification Killers

There are also items to watch out for during the safety certification process. These were learned the same way as the "Key Strategies" listed above. Many of these were lost battles on multiple projects historically in the past. Through multiple products and assessor providers these are the items that will most certainly hinder or even kill your certification effort:

- Failure to recognize safety requirements as real.
- Unclear requirements or requirements never agreed upon up-front.
- Lack of clear evidence of compliance.
- Not doing homework up-front and finding more safety standards that need to be met throughout the development process.
- Lack of dedicated resources, or resources that jump between projects.
- Scope and requirements creep.
- Trying to safety-certify too many things—not developing a boundary diagram and having everyone agree to it.
- Not accounting for enough resources to document the safety case and test based on those requirements.

- Not using a single contact to interface with the assessor (too many cooks!).
- Not being honest with the weaknesses of the proposed system.
- Waiting until the last minute to submit documentation.
- Failure to develop a relationship with the local country where the product will be deployed.
- Failure to properly sequence certification tasks.
- Qualification of software tools and OS to the appropriate safety level.

2 Project-Planning Strategies

The following rules can be applied to help safety-critical software development projects. The strategies listed are typically looked at very early in the project development life cycle before the software is written. These strategies were developed and refined during multiple product certification efforts and following these helps reduce the amount of money and resources spent on the overall effort.

2.1 Strategy 1: Determine the Project Certification Scope Early

Following some of the guidelines and directives listed in Section 1.1, identify which standards your product needs to meet. Determining whether it is a consumer product, has safety implications for the public, and/or which particular certification guidelines satisfy the customer are all part of this step.

2.2 Strategy 2: Determine the Feasibility of Certification

Answer questions up-front whether the product and solution are technically and commercially feasible. By evaluating the top-level safety hazards and the safety objectives, basic defensive strategies can be brainstormed and developed up-front. Involve engineering to determine the type of architecture that is required to meet those defensive strategies, because drastic architectural differences from the base product's current architecture could increase risk and cost.

2.3 Strategy 3: Select an Independent Assessor (if Used)

Find an assessor who has experience with your market segment. Various assessors have specialty industries and areas, so find out if they have experience in certifying products in your industry. Once an assessor becomes comfortable with your overall process and the development procedures it makes certification of subsequent products much easier.

2.4 Strategy 4: Understand Your Assessor's Role (if Used)

The assessor's job is to assess your product with respect to compliance to standards and norms. Do not rely on the assessor to help design your system; the assessor is neither responsible nor obligated to tell you that you are heading down a wrong path! The assessor's role is to determine if the safety requirements and objectives have been met resulting in a report of conformity at the end of the project.

2.5 Strategy 5: Assessment Communication is Key

Having a clear line of communication between your team and the group controlling the standards that need to be met is extremely important. Be sure to document all meetings and action items. Document decisions that have been mutually decided during the development process, so that the assessor and team stay on the same page. Ask for a position on any unclear issues or requirements as early as possible. Insist on statements of approval for each document or artifact that will be used.

2.6 Strategy 6: Establish a Basis of Certification

List all the standards and directives that your product needs to comply with, including issue dates of the documents. In conjunction with your assessment agree, disagree, or modify on a paragraph-by-paragraph basis. Consider placing all the requirements in a compliance matrix, so they can be tracked with the project team. Finally, do not be afraid to propose an "alternate means of compliance" if required.

2.7 Strategy 7: Establish a "Fit and Purpose" for Your Product

Establishing a fit and purpose up-front will prevent future headaches! The "fit" for your product is the space that you plan on selling into. If selling a controller for an overhead crane, then state that up-front and don't incorporate requirements needed for an overhead lighting system. The "purpose" is what the product is supposed to do or how it is going to be used. Clearly define the system boundary and what portions of your overall system and product are included in the certification. Consider things such as user environment, operating environment, and integration with other products. Also, considerations such as temperature and altitude can impact the circuit design, so those should be defined well in advance for successful product certification.

2.8 Strategy 8: Establish a Certification Block Diagram

Generate a hardware block diagram of the system with the major components such as modules and processing blocks. Include all the communication paths as well as a summary of the information flow between the blocks. Identify all the external interfaces including the "certification boundary" for the system on the diagram. The certification boundary shows what is being certified and what is not.

2.9 Strategy 9: Establish Communication Integrity Objectives

Before the system design determine up-front the "residual error" rate objectives for each digital communication path. Defining CRCs and hamming distance requirements for the paths also helps determine the integrity levels required. Also discuss with the assessor up-front how the residual error rate will be calculated, as this could drive specific design constraints or necessary features.

2.10 Strategy 10: Identify All Interfaces Along the Certification Boundary

Generate up-front a boundary "Interface Control Document." From this document identify all the required safety integrity levels for each of the interfaces. At this point research with the potential parties that own the source or destination side of the interface can begin, to make sure they can comply. Quantify and qualify the interface, including definitions of acceptable ranges, magnitudes, CRC requirements, and error checking.

2.11 Strategy 11: Identify the Key Safety-Defensive Strategies

Identify and implement the safety-defensive strategies to achieve the safety objectives for the program. Define key terms such as *fault detection*, *fault accommodation*, and *fail-safe* states. During initial architecture and design keep track of early failure scenarios that could occur. It is difficult to find all of them early in the project, but changes in the architecture and system design are easier made on the front end of the project.

2.12 Strategy 12: Define Built-in-Test (BIT) Capability

Identify the planned BIT coverage including initialization and periodic, conditional, and user-initiated tests. Define a manufacturing test strategy to check for key safety hardware components before shipping to the end user. After identifying each of these built-in-test functions review with the assessor and get agreement.

2.13 Strategy 13: Define Fault Annunciation Coverage

While keeping the system and user interface in mind, define which faults get annunciated. Determine when they should be announced to the operator or put into the log. Determine the level of information that is given to the end user and what is logged. Define the conditions that spawn a fault and what clears that particular fault. Define any fault annunciation color, text, sound, etc. After these are defined make sure the assessor agrees!

2.14 Strategy 14: Define Reliance and Expectation of the Operator/User

Clearly define any reliance that is placed on the operator or user to keep the system safe. Determine the user's performance and skill level, and the human factors involved with safety and vigilance. When placing safety expectations on the user make sure the end customer agrees with the assessment. And, as stated, make sure your assessor agrees with it as well.

2.15 Strategy 15: Define Plan for Developing Software to Appropriate Integrity Level

For each of the formal methods address the compliance with each objective element of the applicable standard you are certifying to. The software safety strategy should include both control of integrity and the application of programming-defensive strategies. The plan should include coding standards, planned test coverage, use of COTS, software development rules, OS integrity requirements, and development tools. Finally, define and agree on software performance metrics up-front.

2.16 Strategy 16: Define Artifacts to be Used as Evidence of Compliance

List all the documents and artifacts you plan to produce as part of the system safety case. List how you plan to cross-reference them to requirements in the system. Make sure any document used as evidence of compliance is approved for release via your configuration control process. Test documentation must have a signature and date for tracking documentation and execution of each test case. Above all, make sure your assessor agrees with your document and artifact plan up-front.

2.17 Strategy 17: Plan for Labor-Intensive Analyses

Plan on conducting a piece part failure mode and effects analysis (FMEA), which is very labor intensive. Also plan on a system level-FMEA and a software error analysis. It is recommended that probabilistic fault trees are used to justify key defensive strategies and to address systematic failures. More information on FMEAs is in Section 3.4.

2.18 Strategy 18: Create User-Level Documentation

Plan on having a user manual that includes the following information: system narrative, normal operating procedures, abnormal operating procedures, emergency procedures, and safety alerts. Also include a comprehensive maintenance manual that contains the following: safety-related maintenance, required inspections and intervals, life-limited components, dormancy elimination tasks, and instructions on loading software and validating the correct load.

2.19 Strategy 19: Plan on Residual Activity

Any change to your certification configuration must be assessed for the impact on your safety certification. There could be features added to the original product, or part changes that need to be made that could affect the safety case. Some safety certifications also require annual independent inspections of manufacturing and/or quality assurance groups. Residual activity (and thus residual cost) will occur after the certification effort is complete.

2.20 Strategy 20: Publish a Well-Defined Certification Plan

Document a certification plan that includes all previous rules, timeline events, resources, and interdependencies. Include a certification "road map" that can be referred to throughout the development process, to have a snapshot of the documentation that is required for the required certification process.

3 Faults, Failures, Hazards, and Risk Analysis

Once the project-planning phase of the project is complete it is important to make an assessment of where the risks may be for the system being designed. To measure the overall risk for the product a risk analysis needs to be performed.

Before getting to the risk assessment a list of safety-critical terms will be explored.

3.1 Faults, Errors, and Failures

A fault is a characteristic of an embedded system that could lead to a system error. An example of a fault is a software pointer that is not initialized correctly under specific conditions, where use of the pointer could lead to a system error. There are also faults that could exist in software that never manifest themselves as an error and are not necessarily seen by the end user.

An error is an unexpected and erroneous behavior of the system, which is unexpected by the end user. This is the exhibited behavior of the system whenever a fault or multiple faults occur. An example could be a subprocess that quits running within the system from a software pointer that is not initialized correctly. An error may not necessarily lead to a system failure, especially if the error has been mitigated by having a process check to see if this subtask is running and restarting it if necessary.

For an embedded system a failure is best described as a system event not performing its intended function or service as expected by its users at some point in time. Since this is largely based on the user's perception or usage of the system the issue itself could be in the initial system requirements or customer specification, not necessarily the software. However, a failure could also occur based on an individual error or erroneous system functionality based on multiple errors in the system. Following the example discussed at the start of this section the software pointer initialization fault could result in a subtask running error, which when it fails causes a system failure such as a crash or user interface not performing correctly.

An important aspect is that for the progression of these terms they may not necessarily ever manifest themselves at the next level. An uninitialized software pointer is a fault, but if it is never used then an error would not occur (and neither would the failure). There may also need to be multiple instances of faults and errors, possibly on completely different fault trees, to progress to the next state. Fig. 1 shows the progression for faults, errors, and failures.

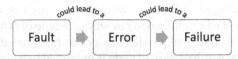

Fig. 1 Faults, errors, and failure progression.

For safety-critical systems there are techniques that can be used to minimize the progression of faults to errors to failures. All these techniques impact the reliability of the system, as discussed in the next section.

3.2   Availability and Reliability

Availability and reliability are related terms but are not the same. Availability is a measure of how much the embedded system will be running and delivering the expected services of the design. Examples of high-availability systems include network switches for voice and data, power distribution, and television delivery systems. Reliability is the probability that an embedded system will deliver the requested services at a given point in time. Even though these terms are related a high-availability system does not necessarily mean the system will also be highly reliable.

An example of a system that could have high availability, but low reliability, is a home network system that has faults. In this example, if every 100th packet is dropped causing a retry to occur, to the user it will seem like a very available system. There aren't any periods of a total outage, but just periods in the background where packets need to be resent. The fault could be causing an error to occur, with delays waiting for processes to restart. The system itself stays up, the browser or whatever user interface stays running, so the user doesn't perceive it as a system failure.

Safety-critical systems are examples of high-reliability systems and, in the case of systems that are monitoring other safety-critical systems, highly available. Systems that are both highly reliable and highly available are said to be dependable. A dependable system provides confidence to the user that the system will perform when they want it to, as it is supposed to. Addressing how to handle faults, which could lead to a system failure, is the source for system dependability.

3.3   Fault Handling

There are four aspects of faults that should be evaluated as part of a safety-critical system. The four types for faults are avoidance, tolerance, removal, and prediction.

Fault avoidance in a safety-critical system is largely an exercise in developing a system that helps mitigate the introduction of software and hardware faults into the system. Formal design and development practices help developers avoid faults. One approach in fault avoidance is designing a software system with a single thread of execution, as opposed to a multitasking preemptive type of task scheduling. This helps avoid issues of parallelism, or timing issues that could occur if a portion of one section of code is impacted in a negative way by another. For this example, it would be unreasonable to include every timing aspect or order that is a part of normal system testing. Safety-critical programming practices that target fault avoidance are listed in Section 5.

Fault tolerance is a layer of software that can "intercept" faults that occur in the system and address them so that they do not become system failures. An important aspect of safety-critical systems is the characteristic that fault-tolerant systems have excellent fault detection. Once an individual hardware or software component has been evaluated as "failed," then the system can take appropriate action. Performing fault detection at a high level in the software should only be done when multiple variables need to be evaluated to determine a fault. One example of good fault detection is evaluation of a temperature sensor. If the sensor has an out-of-range low or high value coming into an A/D converter, the software should clearly not use this value. Depending on the criticality of this input to the system there may be a redundant sensor that should be used (higher criticality). If this input is not critical to the system, then another possible solution could be to use another temperature sensor as an approximation to this one. Architecture, hardware, and software designs all have an impact on a system's ability to be fault tolerant.

Fault removal consists of either modifying the state of the system to account for the fault condition or removing the fault through debugging and testing. Dynamic fault removal is when the system falls back to a less faulty state when a fault is detected. The most difficult aspect of "dynamic" fault removal is to safely determine how to do it. A typical way of doing this is to change the fault state by adopting noncritical data that are part of the safety-critical system. An example of this concept is to make use of a safety-critical system that logs a temperature value for environmental evaluation at a later date. It is not used in any control loops or decisions. If the temperature sensor is an ambient sensor that has failed, it switches over to a less accurate ambient sensor that is integrated into the hardware. For the logs, having a less accurate temperature value is evaluated as being better than having no ambient temperature value at all. Testing and debugging of the system is the primary method for fault removal in safety-critical systems. Systems test procedures cover all the functionality for the safety-critical system, and often require 100% coverage for lines of code in the system as well as different combinations of execution and inputs. Reiterating and addressing faults in this manner is much easier than the complexity involved in dynamically removing the fault.

Finally, fault prediction is an often-missed aspect of safety-critical systems. Being able to predict a fault that may occur in the future and alerting a maintenance person or maintainer is very valuable in increasing the dependability of the system. Examples include sensors that may have spurious out-of-range values, where tossing out those values keeps the system running. However, if the number of out-of-range values increases from a typical rate of one occurrence per day to an unacceptable rate of one occurrence per minute, we are possibly getting nearer to having a failed sensor. Flagging that occurrence and repairing it at a time when the user expects the system to be available is much more dependable than having that sensor fail and cause the system to be unavailable.

3.4 Hazard Analysis

Designing safety-critical systems should address hazards that cause the system to have a failure that leads to tragic accidents or unwanted damage. A hazard is any potential failure that causes damage. Safety-critical systems must be designed where the system operation itself is always safe. Even if an aspect of the system fails, it should still operate in a safe state.

The term fail-safe is used to describe a result where the system always maintains a safe state, even if something goes terribly wrong. A safe state for a locomotive train would be to stop. Some systems, such as aircraft fly-by-wire, do not have a fail-safe state. When dealing with these types of systems multiple levels of redundancy and elimination of single points of failure need to occur as part of system design.

Performing a hazard analysis is key to the design of safety-critical embedded systems. This analysis involves identifying the hazards that exist in your embedded system. It is based on a preliminary design or architecture that has been developed—even if it is just a sketch in its preliminary form.

In this process an architecture for the safety-critical system is proposed and iterated until the architecture could support being highly reliable and available with possible mitigations for the hazards that could be present. Once this is complete, additional hazard analyses will need to be performed on all aspects of the safety-critical system in more detail.

During subsystem design hazard analysis will continue to be performed. When more details are known, one effective way to do this is by performing an FMEA. This is a systematic approach to numerically evaluating each of the failures in the system and provides clarity for the classification of each of the failures. Once the failures are understood with their effects, then mitigations can be performed such as detection, removal, or functional additions to the system to mitigate the condition.

An example work product from an FMEA is tabulated below:

Function	Potential Failure	Potential Effects of Failure	Severity Rating	Potential Cause	Occurrence Rating	Mitigation Plan	Detection Rating	RPN
Vehicle speed sensing	Sensor fails high (out of range)	Cruise control goes off from on	5	Sensor high side shorts high	2	Add overmold to sensor connection	3	30

For an FMEA each function is evaluated for potential failures and each failure condition is evaluated for how often it occurs, the severity of the consequence when it occurs, and how often it can be detected when it occurs. These are typically ranked from 1 to 10, and then an overall score for each failure (risk priority number) is calculated by multiplying these numbers together. This number helps rank the order in which to evaluate the failures, but by no means should any of the failures be ignored! Rules should be set up and put in place for failures that cannot be detected or have serious consequences when they occur. Another important aspect of an FMEA is that it tends to focus on individual failures, where bigger issues could occur when multiple failures happen at the same time.

A fault tree analysis is a top-down approach to doing a hazard analysis. This helps discern how different combinations of individual faults could cause a system failure to occur. The fault tree isn't focused on just software but includes hardware and user interaction that could cause the failures to occur. Its top-down approach starts with the faults themselves and puts a sequence together of logical paths to address how the eventual failure could occur.

Fig. 2 shows a fault tree.

An event tree analysis is done in the opposite way that a fault tree analysis is done, as it is a bottom-up approach to hard analysis. The analysis starts with the event/failure itself and then analyzes how that particular event/failure could occur from a combination of faults. This type starts with the undesired event itself, such as "engine quits running," and then determines how this could possibly happen with individual faults and errors that occur.

Fig. 3 shows an example event tree, with the numbers representing the probability that taking each branch could occur:

In safety-critical systems hazards that can result in accidents, damage, or harm are classified as risks and should require a risk analysis.

Fault tree

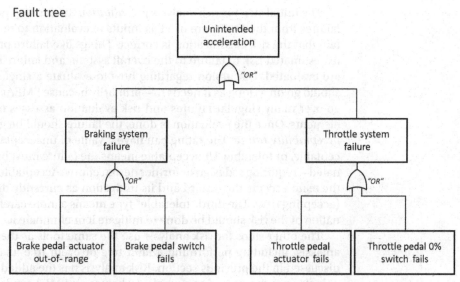

Fig. 2 Fault tree.

Event tree analysis

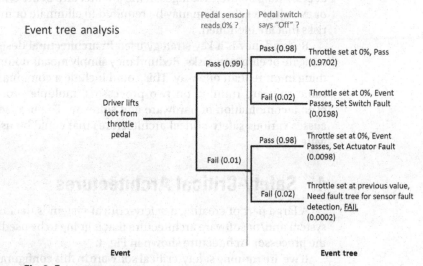

Fig. 3 Event tree.

3.5 Risk Analysis

A risk analysis is a standard method where each of the hazards identified is evaluated more carefully. As part of this process each hazard is evaluated based on the likelihood of the failure occurring, along with the potential for damage or harm when it occurs. A risk analysis helps determine if the given hazard is acceptable, how much risk we are willing to accept, and if there needs to be any mitigation or redesign that needs to occur as a result of that failure.

The initial step for risk analysis is *evaluation*. In this step potential failures from the FMEA are used as inputs to evaluation to make certain that the risk classification is correct. Things like failure probability, estimated risk of failure to the overall system, and failure severity are evaluated. Discussion regarding how to evaluate a single failure should go on as long as it needs to—primarily because FMEAs tend to go over many singular failures and risk evaluation assesses multiple elements. Once the evaluation is done, the failure should be given an *acceptability rating*. The rating can have a value of unacceptable, acceptable, or tolerable. Unacceptable means the failure must be eliminated—requiring redesign or further design efforts. Acceptable means the team accepts the hazard and its mitigation as currently designed (accepting risk). The third "tolerable" type means a more careful evaluation of the risk should be done to mitigate it or eliminate it.

The other steps for risk analysis use the same tools as the hazard analysis, including performing a fault tree or event tree analysis as discussed in the previous section. Risk analysis has the added benefit of looking at the system as a whole—whereas an FMEA tends to look at individual failures. Changes in the architecture of the entire system or even just a subsystem may be required to eliminate or mitigate the risks that are identified.

Redundancy is a key strategy used in architectural design to help mitigate or eliminate risks. Redundancy simply means doing the same thing in more than one way. This could include a combination of the same software running on two processors, multiple processors, or even a combination of hardware and software. The next section discusses various safety-critical architectures that could be used to mitigate risk.

4 Safety-Critical Architectures

A large part of creating a safety-critical system is deciding on the system and/or software architecture that is going to be used. Consider the processor architecture shown in Fig. 4.

If we are running safety-critical software in this configuration, what happens if the processor does something that is unexpected? What if the processor runs variable data out of a bad memory location, or there is a latent failure that only exhibits itself after some period of time?

This processor by itself wouldn't be able to satisfy a truly safety-critical system. Depending on the safety level there may be external components that can be added around the processor to perform the desired safety function in parallel if the processor cannot do so. As the complexity of an interface goes up, replicating with circuitry may not act as a successful mitigation for failures that can happen in your system. This would especially be true if the nature of the critical

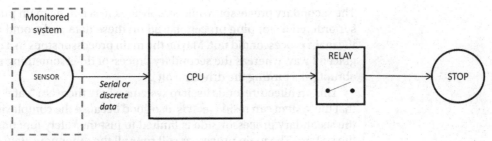

Fig. 4 Single-processor architecture.

data is contained within serial messages or Ethernet frames. When the amount of safety-critical data increases or the number of safety mechanisms increases it is time for a different architecture.

The following sections outline various architectures that could be used for a safety-critical system. For each architecture notes are included to outline various aspects including positives and negatives.

4.1 "Do-Er"/"Check-Er"

In the architecture in Fig. 5 one processor is still performing most of the embedded system work. In this case a second processor is added to look at the safety-related data to make assessments about that data. It then looks at the output of the main processor and determines if that processor is doing what it is supposed to do.

For example, say there is a bit of information in the serial stream that means "Stop" and a separate discrete input signal that also means "Stop." Both processors could be designed to have visibility of both pieces of data. The main processor would process the safety-critical "Stop" logic along with all the other operations it is performing.

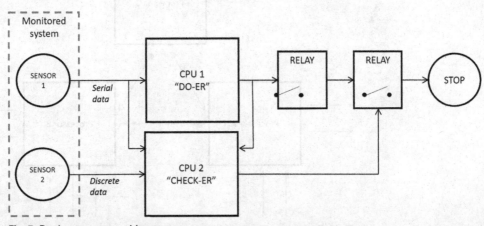

Fig. 5 Dual-processor architecture.

The secondary processor would simply look to see if the main processor ordered a stopping process based on these data and would act if the main processor did not. Maybe the main processor stops in a more graceful way, whereas the secondary processor does something more abrupt (like turning the driveline off).

This architecture lends itself to systems where there is a "safe" state that the system can reside in. It is also good because the complexity on the secondary processor side is limited to just the safety functions of the system. The main processor still runs all the other nonsafety code (the secondary does not).

When the complexity of safety goes up or the safety-critical level goes up, then a different architecture is needed to process data.

4.2 Two Processors

In the architecture in Fig. 6 there are two processors, which could be identical, that are performing the safety aspects of the system. Each of the processors labeled A and B are performing the same operations and handling the same data. The other processor labeled C is performing cleanup tasks and executing code that has nothing to do with the safety aspects of the system. The two safety processors are operating on the same data.

Various tricks can be done on the two processors to make them a little different. First, the memory maps of the processors can be

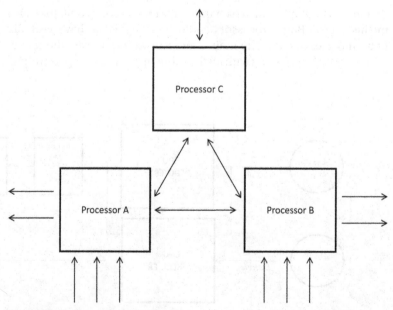

Fig. 6 Triple-processor architecture.

shifted so that a software error dealing with memory on one processor wouldn't be the same memory location on the other processor. They could also be clocked and operated separately—maybe there isn't a requirement to have the processors execute instructions in lockstep with each other. For this architecture, if the processors disagree, then the system would arrive at a safe state for the system. For this and the previous architectures listed the system assumes there is a stop or safe state for the embedded system. If the system must continue to operate, then an even more complex system architecture is needed.

4.3 "Voter"

The architecture in Fig. 7 shows a "voter" type of system.

For this type of system the processors vote on what should be done next. Information is compared between all of them, and the decision with the greatest number of votes wins. Indecision between processors is logged and flagged, so that maintenance can be done on the system. There also needs to be periodic checking of the interpretation of the voting mechanism, so that the voting mechanism itself is known to work and doesn't have a latent failure.

This type of architecture represents a large jump in complexity. There are numerous test cases that need to be performed to evaluate this system—and the number of possibilities greatly increases. Embedded engineers spend their entire lives dealing with the intricacies of systems like this, and development is neither quick nor regular in terms of time.

Selecting the right architecture up-front based on safety requirements is extremely important. Having to shift from one architecture to another after development has started is expensive and complicated.

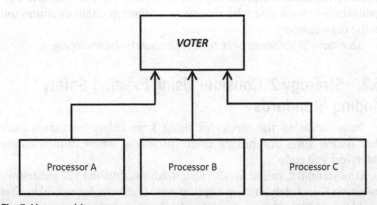

Fig. 7 Voter architecture.

5 Software Implementation Strategies

After the project planning, hazard/risk analysis, and architecture are complete there should be a good understanding of which requirements are safety critical. For software development it is important to treat these as special—even following a separate process to make sure they are designed, coded, and unit-tested correctly.

It is a difficult and unreasonable expectation to have a single process that fits every type of safety-critical application or project. This section's intent is to point out different strategies that should be considered when doing development. The safety requirements for your project may require many of the items listed here, so this provides a good starting point for things to consider. If you are using an independent assessor there may be particular and specific items that need to be included as well.

5.1 Strategy 1: Have a Well-Defined, Repeatable Peer Review Process

A critical part of the development of safety-critical software is having a well-defined peer review process. There must be a process for peer review and consistency regarding what information is provided and the amount of time available to review prior to the review meeting. The reviewers may include people in systems engineering, systems test, safety, and configuration management.

There must also be recognition by the peer review leader that the reviewers may not have had sufficient time to prepare. In this case the meeting should be rescheduled. For safety-critical code development and code sections it is important to have an independent assessment of the source code, so that a single person isn't walking the group through biased code where their opinion could come into play. Such an independent assessment might involve someone external to the organization or someone who reports in a different chain of command in the organization.

An example software peer review process is shown in Fig. 8.

5.2 Strategy 2: Consider Using Existing Safety Coding Standards

In addition to the strategies listed here safety standards exist that define rules for programmers to follow when implementing safety-critical code.

One standard, called MISRA C, initially established 127 guidelines for using C in safety-critical applications. It checks for mistakes that could be made that are entirely "legal" when using the C programming

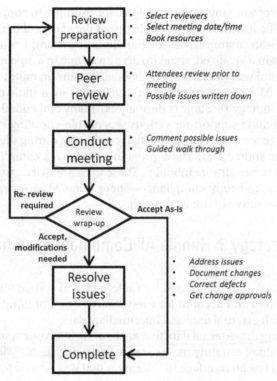

Fig. 8 Review process.

language but have unintended consequences when executed. Based in the United Kingdom the Motor Industry Software Reliability Association (MISRA) felt there were areas of automobile design where safety was extremely important. Their first standard was developed in 1998 and included 93 required rules of the 127 total guidelines. The remaining 34 were advisory.

The MISRA standards were updated in 2004 to include additional guidelines. MISRA increased the number to 121 required rules and 20 advisory rules to bring the total to 141. This newer version of the standard also split the rules into categories such as "runtime failures." Currently, the latest standard released in 2012 added more rules and cross-referenced ISO 26262, a safety standard for electrical and electronic systems. The MISRA C standard document is available at their website (http://www.misra.org.uk). There is also a set of guidelines for C++ in a separate MISRA C++ document.

Let's give an example of a rule: "All code shall conform to ISO 9899 standard C, with no extensions permitted." In simple terms this means using extensions or in-line assembly would be considered nonconformant with this rule. However, accompanying this rule is the following comment: "It is recognized that it may be necessary to raise deviations to permit certain language extensions, for example to support hardware specific features." So, with this caveat the standard permits low-level hardware manipulation or handling of interrupt routines—as long as it is in a localized, standard area and done in a repeatable way. This standard was written with embedded systems in mind!

All the 93 required rules can be checked using a static code analyzer (see Strategy 18: Static code analysis). Many embedded compiler vendors include support for various sets of rule-checking standards. There are also separate programs that can be run during the build for each of the source files. These programs can check compliance and print reports for software modules. These tools are strictly run to cover compliance and only compliance—there is no MISRA certification process that software can go through.

5.3 Strategy 3: Handle All Combinations of Input Data

For data that are processed by a safety-critical system it is important to address and account for every combination of input value including both external data and intermediate data.

Checking the external data that are coming into your system for all possible values certainly makes sense. For example, say your system (System A) has an interface specification that was written to interface with another system (System B). It may state that a data item can only have certain values, but it is important to check system behavior if it receives different values. This could come about later in the life cycle of the product, where System B's baseline is updated with new software, hence new data values and System A are missed. Or it could come about because of a misinterpretation of the specification implementing the interface by either party.

For example, if a data element can have value "0" meaning "Stop," and a value of "1" which means "Go," then what happens if the variable is any other value? Maybe someone adds a new value "2" at a later time that means "Proceed with caution." In such an event logic should be put together not only to specifically check for each case, but also to catch the other case as well. In this situation notifying someone and/or logging the mismatch is important to help correct the situation in the future. An example of this is:

```
if ( input_data_byte == 0 )
{
    Movement = STOP;
}
else if ( input_data_byte == 1 )
{
```

```
   Movement = GO;
}
else
{
   Movement = STOP; // Most restrictive case here
   Log_Error( INP_DATA_BYTE_INV, "Unknown Value" );
}
```

For an intermediate variable declared in your system it is also important to do the same type of checking. Every "if" statement should have an "else" clause, and every "switch" statement should have a default case encompassing the values that are unexpected. More complex combinations of conditions for "if" statements should also have the same "else" condition covered. Having these alternate paths that *should* never be executed helps to better understand the system as a whole. It also helps the programmer explore alternate paths and corner cases that may exist.

5.4   Strategy 4: Specific Variable Value Checking

When writing safety-critical software code it is important to check for a specific value for the permissive condition you are looking for. Consider the following code:

```
if ( relay_status != RELAY_CLOSED )
{
    DO_Allow_Movement(); // Let the vehicle move, everything OK
}
else
{
    DO_Stop(); //The relay isn't positioned correctly, stop!
}
```

In this example the code wishes to look for the relay being open to allow movement. The variable "relay_status" has two values, RELAY_OPEN and RELAY_CLOSED. But, depending on the size of variable that was declared, there are many more values that it can have! What if the memory has a value of something else? With the above code movement would be allowed. This isn't good practice. For the most permissive state always check for a single value (or range when appropriate). The following code is the correct way to write this code block:

```
if ( relay_status == RELAY_OPEN )
{
    DO_Allow_Movement(); // Let the vehicle move, everything OK
}
else if ( relay_status == RELAY_CLOSED )
```

```
{
    DO_Stop(); // It is closed, so we need to stop
}
else // This case shouldn't happen -
{
    DO_Stop(); //The relay isn't positioned correctly, stop!
    Log_Error( REL_DATA_BYTE_INV, "Unknown Value" );
}
```

Another way that the code block could be written based on how the code is structured is to set the most restrictive case in the code at the start of execution. Then specific steps are taken and values are checked to allow movement. For the simple code block above DO_Stop() would be moved outside the conditional "if" and then the code would allow movement if certain checks passed.

5.5 Strategy 5: Mark Safety-Critical Code Sections

For code sections that are safety critical in your code there should be a special way that the code section is marked. The main purpose of this is to carry out maintenance on the code later—or if the code is used later by another group for their project. The real safety-critical sections should be marked with comment blocks that say why it is safety critical and refer to the specific safety requirements that were written. This would also be an appropriate place to refer to any safety analysis documentation that was done as well.

The following is an example of a header that could be used for a safety-critical code section:

```
/*******************************************************
 *******************************************************
** SAFETY-CRITICAL CODE SECTION
** See SRS for Discrete Inputs for Requirements
** Refer to Document #20001942 for Safety Analysis
**
** This code is the only place that checks to make
** sure the lowest priority task is being allowed
** to run. If it hasn't run, then our system is
** unstable!
*********** START SAFETY-CRITICAL SECTION **********/
    // LOW_PRIO_RUN is defined as 0x5A3C
    if ( LP_Flag_Set == LOW_PRIO_RUN )
    {
        LP_Flag_Set = 0;
    }
    else
```

```
    {
        // The system is unstable, reset now
        Reset_System();
    }
    /*********** STOP SAFETY-CRITICAL SECTION ***********/
```

In this example you can see that we are just resetting the system. This may not be appropriate depending on what task your system is performing and where it is installed. Code like this may be appropriate for a message protocol translation device that has safety-critical data passing through it but is likely not appropriate for a vehicle!

5.6 Strategy 6: Timing Execution Checking

For processors that run safety-critical code it is important to check that all intended software can run in a timely manner. For a task-based system the highest priority task should check to make sure that all the other lower priority tasks are able to run. Time blocks can be created for the other tasks such that, if one lower priority task runs every 10 ms and another runs every 1 s, the checking is done appropriately. One method is to check to make sure tasks are not running more than 20% slower or faster than their intended execution rate.

The rate at which the task timings are checked would be dependent on the safety aspect of the code in the task that is being checked.

Another system check is to make sure that the entire clock rate of the system hasn't slowed down and fooled the entire software baseline. Checks like this need to look at off-core timing sources so that clock and execution rates can be compared. Depending on how different timers are clocked on the system it could come from an on-die internal check—but only if what you are checking against is not running from the same master clock input. For example, if the execution timing of a task is running from an external crystal or other input it could be compared with the real-time clock on the chip. When taking this route there may also be a requirement for the source code not to know how (or is not mapped) to change the clock input for the RTC chip.

5.7 Strategy 7: Stale Data

Another safety-critical aspect of a system is performing operations to ensure that we do not have stale data in the system. Making decisions on data that are older than we expect could have serious consequences while running!

One example of this is an interface between a processor and an external logic device such as an FPGA. The logic device is accessed through a parallel memory bus, and the processor uses the interface to read input data from a memory buffer. If the data are used as an input to any kind of safety-critical checking, we do not want the data to be stale as this could impact the safety of the system. In this example this could occur if the process that collects that data on the FPGA stops or has a hardware fault on its input side. The interface should add a counter for the data blocks or have a handshaking process where the memory is cleared after it is read. Additional checks could need to be put in place as well like the processor having a way to check to make sure the memory block was cleared after the request.

There are many ways to get rid of stale data, largely based on how the data are coming in and where they are coming from. There may be a circular buffer that is filled with DMA or memory accesses. In this case it is important to check to make sure data are still coming in. There may also be serial streams of data that are again placed in a specific memory location. It is here that our safety application comes along and operates on these data. Here are some things to consider:

- First, determine if there is a way to delete incoming data once your safety-critical code has run and generated outputs. Clearing out this memory space is a great way to make sure that the data are gone before the functionality is run again.
- Second, when dealing with serial data or sets of data consider using sequence numbers to order the data. This will allow the software to remember which set was processed last time so that an expected increase in sequence number would show the data are newer.
- Third, for large blocks of data where it is impractical to clear the entire memory space, and there is also no sequence number, things are a little more difficult. For these large blocks there should be a CRC or error check of the data themselves to make sure it is correct. After processing these data selectively modifying multiple bytes can help create a CRC mismatch. Although there is a probability of this every effort should be made to avoid a situation in which data are changed and the CRC is still good.

5.8 Strategy 8: Outputs Comparison

Depending on the processor architecture being used, when there is more than one processor the outputs of the safety-critical functions should be cross-checked. This allows each of the processors in the architecture to make sure the other processor is taking appropriate action based on the inputs. There are a variety of ways to check this.

One of the easier ways is for the outputs of one processor to also be run in parallel as inputs on another processor. Again, depending on the architecture, this would be a check to make sure that the other processor(s) is doing what you expect when presented with the same data. For an output that is a serial stream this could also be run in parallel to the intended target as well as fed back into the other processor

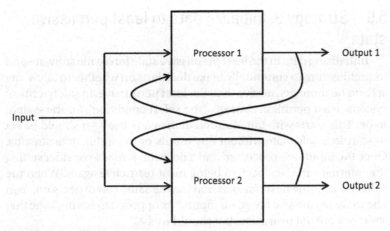

Fig. 9 Processor architecture example.

as an input. A comparison can be done to make sure the other processor is doing the same thing as your processor (as shown in Fig. 9).

Another way this can occur is to send serial or memory-mapped data directly between the two processors. This allows more checking to be done at more granular, intermediate steps in the software process as opposed to when it comes out of the other processor. If one of the safety-critical outputs was "ignite," then it is a little late for another processor to be checking for this. In this case having more checks between the processors would be beneficial before ever getting to the final output case. The latency of the communications channel between them directly corresponds with the regularity and periodicity of the checking. Fig. 10 shows the basics of this serial or memory-mapped communication.

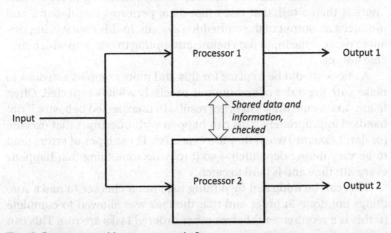

Fig. 10 Processor architecture example 2.

5.9 Strategy 9: Initialize data to least permissive state

Initializing data to the least permissive state forces the software and its architecture to continually make decisions on whether to allow any state to be more permissive than the least permissive. In safety-critical systems least permissive means "the safest condition" for the system to be. This starts with initialization of the code itself—it should be set to start in a safe state without any inputs or incoming data streams. Once the inputs are processed and a decision is made consider setting the internal variables back to being most restrictive again. When the software runs the next time it is making the same sort of decision, "can I be more permissive based on inputs," as opposed to having logic that says, "we are not restrictive, but should we be?"

Architectures that start from a restrictive state tend to be more understandable when the logic is followed than when looking for instances where we should be more restrictive after not being so.

For this case it is permissible to use variables to remember what our last output state was, but that should be used as an input into the logic ("last_output_state") as opposed to the output that the code is generating ("Output_State").

5.10 Strategy 10: Order of Execution

If there are requirements for one code section to run before other code sections, safety checks need to be in place to make sure that this has occurred. This certainly comes into play when software is running in different threads of execution or when tasks and an RTOS may be involved.

For a simple safety-critical application there is a task that takes raw data and converts them to a filtered, or more meaningful set of data. There is then a task that takes that data, performs calculations, and produces an output of the embedded system. In this case it is important to process the input data before attempting to come up with a suitable output.

A check should be in place for this and more complex activities to make sure the order of execution is precisely what is expected. Often failure to execute things in order results in unexpected behavior if not handled appropriately. This can happen with interrupts that execute (or don't execute) when they are expected. These types of errors tend to be very timing dependent—so it may be something that happens every Xth time and is hard to catch.

This can be mitigated by putting together a checker to make sure things are done in order and that the task was allowed to complete (if this is a requirement) before other ordered tasks are run. This can be done using a simple sequence number for the tasks, where it is set to a fixed value to let the next task know that it ran to the appropriate completion point. Then the next task (illustrating the importance of this order) checks the value and proceeds only if it matches what it expects.

Another way to mitigate these types of errors is to use more of the features in the RTOS to help with ordered execution. Use of semaphores and/or flags may be used as an alternate. Be careful with this type of solution—because when it comes to the safety case your dependency on the operating system will go up with the more features you depend on.

Finally, depending on the safety nature of the code another idea is to use a simple timer and run the tasks in frames with your own home-spun scheduler. If all the task priorities are the same and you are comfortable writing interrupts for the code that needs to "run right now," then insuring execution order becomes as simple as function calls.

5.11   Strategy 11: Volatile Data Checking

Any data that are received from another source off board the processor should have their integrity checked. Common ways of doing this involve CRCs of various lengths. Depending on the safety criticality level of the software a different CRC other than an established standard may be needed, as described below.

In embedded networks the parameter that is the most looked at using CRCs is the hamming distance. This property specifies the minimum number of bit inversions that can be injected into a message without it being detected by CRC calculation. For a given message bit length with a hamming distance of 4 that means there exists no combination of 1-, 2-, or 3-bit errors in that message that would be undetectable by CRC calculation.

The use of CRCs and other types of data checking is ideal for streams of data that are coming into your system. When data arrive a check can be performed to make sure it is a good message before it is placed into memory. Additional, periodic checks should be made on the data after they are placed in memory to make sure they aren't altered by errant pointers or memory conditions.

As a strategy all volatile data considered safety critical, meaning they can alter the safety mechanisms of the system depending on its value, should be checked. Data updated by calculation should either be set to the least permissive state before calculation or have a check performed to make sure the variable is updated during execution when we expect it to. This could involve setting the variable to an invalid state and then checking at the end of the code to make sure it is not in an invalid state.

Data that are considered safety critical but cannot be updated via calculation and their associated variables should be checked using a CRC.

For example, let's say there is a test variable that is set to "on" to output data to a maintenance port. A remote tool can set this variable to "on" or "off" as requested. What happens when this volatile memory region becomes corrupted? With no calculation to change it back to the correct value we could start sending data out the maintenance port. Again, if this variable is safety critical in nature, we need to have a check to keep that from happening. Including this variable with others and having a CRC for the set is a good way to see if this memory has become corrupted from an errant pointer or other situation. Then our program can periodically check the CRCs for these data blocks to know that they are set correctly. Having these CRC calculations on data blocks is, as discussed, especially important for data that are not updated continuously.

Lastly, safety-critical data should be sanity-checked before they are used in calculations throughout the code. This wouldn't include a variable that was set to one value or another in the previous statements, but rather variables that could have influence outside the current function. This certainly would be the case if the variable is modified by other threads of execution. For example, say we want to execute a section of code every six times a function is called. This can be done by setting a maximum of five or six (depending on decrement logic), and then decrementing when the function is called. If we are executing code that decides whether we should perform our task (value of zero), what should we also check? Does it make sense to remember the "last" value this variable had to make sure it is different from the "current" value? It makes sense to make sure the variable is currently set to no higher than six!

A large part of how the volatile data are checked depends on the safety criticality level of the application. Keep these strategies in mind to lower the chance of dealing with stale, corrupted, or invalid data.

5.12 Strategy 12: Nonvolatile Data Checking

Nonvolatile data are a little easier to check because they aren't supposed to change. A useful strategy is to consider having your makefile calculate a CRC for the program image at build time. If a single CRC for the entire image doesn't provide enough bit error checking for the length of the image, then use multiple CRCs for various sections of the code space. One approach could be to have a CRC cover the first third of the image, another to cover the first two-thirds, and another to cover the whole image. Different variations of this could be used as well.

The primary reason for using multiple CRCs like this is to be able to keep the CRC length the same as the atomic size of the processing unit. This will help speed CRC processing.

The safety case will drive how often the image is checked. Inputs to the safety case include the MTBF data for the hardware involved, how often the system is executing, and the safety criticality of the code on that single processor itself. If image checking is assigned to the lowest priority task (which is typical), then there should be a check at a higher priority to make sure that it is able to run and that it completes in the time expected.

Another point of nonvolatile data checking is to check the image before running it. A bootloader or initial program function should check the CRCs upon initialization and only run the image if the integrity is verified.

5.13   Strategy 13: Make Sure Entire System Can Run

For a safety-critical system it may not make sense for a real-time operating system to be running. Depending on the safety requirements for the software that is being written it may be too cost prohibitive or complicated to include an RTOS where additional complexities are introduced. Maybe a simple scheduler could also meet the needs, with interrupts to handle time-critical data that are coming in or out. Regardless of what tasking type of system is being used the system needs to be checked to make sure everything is running correctly.

For a task-based RTOS-type system this involves making sure that all the tasks are being allowed to run. It is straightforward to have a high-priority task to make sure the lowest priority task is running. It gets a little more difficult to make sure that all the tasks in the system are running correctly, have enough time to run, and are not hung and doing something unexpectedly. More run-type checking will need to be performed with tasks that have safety-critical code within them. Tasks that contain code that is not safety critical or part of the safety case probably don't need as much checking.

For a simple scheduler system, where the code runs in a constant loop with some sort of delay at the end of the loop waiting for the frame time to end, checking whether everything was able to run is a little easier. If function pointers (with "const" qualifiers for safety-critical systems) are used, then the checking does become a little more difficult. Since this type of software architecture can get held up at a code location in the main loop forever it is important to have a periodic interrupt check to make sure that the main code can run.

For both types of systems it is always good to have an external watchdog circuit that can reset the system (or take other action) if the code appears to quit running altogether.

Another aspect of execution is making sure that the timing you expect is real or not. In other words, if you are running an important sequence of code and it needs to run every 10 ms, how do you know it is really 10 ms or plus/minus some margin? For this type of case it is a good idea to have an external timing circuit that provides a reference that can be checked. A waveform that is an input to your processor could be checked to make sure there is a match. For instance, if you have a simple scheduler that runs every 10 ms, you could have a signal with a period of 10 ms. A mathematical calculation can be done, based on the acceptable margin, of how many "lows" or "highs" are read in a row at the start of the loop for it to be "accurate enough." When the input is read it should be either low or high for several consecutive samples, and then could shift to the other value for a consecutive number of samples. Any condition where the input is changing more often than our consecutive samples could constitute a case where a small-time shift is required because the clock input is synced with our loop timing.

If timing accuracy has some flexibility, using an output at the start of the main loop to charge an RC circuit could also be used. Based on tolerances and accuracy, if it isn't recharged through the output, then an input could be latched showing time expiration like an external watchdog circuit without the reset factor.

Any of these or other methods could be used. However, checking to make sure all the code can run and that its execution rate matches what is required is important.

5.14 Strategy 14: Remove "Dead" Code

Another strategy is to remove any code and/or functions that are not currently being called by the system. The reason for this is to ensure that these functions cannot start executing accidentally; they are not covered by testing that is being done so it could certainly lead to unexpected results!

The easiest way to remove "dead" code that is not currently executed is to put conditional compiles around the block of code. It is possible there is special debug or unit test code that you want to include for internal builds, but you never intend for this code to be released in the final product. Consider the following block of code:

```
#if defined (LOGDEBUG)
    index = 20;
    LOG_Data_Set( *local_data, sizeof( data_set_t ));
#endif
```

This code block is created whenever a debug version is created, whereas the conditional definition "LOGDEBUG" is defined at build time. However, a situation could arise where a developer defines this elsewhere for another purpose and then this code gets included unexpectedly! In situations where there are multiple conditional compile situations associated with internal releases consider doing something like the following code block:

```
#if defined (LOGDEBUG)
#if !defined(DEBUG)
    neverCompile
#else
    index = 20;
    LOG_Data_Set( *local_data, sizeof( data_set_t ));
#endif
#endif
```

This code block helps when multiple conditional compiles exist for different features. If "LOGDEBUG" gets defined and is not part of an overall "DEBUG" build, then there will be a compiler error when it gets compiled. A good way to make sure that code segments do not end up in the final deliverable is if "DEBUG" is never allowed to be defined in external release software deliverables. This is an excellent way to add extra protection to conditional compiles.

5.15 Strategy 15: Fill Unused Memory

For nonvolatile memory that contains program code filling unused memory with meaningful data is a good idea. One older processor family decided to have the machine opcode "0xFF" equate to a "no operation" instruction, where it would use a clock cycle then go on to execute the next instruction! For any processor architecture it is good to protect yourself in case there is an unexpected condition where the program counter gets set to an invalid address.

When the program image is built and linked it is a good strategy to fill the memory with instructions that cause the processor to reset. There are opcodes that can be used to cause an illegal or undefined interrupt. When the interrupt routine receives such an interrupt it does a reset because the unexpected interrupt has code that executes this. Or you could use instructions that do a software reset depending on the processor core. Executing in invalid locations isn't a good situation—so for your safety case determine the best course of action!

5.16 Strategy 16: Static Code Analysis

The last strategy to use with safety-critical code development is to run a static code analyzer when the code is compiled. Many different static code analysis packages exist for C and C++ that also conform to published standards such as MISRA C (discussed in Strategy 4: Specific variable value checking). Irrespective of the checking done as part of static code analysis there shouldn't be any warnings when the analysis is complete.

Static code checkers typically include a way to "ignore" certain warnings in the code. Since many checks that are done can be considered "optional" or "good practice" there may be instances that the code written is really intended to be the way it is and fixing it to match the checking standard is not optimum or feasible. In these situations it is important to document in the code exactly why you are doing it the way you are doing it and then include the appropriate "ignore" directive immediately preceding the offending line of code.

Exercises

1. **Q:** Describe the progression leading to system failures.
   A: Faults could lead to errors which could lead to failures.
2. **Q:** Why would you want to add an additional processor to a safety-critical system's architecture?
   A: To provide a way for functionality to continue in case the primary processor becomes unusable or corrupted.
3. **Q:** What does the word "fail-safe" mean?
   A: The term "fail-safe" is used to describe a result where the system always maintains a safe state, even if something goes terribly wrong.

12

NETWORKING SOFTWARE

Sandeep Malik, Shreyansh Jain, Jaswinder Singh

Digital Networking, NXP, Delhi, India

1 Introduction

In the past few decades the semiconductor industry has undergone a revolution. Initially, embedded devices were thought to have very limited capability, both in terms of processing power and memory.

Most embedded devices were designed using microcontrollers rather than actual processors.

However, with an increased number of users exchanging data over the internet and with the increased availability of high-speed networks, the demand for high-processing embedded networking devices increased. This demand was also fueled by advances in semiconductor technology, which led to the computational power that is present in today's devices being affordable.

This transition of embedded network products using microcontrollers to embedded network processors has gone through multiple phases. This evolution did not only focus on hardware but also on the complete ecosystem. Earlier embedded devices started by having most processing occur in the core. Then came an era in which network processing was subdivided into hardware and software. A subsequent stage introduced intelligent hardware accelerators that could be programmed for autonomous packet forwarding, post configuration. As technology continued to progress, processing power became cheaper, such that designs with custom accelerators became too expensive to design and develop.

These days, the industry focus is mainly on producing devices with raw processing power that can be used as general-purpose devices or as embedded networking devices.

2 Embedded Linux Networking

In the wake of technological advances, communication has become a vital requirement for a fully functional system. The requirement to have a communication network is now not just limited to legacy network infrastructure, involving core devices such as switches, routers, firewalls, or gateways; rather such networking functionality is becoming integrated in devices belonging to various new domains in a variety of forms.

For example, in the automotive world, the market is shifting toward **autonomous cars** with advanced connectivity requirements. These requirements are pushing the limits at which data is exchanged. To cater to such needs, connected cars are being designed with security gateways as a core component, responsible for controlling secure data exchanges between various components as well as with external cars on an as-needed basis. Since the use case of cars is a real-time use case, where data needs to be processed in a time-bound fashion, along with making sure that it is not compromised, having a security gateway as an integrated part is essential to the smooth operation of NextGen connected cars. This has opened up an entirely new domain with challenges involving real-time deadlines, data processing at high speeds with accuracy, and having security as a key component to ensure that malicious users do not hack into the internal vehicle network with devastating effects.

Another such requirement comes from the IoT domain. Due to the increased usage of the Internet and the availability of high-speed infrastructure, the number of devices and the amount of data that these devices have to exchange among themselves has increased multifold in last couple of years. Also, the adaptability of the IoT is acting as a catalyst for this. With the IoT coming into the picture, the estimated number of devices connected over the internet is in the order of billions. With so many devices in place, the need for embedded Linux networking is becoming more and more crucial. One important area in the IoT domain, where embedded networking plays a critical role, is "**Edge Routing.**" As IoT technology is evolving, the need for distributed computing power is coming to the forefront, wherein instead of relying completely on the cloud, some computational processing would be offloaded to edge routers which would eventually communicate on the north side with various sensors, actuators, and other devices, and on south side connect to the cloud network.

This chapter will start with an introduction into the core component of networking, namely the network stack in Linux. The introduction will cover the layered architecture of a network stack including a glimpse of both OSI and TCP/IP models. After the introduction, the chapter will consider various embedded network device use cases. Then it will shed light upon the tools that can be used in Linux to configure the networking aspects of the kernel. The chapter concludes by discussing the issues with Linux kernel–based networking.

2.1   Network Stack

The network stack in the Linux kernel is a core module which allows two systems, connected over a network, to communicate with one another. The network stack primarily defines a set of protocols which allows two entities to communicate with each other by following a set of rules. These rules primarily govern how the basic element flowing in the network, primarily known as a frame/packet, shall appear. The overall networking stack is implemented in the form of discrete layers where each layer is assigned a clear task. For the packets originating from the system, the task of a layer is primarily to encapsulate the information received from the upper layer, with an appropriate header, and pass it to the next layer until the packet is placed on a physical medium. In the reverse direction, each layer would extract the respective header before handing over data to the layer above.

There are mainly two models used to represent the layered architecture of the Linux network stack. The first is called the OSI model and the second is popularly known as the TCP/IP model.

Fig. 1 depicts the OSI model and the protocols functioning in each of its layers.

Application Layer (HTTP, TFTP, STUN, SMTP, Telnet etc.)
Presentation Layer (TLS, SSL)
Session Layer (RTCP, SSH, NFS)
Transport Layer (ESP, AH, TCP, UDP)
Network Layer (IPv4, IPv6, IPsec, ARP)
Data Link Layer (Ethernet, IEEE802.1Q, MAC, ATM)
Physical Layer (Ethernet Physical Layer, CSMA/CD)

Fig. 1 OSI seven layers.

Now let us talk about the other model, namely the TCP/IP stack. This model envisions that complete functionality can be achieved even if we have five layers instead of seven, as in the OSI layered model. The TCP/IP model combines the highest three layers, namely the Application, Presentation, and Session Layers into one Application Layer. The designers of the TCP/IP model considered that the OSI model was overly layered and that a few layers could be combined to achieve a clearer, simpler design. Fig. 2 provides an overview of the TCP/IP model along with the protocols executing in each of its layers.

It is worth mentioning here that even though Fig. 2 shows only four layers, the numbering is such that the Application Layer is called L5 or Layer 5. Similarly, the Transport Layer is called L4 or Layer 4, the Network Layer is called L3 or Layer 3, and the Data Link + Physical Layer is called L2/L1, interchangeably.

The main advantage of the layered approach is that the task of packet handling has been distributed among different layers. Due to this, the maintainability of the code has increased and an issue originating in

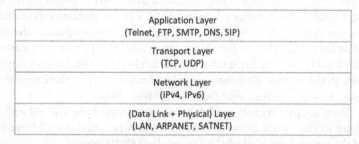

Application Layer (Telnet, FTP, SMTP, DNS, SIP)
Transport Layer (TCP, UDP)
Network Layer (IPv4, IPv6)
(Data Link + Physical) Layer (LAN, ARPANET, SATNET)

Fig. 2 Five layer TCP/IP model.

a particular layer will remain confined to that layer. This eventually helps with isolating software issues a little faster than a monolithic implementation. Details about the function of each layer are provided in great depth in both the online and offline literature and are beyond the scope of this chapter.

2.2 Embedded Network Devices

This section focuses on various use cases of embedded network devices. As mentioned in the Section 2.1, each layer has discrete responsibilities. This section touches on the family of network devices which are commonly used and also covers details such as which layer the device operates on or belongs to. In this section the layered architecture discussion is focused only on Ethernet devices.

Let us start with L1 or Layer 1 devices. These devices primarily work on the physical layer and usually are dumb devices without any frame-parsing capability. The data received on one port is religiously forwarded to other ports. There is generally no intelligence applied by devices operating on this layer. A **Hub** is one such device which operates at this layer.

Next comes L2 or Layer 2 devices. These devices focus on the L2 information which is primarily comprised of MAC addresses. These devices permit frames to allow or drop on the basis of the MAC address associated with that frame. A **Switch** is one of the commonly used network devices operating at Layer 2. A switch can be thought of as an intelligent hub which can educate itself based on the MAC addresses learned from the packets received on its ports and post awareness will not flood traffic to other ports. A switch usually maintains forwarding database tables where the mapping of MAC address and ports is stored. This mapping helps in deciding upon the destination of a particular stream.

Layer 3 devices focus on the destination IP address in the packets to decide on the action to be taken for a frame. The action depends on whether the destination IP address is a self IP or some other IP. In the case that it is a self IP, the packet gets delivered locally to a host. In the case of an IP address not matching the self IP, the device tries to find a route for this traffic, including if the destination belongs to a subnet of any interface IP. In the case that a match is found the packet is forwarded to that interface or else the system checks whether there is a default route configured in the system. If a default route exists, the packet is forwarded, otherwise the packet is discarded. Sometimes the routing decision may involve modified IP addresses in the case that a NAT (network address translation) is configured. A NAT is not limited to Layer 3, rather it spreads from Layer 3 to Layer 5 depending on the configuration. Since this layer is primarily involved in finding the route for the next hop, devices which operate in this layer are called **Routers**.

Layer 4 devices operate on five-tuple information, including IP source address, IP destination address, protocol, source port, and destination port. The packet undergoes various processing stages/lookups before coming out of a particular port. The stages typically help with the acceptance or rejection of the flow based on the firewall rules configured in the system. Network devices that operate at this layer are called **Firewalls**.

Layer 5 devices operate on the complete packet, including the payload, to make a decision. These devices are highly intelligent devices and are typically deployed in areas of deep-packet inspection or work as proxies. Such devices, if working as proxies, are typically called **Application Layer Gateways** or if working in areas of deep-packet inspection are called **Next Generation Firewalls**.

2.3 Network Configuration and Analysis Utilities

This section considers the various utilities which help configure networking services on a Linux-based embedded network device. The utilities discussed in this section are all open-source utilities and a plethora of literature about them is already available online. Some of the references about useful resources are also available in the "Further Reading" section of this chapter.

ethtool: This utility allows the user to view and update the properties of a network device and a few properties supported in the device driver. This utility is widely used across the community to peep into the network device. A sample output of the ethtool command follows.

When the device link is not up:

```
ethtool eth0
Settings for eth0:
    Supported ports: [ ]
    Supported link modes: Not reported
    Supported pause frame use: No
    Supports auto-negotiation: No
    Advertised link modes: Not reported
    Advertised pause frame use: No
    Advertised auto-negotiation: No
    Speed: Unknown!
    Duplex: Half
    Port: Twisted Pair
    PHYAD: 0
    Transceiver: internal
    Auto-negotiation: off
```

```
    MDI-X: Unknown
    Current message level: 0xffffffff (-1)
                drv probe link timer ifdown ifup rx_err tx_err tx_queued intr tx_done
rx_status pktdata hw wol 0xffff8000
    Link detected: no
```

When the device link is up:

```
ethtool eth0
Settings for eth0:
    Supported ports: [ MII ]
    Supported link modes: 10baseT/Full
                100baseT/Full
                1000baseT/Full
    Supported pause frame use: Symmetric Receive-only
    Supports auto-negotiation: Yes
    Advertised link modes: 10baseT/Full
                100baseT/Full
                1000baseT/Full
    Advertised pause frame use: Symmetric Receive-only
    Advertised auto-negotiation: Yes
    Speed: 10Mb/s
    Duplex: Half
    Port: MII
    PHYAD: 28
    Transceiver: external
    Auto-negotiation: on
    Current message level: 0xffffffff (-1)
                drv probe link timer ifdown ifup rx_err tx_err tx_queued intr tx_done
rx_status pktdata hw wol 0xffff8000
    Link detected: no
```

More details about ethtool command parameters form part of the
Linux man pages.

ifconfig: Another popular utility, widely used by network admin-
istrators as well as individual users for configuring network device
parameters, is ifconfig. This utility allows the configuration of the network interface, specifically the IP address, MAC address, and devise
status, to be either up or down. A sample output of the ifconfig command follows.

```
root@t1040qds:~# ifconfig
eth0  Link encap:Ethernet HWaddr 00:e0:0c:00:58:00
      inet addr:192.168.2.108 Bcast:192.168.2.255 Mask:255.255.255.0
      inet6 addr: fe80::2e0:cff:fe00:5800/64 Scope:Link
      UP BROADCAST MULTICAST MTU:1500 Metric:1
      RX packets:0 errors:0 dropped:0 overruns:0 frame:0
```

```
                          TX packets:3 errors:0 dropped:0 overruns:0 carrier:0
                          collisions:0 txqueuelen:1000
                          RX bytes:0 (0.0 B) TX bytes:238 (238.0 B)
                          Memory:fe4e0000-fe4e0fff
              lo    Link encap:Local Loopback
                          inet addr:127.0.0.1 Mask:255.0.0.0
                          inet6 addr: ::1/128 Scope:Host
                          UP LOOPBACK RUNNING MTU:65536 Metric:1
                          RX packets:13 errors:0 dropped:0 overruns:0 frame:0
                          TX packets:13 errors:0 dropped:0 overruns:0 carrier:0
                          collisions:0 txqueuelen:0
                          RX bytes:1456 (1.4 KiB) TX bytes:1456 (1.4 KiB)
              root@t1040qds:~# ifconfig eth0 down
              root@t1040qds:~# ifconfig
              lo    Link encap:Local Loopback
                          inet addr:127.0.0.1 Mask:255.0.0.0
                          inet6 addr: ::1/128 Scope:Host
                          UP LOOPBACK RUNNING MTU:65536 Metric:1
                          RX packets:13 errors:0 dropped:0 overruns:0 frame:0
                          TX packets:13 errors:0 dropped:0 overruns:0 carrier:0
                          collisions:0 txqueuelen:0
                          RX bytes:1456 (1.4 KiB) TX bytes:1456 (1.4 KiB)
```

The ifconfig utility is part of the **nettools** package.

Note that there are other useful utilities which are part of the **nettools** package, including **arp**, **netstat**, and **route**. These utilities are also worth exploring.

ip: This section considers another utility named ip. This allows the user to play with not only the interface related parameters but also some advanced functions like route configuration. This utility is a part of the **"IPROUTE"** package.

This utility provides features of multiple **nettools** utilities, such as **arp, ifconfig**, and **route**, using a single interface. Some sample commands executed using the **ip** utility follow.

```
root@t1040qds:~# ip -s neigh
root@t1040qds:~# ip neigh add 192.168.2.10 lladdr 1:2:3:4:5:6 dev eth0
root@t1040qds:~# ip -s neigh
192.168.2.10 dev eth0 lladdr 01:02:03:04:05:06 used 2/2/2 probes 0 PERMANENT
root@t1040qds:~# ip route
192.168.2.0/24 dev eth0 proto kernel scope link src 192.168.2.108
root@t1040qds:~# ip route add default via 192.168.2.1
root@t1040qds:~# ip route
default via 192.168.2.1 dev eth0
192.168.2.0/24 dev eth0 proto kernel scope link src 192.168.2.108
```

This section lists the two most popular utilities used by hackers and network programmers for sniffing packets in the network.

tcpdump: This utility is very useful to debug and to root out the causes of problems related to traffic entering the Linux box. This utility allows the user to monitor packets on a specific interface in real time, irrespective of whether the interface is wired or wireless. This also allows the user to redirect packet captures in a file which can later be viewed using any tool, like wireshark, or played using tools like tcpreplay. The following snapshot shows a typical view of a system running the tcpdump utility to capture packets when ping is initiated on an interface.

```
tcpdump: verbose output suppressed, use -v or -vv for full protocol decode
listening on nc_eth1.10, link-type EN10MB (Ethernet), capture size 262144 bytes
Added new mac 4e:80:65:d4:2f:7 serial=n_p1_m0
00:47:53.298267 ARP, Request who-has 192.168.1.100 tell 192.168.1.10, length 50
00:47:53.298293 ARP, Reply 192.168.1.100 is-at 18:03:73:14:00:11 (oui Unknown), length 28
00:47:53.331544 IP 192.168.1.10 > 192.168.1.100: ICMP echo request, id 594, seq 1, length 64
00:47:53.331565 IP 192.168.1.100 > 192.168.1.10: ICMP echo reply, id 594, seq 1, length 64
00:47:56.258757 IP 192.168.1.10 > 192.168.1.100: ICMP echo request, id 594, seq 2, length 64
00:47:56.258776 IP 192.168.1.100 > 192.168.1.10: ICMP echo reply, id 594, seq 2, length 64
00:47:58.387620 ARP, Request who-has 192.168.1.10 tell 192.168.1.100, length 28
00:47:58.390463 ARP, Reply 192.168.1.10 is-at 4e:80:65:d4:2f:07 (oui Unknown), length 50
00:47:59.179691 IP 192.168.1.10 > 192.168.1.100: ICMP echo request, id 594, seq 3, length 64
00:47:59.179707 IP 192.168.1.100 > 192.168.1.10: ICMP echo reply, id 594, seq 3, length 64
00:48:02.092884 IP 192.168.1.10 > 192.168.1.100: ICMP echo request, id 594, seq 4, length 64
00:48:02.092902 IP 192.168.1.100 > 192.168.1.10: ICMP echo reply, id 594, seq 4, length 64
00:48:05.011308 IP 192.168.1.10 > 192.168.1.100: ICMP echo request, id 594, seq 5, length 64
00:48:05.011327 IP 192.168.1.100 > 192.168.1.10: ICMP echo reply, id 594, seq 5, length 64
00:48:07.920985 IP 192.168.1.10 > 192.168.1.100: ICMP echo request, id 594, seq 6, length 64
00:48:07.921005 IP 192.168.1.100 > 192.168.1.10: ICMP echo reply, id 594, seq 6, length 64
00:48:10.837676 IP 192.168.1.10 > 192.168.1.100: ICMP echo request, id 594, seq 7, length 64
00:48:10.837695 IP 192.168.1.100 > 192.168.1.10: ICMP echo reply, id 594, seq 7, length 64
00:48:13.755429 IP 192.168.1.10 > 192.168.1.100: ICMP echo request, id 594, seq 8, length 64
00:48:13.755449 IP 192.168.1.100 > 192.168.1.10: ICMP echo reply, id 594, seq 8, length 64
00:48:16.743616 IP 192.168.1.10 > 192.168.1.100: ICMP echo request, id 594, seq 9, length 64
00:48:16.743645 IP 192.168.1.100 > 192.168.1.10: ICMP echo reply, id 594, seq 9, length 64
00:48:19.659724 IP 192.168.1.10 > 192.168.1.100: ICMP echo request, id 594, seq 10, length 64
00:48:19.659744 IP 192.168.1.100 > 192.168.1.10: ICMP echo reply, id 594, seq 10, length 64
```

More information about the tcpdump utility can be found on the main pages in Linux.

wireshark: Another very widely used utility, similar to tcpdump, is known as wireshark, aka ethereal in its early days. The wireshark/ethereal utility is usually installed in Linux box by default but can be installed easily in cases when it is not present since almost all Linux distribution offers this utility's installer. It provides an easy-to-use graphical interface which allows the user to add filters in order to select the display of a specific packet stream or a specific protocol from all captured packets. Below is a snapshot of the wireshark tool capturing packets on a live network interface.

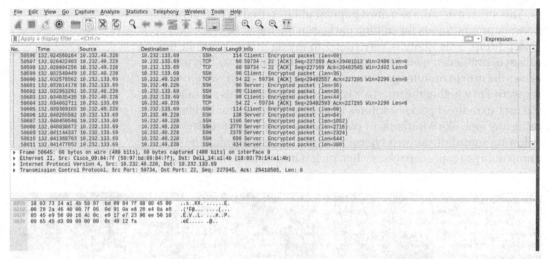

Further details about the usage and capturing capability options of wireshark can be found on the wireshark website.

3 Moving From the Linux Kernel to User Space

The Linux networking stack is a very large block that enables almost all the well-known protocols active today. It enables the Kernel to read a packet, process its various headers, and perform operations like handing over a user space application or forwarding ahead. However, being a general-purpose operating system, the Linux network stack includes a much larger set of functionalities than generally used. The design tenets which Linux attempts to achieve impacts the design of the network stack as well:

1. *Inherent support for a large set of protocols and devices.* The use cases which the Linux kernel supports are very wide ranging—from embedded devices to large supercomputers. With such a large set comes the need to support an even larger number of protocols, irrespective of the density of deployment. The Linux networking stack supports a huge number of protocols—some of which enable the Internet (TCP) and others that are specific to domains like the IoT (e.g., 6lowpan).

2. *Compliance and quality assurance.* With such a large user base comes the responsibility to be compliant to industry standards. This requires rigorous verification, both in implementations and deployments. This makes the kernel reliable and its network stack standardized. This also leads to long development and mainline timelines for new devices or protocol support.

Performance as a criterion sits on a lower priority strata than flexibility, scalability, and adaptability, which the kernel strives to achieve. Though the Linux kernel can be used in innumerous generic scenarios, it is not an adequate solution for some specific problems—more precisely, problems around achieving line rates or an adequate utilization of a network fabric.

Upcoming use cases and subsequent advancements in network infrastructure have long out-paced the development of the Linux kernel network stack's ability to keep up with the available bandwidth and latency requirements. For example, in the case of high-frequency trading (HFT), the available margin of packets is much lower than the performance achieved by the Linux kernel-based environment. The focus on reliability and compliance is not a primary criterion.

The general reasons for Linux network stack's unsuitability are:

1. Its performance is not able to keep up with the network fabric available in the market. It cannot reach line rates for raw packet performance on many of the hardware NICs (like Intel XL710 with 40G interfaces).
2. Its lack of flexibility to adapt to new protocols or packet types—at least through mainline support. It is always possible to hack into the kernel, but that is a risky approach given that it impacts system stability.
3. Its inability to keep up with the lightning fast times to market required by organizations and their eventual customers, associated with rapid advancements in hardware technology in compliance with an increasing number of use cases.

3.1 Analyzing the Expected Packet Rates

For a 1-G line card, the maximum theoretical throughput that can be achieved is 1.448 million packets per second (Mpps). This is based on the fact that the smallest packet would be 84 bytes, including the preamble and IFG (interframe gap). This implies that to achieve line rate, the CPU (or the packet processing engine, in its entirety) should be able to process 1.44 million packets per second. For a 10-G link, that would be 14.88 Mpps.

1 Gbps = 1 x 1000 x 1000 x 1000 bits per second = 1000,000,000/8 bytes per second = 125,000,000 bytes per second

With minimum packet size of 84 bytes = 64 bytes + 12 bytes preamble + 8 bytes IFG = 67.2 ns processing time to achieve line rate

For processing that number of packets, about 67.2 ns are available for the packet processing engine to perform processing on the packet.

For a NIC card capable of a 10-G line rate, the average time available for processing each packet, so as to achieve line rate, is about 67.2 ns.

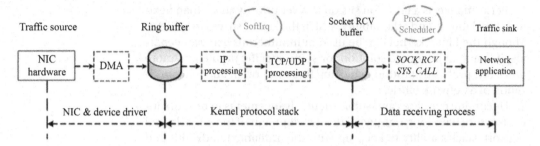

In the above image, extracted from https://people.cs.clemson.edu/~westall/853/tcpperf2.pdf, the RX path of a packet in a Linux stack has been described.

The costliest operation, shown in this figure, is the "kernel protocol stack"—which parses each packet and finds the appropriate application to hand it over to, using the socket interface. This operation is expensive because it supports a very large number of protocols, most of which are not used in the majority of use cases. This makes the Linux Kernel an ideal stack for deployment in a large number of use cases—however, performance is sacrificed. Usually, over a 10-G NIC, approximately 1–2 Mpps can be processed by a kernel stack on a general 2-GHz processor.

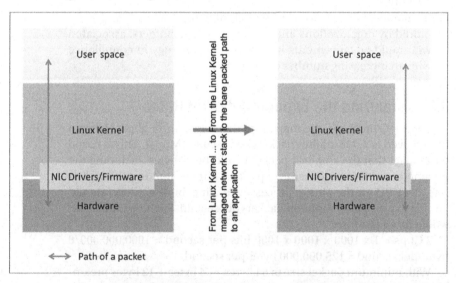

In case the number of protocols supported is not required and the complexity of the stack does no't add any value to the use case, it is possible to completely bypass the stack. There are proven solutions which showcase performance at line rate for supported NICs. These solutions either completely bypass the Linux kernel network stack or just use it as a secondary or fallback option—performing protocol-specific operations.

The following text outlines what is required to achieving the best performance outside of the Linux network stack.

3.2   Direct Access to the Hardware

The primary reason the network path of a packet through a Linux kernel is not considered optimal is the thickness of the layers between the actual hardware receiving the packet and the application consuming the packet.

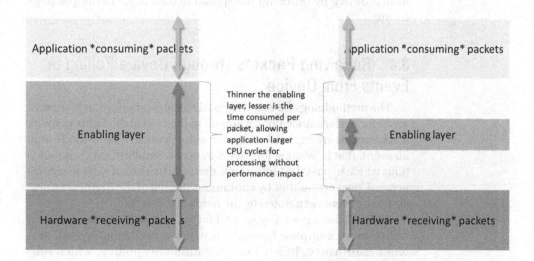

Various solutions propose thinning this layer of its "enabling software or layers," allowing more CPU time for applications to write their custom logic. Some solutions work directly with hardware registers, configuring them with optimal values based on use cases. Other solutions work with very thin layers of abstract APIs, which hide hardware complexities. However, in both cases, the challenge is to expose the hardware to the application.

Solutions like DPDK and ODP rely on the user space access of the hardware through memory mapping. Whereas, solutions like XDP and Netmap use a thin set layer in the Linux kernel (kernel module) for enabling access. Further, solutions like eBPF go a step further and provide in-kernel execution of custom code within the kernel scope.

3.3   Virtual I/O Layer (VFIO/UIO) and HugePages

Two of the Linux kernel's concepts, VFIO/UIO and HugePages, play a big role in enabling the user space access to devices.

Virtual function I/O or VFIO and user space I/O or UIO represent the Linux kernel's ability to expose a device to the user space through file-based interfaces from sysfs. A device, if bound to a

VFIO/UIO driver, exposes its configurable memory space to the driver. Thereafter, the VFIO/UIO driver would allow any user space application to read the memory space and map it in its own address space. All this is done through the standard file open/close/read/write/ioctl interface.

At the same time, HugePages support in the Linux kernel allows the application to pin to memory large chunks of address space, which have larger than usual page sizes (2 MB, 4 MB, 1 GB, etc.). This increases the performance of accessing an address for I/O from a device by reducing the spread of data across multiple page counts.

3.4 Receiving Packets Through Device Polling or Events From Device

The methodology for receiving packets by an application is another arena for performance improvement. In the case of the Linux network stack, sockets are exposed which can either work in polling mode or in an event, that is, waiting for events to occur. Similarly, for those solutions which bypass the Linux stack, there are two broad ways to receive network packets—either by continuous polling of the hardware interface or by events generated by the hardware interface.

As CPUs become cheaper and faster, network applications can utilize more compute bandwidth without impacting overall system performance. In such cases, continuous polling, which otherwise is expensive, proves to be a performant alternative. Yet, in many small or edge devices hosting a network application, where CPUs are a precious resource, event-based packet reception models work well.

The following section considers a few of those solutions, that is, ODP, Netmap, DPDK, XDP, and BPF. Of these, ODP, Netmap, and DPDK are primarily used for performance extraction but XDP and BPF are used for enhancing Linux network stack ability through external plugins and programmability.

3.5 ODP—Open Data Plane

ODP or OpenDataPlane is an open-source project under the Linaro umbrella which aims to provide a cross-platform set of application programming interfaces (APIs) for the networking data plane. It leverages vendor-specific hardware blocks and their implementation and layers a generic set of APIs exposed to the applications. The intent is to create network data plane applications which are agnostic to underlying hardware and yet able to extract the best hardware support.

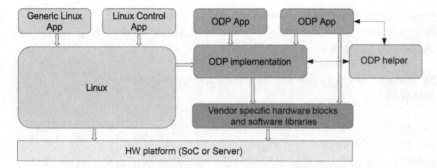

From: https://opendataplane.org/index.php/service/technicaloverview/

In the image above, green blocks represent the ODP high-level components compared with the Linux stack and vendor hardware.

ODP is a C-based framework. A typical ODP application would be similar to the following snippet:

```
int main(int argc, char *argv[])
{
    odp_init_t init;
    odph_odpthread_t thr;
    /* Install some kind of signal handler as thread would be
     * infinite loop */
    signal(SIGINT, sig_handler);
    ...
    /* Global initialization */
    if (odp_init_global(&instance, &init, NULL)) {
        LOG_ERR("Error: ODP global init failed.\n");
        exit(EXIT_FAILURE);
    }
    /* Initialization for a local thread, to be done for each
     * thread created. */
    if (odp_init_local(instance, ODP_THREAD_CONTROL)) {
        LOG_ERR("Error: ODP local init failed.\n");
        exit(EXIT_FAILURE);
    }
    ...
    /* Create a pool of buffers which would be used by
     * hardware for packets - ethernet, crypto, etc */
    odp_pool_t pool = odp_pool_create("packet pool", &params);
    ...
    /* Create a pktio device - essentially representing a hardware device
     * which would Rx and Tx data packets. Like Ethernet or Crypto.
     * This function also initializes any queues attached to the pktio
     * device. */
    if (create_pktio(dev, i, num_rx, num_tx, pool, grp))
        exit(EXIT_FAILURE);
    ...
```

```
        /* Create an ODP worker thread - which would essentially do I/O.
         * 'thr_params' would contain thr_params.start = <some_IO_func> */
        odph_odpthreads_create(thr, &thd_mask, &thr_params);
        ...
        /* At this point, the workers threads are awaiting device start */
        odp_pktio_start(pktio)
        ...
        /* Collect some stats */
        /* When done, stop the Pktio device */
        odp_pktio_stop(pktio);
        ...
        odp_pool_destroy(pool);
        odp_term_local();
        odp_term_global();
        return ret;
}

static int create_pktio(char *dev_name, odp_pool_t pool, int num_tx_queue,
                        int num_rx_queue)
{
        odp_pktio_t pktio;
        pktio = odp_pktio_open(dev, pool, <some params>);
        ...
        /* Initialize the parameters which would be used for configuring the
         * Rx and Tx queues. */
        odp_pktin_queue_param_init(<Rx queue params>);
        odp_pktout_queue_param_init(<Tx queue params>);
        ...
        /* Configure the queues */
        odp_pktin_queue_config(pktio, <Rx queue params>);
        odp_pktout_queue_config(pktio, <Tx queue params>);

        /* Eventually, enable the queues, ready when odp_start is called */
        odp_pktin_queue(pktio, <Queue configured>);
        odp_pktout_queue(pktio, <Queue configured>);
        ...
}

static int some_IO_fun(void *arg)
{
        odp_queue_t queue;
        /* Queue would have been initialized to */

        while (<signal not received>) {
                /* Receive packet on a queue */
```

```
        pkts = odp_pktin_recv(rx_queue, pkt_buffer, <num of buffers>);
        /* Send it out of another queue */
        sent_pkts = odp_pktout_send(tx_queue, pkt_buffer, pkts);
        /* Release the packets which were sent */
        odp_packet_free(pkt_buffer[sent_pkts]);
    }
}
```

In the above snippet, there are various functions represented by the odp_*   naming convention. These are ODP APIs which define an abstract layer over a complex hardware-dependent code. These APIs can be broadly categorized into following sets based on their functionality:
1. Initialization APIs.
2. Packet and packet I/O for representing a packet and an interface over which packets are either received or transmitted.
3. Buffers and buffer pool for representing a block of memory for holding packets or other metadata, arranged in a pool for efficient allocations and deallocations.
4. Queue for representing the serialized packet interface connected to an I/O device.
5. Scheduler which represents a software logic which controls the next queue to serve, based on prioritization or order.
There are various other blocks like Crypto, Hash, and Traffic Management, used for tying together functionalities to create a solution.

ODP's primary aim is to create a uniform set of APIs. All the implementation is left for hardware owners to handle, including any specialized libraries like hashing and distribution. These allow hardware owners to focus on their implementation without integration issues. This also enables gluing together two or more underlying bits of hardware (if supported layers are present) with a high degree of interoperability. Furthermore, this strategy also enables a fast time-to-market for hardware vendors as they only have to integrate their implementations with a uniform set of APIs.

3.6   DPDK—Data Path Development Kit

DPDK was originally an Intel project but was open sourced (BSD) in 2010 as a stand-alone project. In the year 2017, it merged with the Linux Foundation. As mentioned online (https://www.dpdk.org/about/in), "Since then, the community has been continuously growing in terms of the number of contributors, patches, and contributing organizations, with 5 major releases completed including contributions from over 160 individuals from 25 different organizations. DPDK now supports all major CPU architectures and NICs from multiple vendors, which makes it ideally suited to applications that need to be portable across multiple platforms."

Whereas ODP focuses on abstraction of APIs, DPDK focuses on performance through optimization of general-use functionalities. ODP allows hardware owners to completely focus on their implementation, with all their support libraries—whereas, DPDK architecture allows hardware owners to off-load generic operations to an already available optimized set of libraries, allowing them to focus completely on core I/O paths.

However, both solutions are classic examples of frameworks for user space acceleration packets from network and nonnetwork (crypto, compression, etc.) blocks.

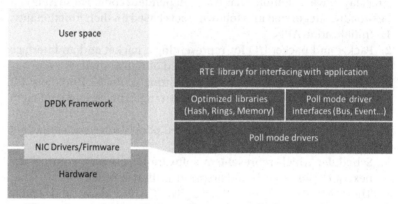

The primary logic of controlling network devices is contained within the "DPDK Framework" layer. This in turn is composed of multiple libraries, each for a specific function, such as Hash Table, Ring, GRO, GSO, Fragmentation, and Classification. This is glued together with NIC specific drivers in user space for enabling and devices. The library is also referred to as RTE and EAL, which stands for RunTime Environment (RTE) and Environment Abstraction Layer (EAL), respectively.

Just like ODP, DPDK is also a C language-based framework. A typical program in DPDK would be similar to the following snippet:

```
int main(int argc, char *argv[])
{
    ...
    /* Initialize the EAL framework; This initializes all the necessary components like
    memory, threads and command line arguments for device parameters. */
    ret = rte_eal_init(argc, argv);
    ...
    /* Those arguments which are consumed by rte_eal_init are the ones passed to the
    framework. Separated by a '--', all the application specific arguments can be
    provided which would remain untouched by EAL framework. */
```

```
ret = parse_args(argc, argv);
...
/* Find the number of ports which were detected by the EAL layer. */
nb_ports = rte_eth_dev_count_avail();
if (nb_ports == 0) {
    printf("No ports were detected\n");
    return -1;
}
/* Applications can also find a port using its name. This API would return a port
ID number which can then be used to reference the port. */
ret = rte_eth_dev_get_port_by_name(name, &portid);
...
/* Create a pool for Rx/Tx of packets; This pool would be attached to the device/
port and when driver Rx's packets, memory would be allocated from this pool.
Similarly, the application can use this pool to allocate memory (rte_mbuf) from
this pool. */
pktmbuf_pool = rte_pktmbuf_pool_create("mbuf_pool", nb_mbufs,
            MEMPOOL_CACHE_SIZE, 0, RTE_MBUF_DEFAULT_BUF_SIZE,
            rte_socket_id());
if (l2fwd_pktmbuf_pool == NULL)
    printf("Error; Not enough memory to work with\n");
...
...
/* Then, for each port, configure it. This can also be replaced with a logic for a
specific port fetched from rte_eth_dev_get_port_by_name(). */
RTE_ETH_FOREACH_DEV(portid) {
    struct rte_eth_dev_info dev_info;
    /* Get the port information, which includes its name, limits (number of
    queues, buffer and ring sizes, capabilities and other information based on
    which application can define its configuration parametres. */
    rte_eth_dev_info_get(portid, &dev_info);
    /* Creating a local set of configuration based on fetched set to configure
    the device. If the values of configuration exceed those specified in the
    rte_eth_dev_info_get API, the configuration can return an error. */
    struct rte_eth_dev_info local_info;
    /* for example, for enabling VLAN stripping on Rx'd packets, after checking
    if that is supported or not... */
    if (dev_info.rx_offload_capa & DEV_RX_OFFLOAD_VLAN_STRIP)
        local_info.rxmode.offloads |= DEV_RX_OFFLOAD_VLAN_STRIP;
    ...
    /* Configure device with n_rx and n_tx number of Rx and Tx queues, respectively. */
    ret = rte_eth_dev_configure(portid, n_rx, n_tx, &local_conf);
    ...
    ret = rte_eth_rx_queue_setup(portid, 0, n_rx, <Socket ID>, rxq_conf,
    pktmbuf_pool).
```

3.7 BPF—Berkley Packet Filter

Most packet processing software stacks focus on working on the packets either in Kernel space (drivers, Linux TC, XDP) or in user space (DPDK, ODP). The eBPF, or Berkley Packet Filter, is a novel approach which focuses on reducing the overhead of copying packets from the Kernel to user space by using "Packet Filters" which are capable of processing packets as early as the software queue but without the complexity associated with a Linux Kernel network driver.

This method was first proposed way back in 1992 (http://www.tcpdump.org/papers/bpf-usenix93.pdf). The original proposal talks about providing a limited set of instructions, called a "filter machine," which are available to be plugged and configured from user space. The authors implemented a register-based RISC emulator which traps the limited instruction set of the binary blob used for process packets. This filter is executed within the context of the kernel. Such filters are invoked on each arriving packet, and the result of the filter is passed to some user space application to make a decision (if not already made by the filter logic).

This approach of creating a pseudo virtual machine, with a limited instruction set, within the Linux kernel was unique to the packet-processing scene. In recent times, it was improved with the introduction of eBPF or enhanced BPF. The virtual machine now employs instructions closer to the hardware for data movement in register-sets. Consequently, the filter sets written can use more advanced compare, load, and store operations.

4 Life of a Packet in a Native Linux Network Stack

This section discusses the details of the life of a packet in a Linux-based system. A packet first enters the NIC card that usually has a DMA engine responsible for dumping packet data into the buffers that belong to its internal memory in the case that it is an autonomous IP acting as an accelerator, or the DDR in the case that packet processing is done at the CPU. For the sake of simplicity and explanation this section covers the latter case, where buffers are from DDR and packet processing is done at the CPU. To be more precise this section discusses the receive path where a packet is delivered to the application running on the core after being received by an Ethernet driver.

The complete receive path can be thought of as consisting of three major components. First, the NIC and the Ethernet driver; second, the kernel network stack; and third, the application or the consumer of the packet.

In the first stage, the NIC card receives a packet which will eventually be processed by the network driver and performs the functions of Layer 1 and Layer 2 of the OSI layer architecture. The general architecture for the NIC consists of a DMA engine and a set of rings which have buffers to receive the packets. The following figure is a high-level view of the system. The NIC card DMAs the data received from the network into the buffers attached to the buffer descriptor ring and sets the status to FULL. In above diagram the filled buffers have been grayed out. At the same time, it also raises an interrupt to the core—a hardware interrupt. This interrupt eventually triggers the invocation of the NETIF_RX softirq which is responsible for dequeuing the buffer from the ring and replenishing it with a new empty buffer. These buffers then are handed over to a higher layer for further processing. In the case that there are no free buffers available, the incoming packets are discarded.

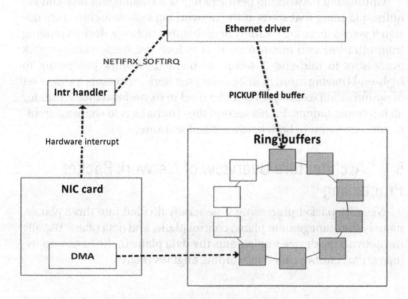

In the second stage, the buffer is processed by the Linux network stack where the packet is processed by different protocol layers such as IP, IPsec, TCP, or UDP. The driver invokes the standard Linux network stack API to hand over the packet to the kernel. Initially, the IP layer completes integrity checks on the IP header. Once the packet is found to be valid, it is then checked against the forwarding database to see if it should be forwarded to another hop or handed over to some application running on the system. The application could be listening on a TCP or a UDP socket in which case the packet would be picked by the application using an interface.

In the third and final stage, the data is received by the application from a socket's receive buffer and is copied to user space memory with the help of standard Linux interfaces like *struct iovec* and other related calls like copy_to_user. Based on the protocol used for communicating with user space, processing may also start in the context of process dequeuing of the buffers. However, if packets are targeted for forwarding to the next hop, all processing from dequeuing the buffer from the receive buffer descriptor ring to finally putting the packet back into the transmit buffer descriptor ring for transmission, is done in the same context as NETIFRX_SOFTIRQ.

5 Networking Performance Optimization Techniques

Optimizing networking performance is a challenging task and requires planning that starts at the networking system architecture design level. As there are a variety of embedded network devices ranging from ultra-low-end microcontrollers to low-end single-core network processors to mid-end 2–4-core embedded network processors to high-end (having more than 24 cores) network processors a single set of optimization techniques can't be used to fix performance issues for all implementations. In this section the emphasis is to share an architectural overview of both hardware and software.

5.1 Architecture Overview of Network Packet Processing

Network packet processing is generally divided into three planes, namely the management plane, control plane, and data plane. Usually the networking device implements the data plane in the hardware using various entities called forwarding engines (Fig. 3).

Management plane				
Control plane				
Control engine 1	Control engine 2	Control engine 3	......	Control engine m
Data Plane				
Forwarding engine 1	Forwarding engine 2	Forwarding engine 3	......	Forwarding engine n

Fig. 3 Packet processing components.

These forwarding engines can be designed for dedicated processing, for example, one forwarding engine doing the L3 forwarding and another engine dedicated to packet inspection and the firewall. Similarly, there can also be a forwarding engine dedicated to IPsec processing. These forwarding engines usually maintain flow contexts to provide respective services. Each of the forwarding engines caters to multiple flow contexts. These flow contexts are usually programmed by the control plane.

The creation of a flow context can be done either on need basis; when a flow enters system and the forwarding engine doesn't have a flow context entry. The packet is given to the control plane, which initiates the process of the creation of a flow context entry in the forwarding engine. Once the flow context is created, the next packet is served by the forwarding engine.

Another way to create a flow context is to initiate the process even before the packet enters the system. In this case, the control plane creates flow contexts and whenever a packet enters it finds the matching flow context and is handled by the forwarding engine without intervention of the control plane.

The information stored in the flow context varies based on the service it is catering to. For example, in the case that the forwarding engine is catering to the L2 bridge or switch service, the flow context usually stores the destination MAC address and outbound port for that flow. The flow context usually has a lifetime associated with it, meaning that if no packet matching a flow enters the system, the flow context will be deleted, post lifetime expiry, to make space for a new flow. Whenever a matching packet for a flow enters the system, the flow context timer is refreshed, allowing the entry to be alive.

The control plane has multiple control engines which are usually implementing protocols responsible for interacting with other networking endpoints. The control plane is responsible for handling the exception packets coming from the forwarding engine. The exception packets are usually packets for which a matching flow context is missing in the forwarding engine. On receiving an exception packet, the control plane initiates the process of flow context creation in the forwarding engine.

The outcome of the interaction between the two entities eventually triggers flow context creation in the forwarding engine.

5.2 Network Packet Processing Implementation

Now that we have learned about the basic building blocks involved in network packet processing, let's consider the methods that can be used to implement network packet processing solutions. Since the processing of networking traffic involves multiple components, software can be designed to work either to have all the processing done in a single core or break packet processing across multiple cores, such that each core does a certain part before handing over the packet for further processing to the next core, and so on.

A programming model where packet processing is divided into multiple components being executed to different cores is called pipeline processing. This kind of implementation has severe implications on aspects of performance. With such an implementation, the overall performance of a system is limited by the slowest component involved in the complete processing. Also, this kind of implementation induces lots of delays in packet processing. Complexities may arise in cases where there is a mismatch between processing modules and the number of cores available. This would require the identification of components that can be further broken down or can be merged, based on core availability. Identifying and breaking or clubbing the components is not an easy task in most instances.

The other model involves complete packet processing done on a single core. This processing model gives flexibility in terms of taking advantage of the multiple cores available in the system to deliver best performance. However, even in this case, based on the nature of traffic tweaking, there is a requirement that packets belonging to a particular flow are processed by a particular core, otherwise packets belonging to a particular flow may start going out of order.

5.3 Considerations for Optimized Network Packet Processing

Network packet processing can be optimized by having a dedicated hardware component, wherein complete processing can be off-loaded. The off-load to hardware can be controlled by software. Whenever a new flow enters the system, the control plane programs the underlying hardware with details about the flow, such that going forward the hardware block can autonomously handle the packets. A high-level diagram showing such an arrangement is provided below.

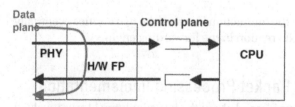

Such dedicated bits of hardware are usually called network coprocessors or packet processing engines. These can be designed in such a way as to be programmed via standard interfaces or may provide a proprietary interface. In an advanced multicore network processor there could be multiple hardware blocks handling a variety of traffic. These hardware blocks are primarily required for two reasons: first, they help in off-loading packet processing otherwise done on the CPU; and second, they help in achieving line rate without consuming CPU cycles. These acceleration IPs are usually dedicated to a specific task— one IP only doing network-related processing with another involved in security-related processing.

Another approach to achieve optimized network packet processing is to have a software module designed to imitate the dedicated hardware coprocessor. What this means is that the software module would provide APIs that can be called by the control plane to off-load flow-related information.

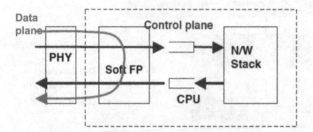

Based on this information the software module indigenously starts processing packets. This avoids lots of checks that every packet undergoes if processed by the normal flow in Linux. Based on the capabilities provided by this software, it needs to register for hooks at different points in the Linux stack. For example, if the module provides only firewall/router functionality, it needs to be aware of any flow creation or any firewall rule change happening in the Linux subsystems. Similarly, if the module is also providing IPsec-related processing, it registers hooks in the Linux XFRM infrastructure to be aware of the changes happening in the security policy database (SPD) and security associations (SA). In the next section we discuss at a higher level the implementations used in actual deployments to achieve the goal of fast packet processing.

5.4 Application-Specific Fast-Path (ASF) for Linux

This section captures some high-level detail about ASF implementation for Linux by NXP. The intent of ASF is to accelerate data plane packet processing for the most commonly used functionalities, such as IPv4 forwarding, NAT, firewall, and IPSEC. Here, control plane processing is still handled by the Linux networking stack, while fast data path packet processing is completed in ASF. The figure below gives an overview of the layered architecture of ASF.

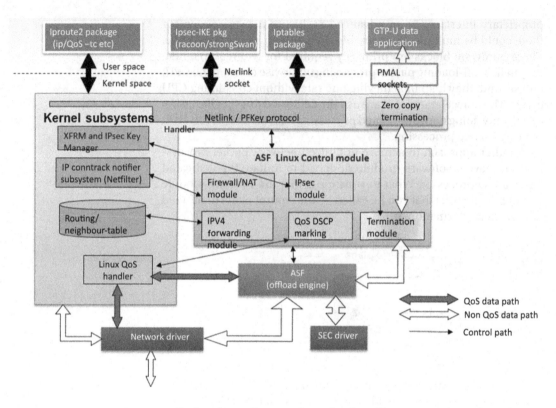

In the above diagram the red-colored boxes constitute the software module providing the packet acceleration capability, in its entirety called the ASF. So, there is a control logic which sits in the Linux network subsystem and there is fast-path processing logic which can either be in the form of a loadable kernel module or can be built as a static module.

All the packets entering the system are forwarded from an Ethernet driver to the ASF module. In the ASF module, a given packet flow (based on L2/L3/L4 header information) is checked in ASF lookup tables, which become populated through the ASF control module. If a matching flow is found for a received packet, it is processed as per the action configured for that flow and is forwarded to the configured egress interface or terminated locally.

ASF can cater to various use cases ranging from IP forwarding, firewall, NAT, QoS, and IPSec. Let us consider an example and delve into the details of the IPsec use case. For IPsec, ASF control module registers hook into the IP XFRM framework of Linux, via Key Manager, to receive notifications about any updates happening in SPD or SA databases. Whenever a new security policy or a new security association is created in Linux, an ASF control module callback notifier function is invoked to make any required updates. This update notification triggers an update of the database maintained in the ASF module.

This notifier is invoked for every addition/deletion/modification to the database. The following is a snippet showing how to register for an event notification in case of any change in the database.

```
static struct xfrm_mgr fsl_key_mgr = {
        .id                                          =    "fsl_key_mgr",
.notify        = fsl_send_notify,
        .acquire                                     =        fsl_send_acquire,
.compile_policy = fsl_compile_policy,
        .new_mapping                         =        fsl_send_new_mapping,
.notify_policy                       =        fsl_send_policy_notify,
.migrate                             =        fsl_send_migrate,
};
..........
xfrm_register_km(&fsl_key_mgr);
```

The functions specified in the *xfrm_mgr* are the notification functions which are invoked on any state change. These notification func-

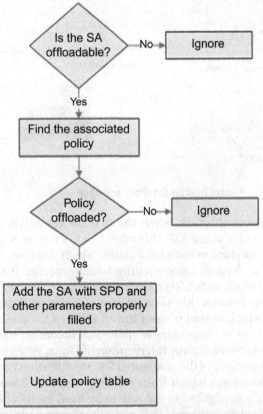

Off-loading the SA/SPD in the ASF.

tions will eventually populate the flow database used in the ASF for the lookup of flows for incoming packets. The following figure is a flow chart indicating how the SA/SPD is updated in the ASF.

The diagram below indicates the flow of packets entering the ASF and various modules participating in the creation of flow entries in the ASF database for the IPsec use case. This flow considers that IPsec is dynamically configured using the IKE tool.

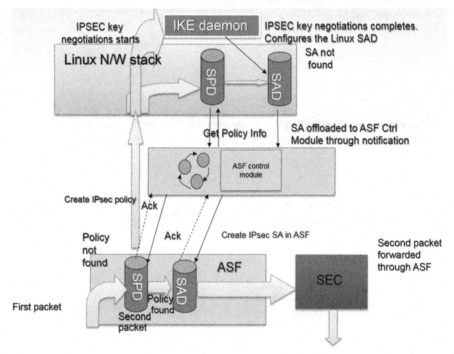

Packet flow for the IPsec use case.

In the above diagram, when the first packet enters the system there is no policy in the ASF SPD table and the packet is given to the Linux network stack to fetch information about this flow. In this case the packet undergoes some security transformation, it triggers the IKE process to set up the SAs used for this flow. Whenever there is an update in the Linux XFRM security database, the notification comes to the ASF which in turn triggers the addition of the flow to the ASF database. Once the flow entry is created all incoming traffic is handled by the ASF without any intervention of Linux. Whenever there is a soft lifetime expiry of the configured SA, a notification is given to the Linux network stack which starts fetching new SAs. These new SAs are eventually configured in the ASF to be used for incoming traffic. Such a lightweight implementation usually gives a multifold increase in IPsec throughput compared with native Linux IPsec processing, depending upon the packet size.

Another such implementation recently added to Linux is XDP. The next section provides an overview of XDP.

5.5   eXpress Data Path (XDP) for Linux

XDP, popularly known as eXpress Data Path, has been a recent addition to the Linux kernel, facilitated by the Linux community. Initially, there were doubts and concerns about adding this to the mainline kernel, however, after lengthy discussion it was agreed to include it within Linux. This section provides a high-level overview of XDP. Most of the information included here is available online—this is an open-source utility which is part of the Linux kernel. This has been developed as a part of the IO Visor Project (details about the project are available at https://www.iovisor.org/technology/xdp).

To start, let's discuss what is behind the motivation for having such a solution added to Linux. In Linux, the network packet processing path has a significant overhead due to the generic nature of the software. In the general network packet processing path of Linux, when the packet enters a system it needs to be attached to a sk_buff which is a core entity traversing various layers of network stack before deciding on the fate of the packet—either to hand it over to some user space application, forward it, either as it is or post some transformation, or discard it, depending on system configuration. Due to all such overheads, the overall performance of the network data path is very slow when using a vanilla network stack. However, most of the time this repeated processing is not required for all the frames of a particular flow. There have been demonstrations by various implementations that the performance of the network data path can be enhanced significantly, as measured against standard Linux network performance. One such implementation is covered in the previous section under the Application-Specific Fastpath. However, there are other implementations in user space as well, like DPDK, which show that network packet performance can be enhanced multifold compared with standard Linux. All these different implementations triggered a need to have an implementation which can demonstrate good performance with Linux network stack.

XDP is a specialized programmable application that can deliver high performance for networking workloads in the Linux network data path. It introduces hooks in the software stack at the lowest level, basically the Ethernet driver level, to pull packets. These packets are processed in the XDP instead of being given to the Linux network stack. Due to such a design, there is no need to create an sk_buff and complete network stack processing can be avoided. The figure below (extracted from: https://www.iovisor.org/technology/xdp) is a high-level overview of XDP architecture.

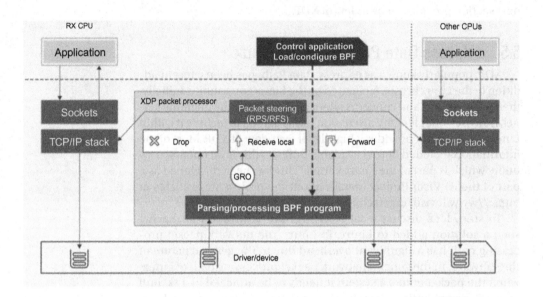

Since, XDP extracts the frame at an almost bare-metal level it is ideal for speed benchmarking without compromising programmability. Also, adding new features will not impact the existing packet flow of the Linux kernel. Another benefit is that it is not disintegrated fully from the Linux network data path, meaning that on a need basis it can either handle the packet independently or can enqueue the packet to the Linux network stack. It is also not a replacement for the existing TCP/IP stack but rather is an augmentation to the existing stack, working in concert. It also utilizes various performance techniques, such as lockless and batched I/O operations, busy polling, direct queue accesses, page recycling to avoid overhead of allocation and freeing up, avoiding overhead of sk_buff maintenance, optimization of lookups of flow state tables, packet steering, flow hashing, and NIC off-loads. The NICs are expected to have multiple queues, support for common offloads like checksum offloads, TSO and RSS to name a few. To avoid locking, it assigns one CPU to each NIC RX queue and can work in either busy polling mode or interrupt mode. The packet processing for XDP is usually done by the BPF program which parses the packets, does lookups, and performs packet transformations. This will return the action to be taken on such frames.

Now that a high-level overview of two different implementations has been shared, the user is advised to try to benchmark the system with XDP and without XDP for the IP forward use case. While setting up the IP forward use case, the expectation is to configure the packet streams destined to be forwarded from one network port to another and look for differences between the two use cases.

5.6 General Techniques for a Better Performance Using Efficient Resource Utilization

Apart from the techniques mentioned above there are also generic techniques which improve resource utilization and eventually overall system performance. The most widely used concept in this domain is *RSS*, popularly known as *Receive Side Scaling*. This concept ensures that the multiple packet processing queues in the system get evenly loaded, based on some distribution attribute, and eventually processed by multiple CPUs. This helps in distributing the packet processing load evenly across multiple CPUs. However, this solution only works well when there are many flows leading to an even distribution across cores. In the case that there is only a single stream of data, the RSS will be unable to distribute traffic evenly across multiple cores.

There is another concept called *RPS* (*Receive Packet Steering*) which is generally implemented in the software. This has multiple advantages, including that its usability is hardware independent and that it helps in adding software filters, giving flexibility for filtering packets in terms of a variety of parameters. This method primarily relies on interprocessor interrupts to distribute traffic. RPS is beneficial in cases where the number of queues is less than the number of available CPUs and when the CPUs involved in packet processing belong to the same domain.

There is another similar concept called an *RFS* (*Receive Flow Steering*) which again is mostly implemented in the software. This allows traffic coming to a particular core to be steered away to another core running the required application for processing incoming traffic, thereby helping distribute traffic across various cores. With the help of RFS, the processing of even a single stream of data traffic occurs on multiple cores.

Apart from the techniques mentioned above, software programmers need to ensure that the code they write is highly optimized and uses cache in an optimal way to ensure minimal cache misses during packet processing. The idea is to exploit spatial locality to the fullest to avoid cache misses as much as possible. The following example shows how to rewrite code to have better spatial locality and thus a smaller number of cache misses, eventually leading to better performance.

Another aspect to keep in mind while planning optimal usage of memory is that cache sizes are usually limited. Moreover, the closer the memory is to the CPU the better speed it provides, however, the costlier it will be. The following diagram captures details about different

Example

This example shows how adding the array elements in two different ways can lead to different behavior with respect to cache utilization. In C, arrays are allocated in row-major order which means each row element is contiguous. This means if each element of a column for a particular row is accessed it uses its spatial locality advantage leading to fewer cache misses compared with the case where the elements are accessed in reverse. The following figure captures data from the cache miss for the abovementioned use cases, considering that the block size is 4 bytes. From the diagram it can be deduced that the miss rate changes significantly because of the way elements in an array are accessed.

```c
int sumarrayrows (int a[M][N])
{
    int i, j, sum = 0;

    for (i = 0; i < M; i++)
        for (j = 0; j < N; J++)
            sum += a[i][j];
    return sum;
}
```
Miss rate = 1/4 = 25%

```c
int sumarraycols (int a[M][N])
{
    int i, j, sum = 0;

    for (j = 0; j < N; J++)
        for (i = 0; i < M; i++)
            sum += a[i][j];
    return sum;
}
```
Miss rate = 100%

aspects like speed versus cost and the standard supported sizes of memory at different distances from the core. So, it is very challenging to optimize code such that the critical and most frequently used code resides in the L1 cache most of the time, for best performance.

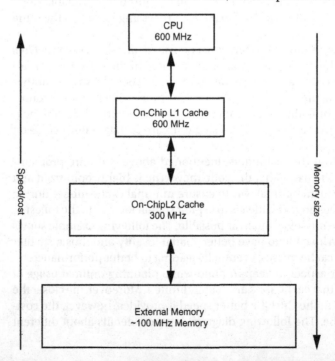

6 Case Studies: Covering Microcontrollers to Network Processors

6.1 IoT Subsystem

The Internet of Things (IoT) was initially coined in a Finnish article (https://en.wikipedia.org/wiki/Internet_of_things#cite_note-TIEKE-28) as "... *an information system infrastructure for implementing smart, connected objects.*" With an exponential increase in number of sensors, packaged well enough to be plug-and-play and with integrated networking, IoT as a technology is rapidly gaining popularity. There has always been a requirement to connect with sensors and consider the information they provide—as far as back the early 1900s, large railway signaling systems were commissioned across the United Kingdom. However, the need to assimilate data to make smarter decisions is a new phenomenon, not possible in the past because of a lack of data processing points.

Initially, sensors were stand-alone devices—a temperature sensor would provide discrete data about temperatures at specific locations. Such a sensor required human intervention to be read, and its data needed to be copied over manually to other machines in order that they could make decisions. For example, let us consider a large power house using steam for power generation. Turbines and steam boilers are two of many important systems, which have inherent linkage—steam boiler output is input to turbines. As long ago as 1950s, enough sensors were available to capture data from various points across the boiler and turbine. But, it required human intervention to read and make logical decisions on that data. Based on steam output, turbine input was manually regulated.

Then, with the use of microcontrollers, one was finally able to connect directly to data, and data could be directly connected to other devices. For example, microcontrollers meant that smart decisions could be made about whether boilers were underheating or overheating and consequently make decisions about releasing excess steam. Microcontrollers read data from temperature sensors and made sense of whether servo motors, attached to boiler valves, should be opened or closed. All this was possible because microcontrollers could make sense of larger sets of data, beyond what simple temperature sensors could handle, and then find a correlation before producing another set of data on which an action could be based.

Multiple microcontrollers could be chained to create a workflow based on this stream of data. Once advancements in memory and SoC were achieved, it became possible to create larger streams to make better or "smarter" decisions. Thus completely automated power stations became a contemporary reality—one in which millions of sensors could feed their discrete data into an array of controllers which in turn could cascade into another set of controllers leading to an output interpreted as processed data or information worthy of decision making.

Power plants, like the steam turbine example used above, are capital intensive industries where enough capital is available for standalone research for sensors and their interaction to gradually move towards automation. Such research over the years moved them towards earliest example of "IoT Networks." Closer to home, cars too have been through a drastic transition over the past few decades. Automotive domains were always a focal point for combustion technology in terms of generating power—almost everything else represented a mechanical engineering achievement. However, with a network of sensors and controller arrays, it is now possible to set aside the mechanical engineering marvel that is the car, and shift our focus to make it a device with which humans can interact on a daily basis through smarter decisions, making travelling more efficient and safe.

Presently, cars are huge compute nodes—processing millions of sensor inputs across their infrastructures, making critical decisions about braking, and less critical decisions about specific climate controls for users. In fact, networks of sensors expand well beyond the physical zone that is the car. Cars now interact with various surrounding infrastructure items such as other cars or even buildings and fixtures. With enough processing power, a car should eventually be able to understand road conditions by sensing the presence of all other traffic. A car should be able to communicate with a traffic light and decide to reduce its engine output for a specific duration to conserve fuel. It should also be possible for a car to interact with fixtures like building to understand the density of incoming traffic at a blind turn or crossing. It should also be possible for a car to understanding the traffic or environmental conditions that lie ahead on its journey—snow, rain, and air temperature—by communicating with cars coming from that direction, having previously encountered those conditions.

6.1.1 Choosing the Right Device

The present-day market has numerous devices spanning use cases from homes to industrial plants. However, when it comes to choosing the right device, or SoC, various factors come into play, such as compute power, power consumption, peripherals that can be mounted, form factor, and connectivity options. Based on use case, it is important to define the values of each parameter before choosing a prototyping and eventual product device. Ease of programmability is another criterion which helps reduce time to market.

For the purpose of this case study, an NXP i.MX 1060 RT SoC package was chosen. This SoC is a microcontroller platform based on an ARM Cortex M7 core, having the capability to run real-time loads (if backed by adequate OS capabilities). Being a microcontroller board has advantages and disadvantages—the primary advantage being that it has a lower cost than microprocessor-based SoC packages; the primary disadvantage being that it has limited support for generic application stacks. The i.MX

1060 RT can offer a significant number of applications—being a real-time board, it is a platform capable of controlling devices such as industrial robots or servo motors. Having a low power consumption means that it can be used as an edge device for "always on" use cases. NXP MCUXpresso IDE provides an easy and convenient way to enable the board and deploy applications in the least possible time.

Another processor chosen was the LS1012, which is a low-end ARM-based network processor from NXP. This SoC is a low power communication processor offered under the LayerScape QorIQ family of ARM-based processors. It uses a single ARM Cortex A53 core which can be clocked as high as 1 GHz, with a hardware packet forwarding engine and high-speed interfaces to deliver line-rate networking performance in an ultrasmall size envelope with a typical power dissipation of 1 W.

In this use case the standard release code from SDK was used without doing any local optimizations and the same workload was offered to both SoCs. Their performances were measured. The workload was an IPERF client/server application running a performance benchmark for a TCP use case.

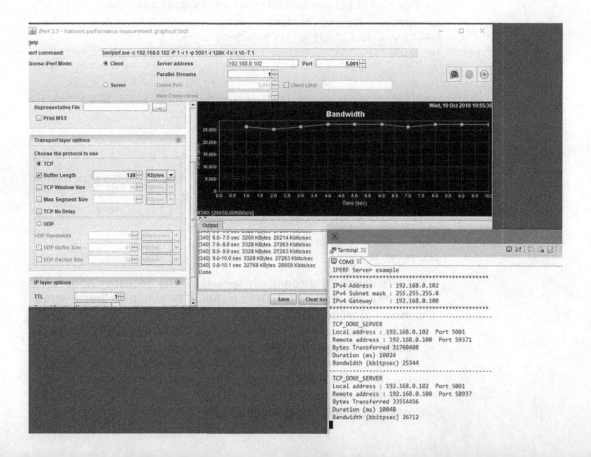

462 Chapter 12 NETWORKING SOFTWARE

First, the data was shared with the i.MX 1060RT SoC running at 600 MHz. The snapshot below captures the data which clearly shows that the maximum performance achieved for an iperf application offering the TCP workload is around 27 Mbps on the i.MX RT1060. Please note that the performance measurement was done for the stock released code for the SOC without any local changes being made to it. Performance improvement or degradation may occur if changes are made to the code.

The same workload was run on the LS1012 network processor running at 1 GHz. This SoC was able to saturate the 1-Gbps link with ease.

```
IPERF Server
------------


Ubuntu:~# iperf -s
------------------------------------------------------------
Server listening on TCP port 5001
TCP window size: 85.3 KByte (default)
------------------------------------------------------------
[ 4] local 192.168.3.11 port 5001 connected with 192.168.3.1 port 58882
[ ID] Interval     Transfer    Bandwidth
[ 4] 0.0-10.0 sec  1.10 GBytes   939 Mbits/sec
Ubuntu:~#



IPERF Client
------------


Ubuntu:~# iperf -c 192.168.3.11
------------------------------------------------------------
Client connecting to 192.168.3.11, TCP port 5001 TCP window size: 85.0 KByte (default)
------------------------------------------------------------
[ 3] local 192.168.3.1 port 58882 connected with 192.168.3.11 port 5001
[ ID] Interval     Transfer    Bandwidth
[ 3] 0.0-10.0 sec  1.10 GBytes   942 Mbits/sec
Ubuntu:~#
```

To conclude, theoretically speaking if the frequency of the i.MX 1060RT was clocked to 1 GHz, it should be able to deliver a performance of around 41 Mbps, which is still much lower than the performance that can be achieved on the LS1012 platform. Hence, it can be deduced that core frequency cannot be used as a benchmark when deciding which SoC to use for a particular use case. It can be seen from the above case study that even though both the processors are ARM-based low-cost/power products, if the requirement is to design a networking product which can deliver high networking performance with low cost then the LS1012 is preferable to the i.MX Rt1060. However, if the use case requires networking capabilities along with other capabilities like connectivity with external sensors in order to collect data along with an interface to connect to external input devices like a camera with a real-time processing capability, then the i.MX RT would be the obvious choice.

Exercises

Please create programs for the following problem statements:
1. Create two namespaces, net_ns1 and net_ns2. Configure the system to allow forwarding of packets from one namespace to another.
2. Design a packet sniffer which sniffs all packets coming on a network device.
3. Design a packet sniffer using raw sockets which will sniff packets for a particular vlan on a network device. Assume that there are three vlans configured on eth0, namely, eth0.10, eth0.20, and eth0.30. Write a sniffer using raw sockets to capture the packets coming on eth0.20.
4. Write a program to print the state change of an NF_CONTRACK entry. The program shall print the details of the entry created, deleted, or updated for each flow.

Further Reading

[1] A. Carrion, Very fast money: high-frequency trading on the NASDAQ, J. Financ. Mark. (2013).
[2] DPDK | About Us. (n.d.). Retrieved from http://www.dpdk.org/about.
[3] Open Data Plane. (n.d.). Retrieved from https://www.opendataplane.org/.
[4] V.J. Steven McCanne, The BSD Packet Filter: A New Architecture for User-level Packet Capture, Lawrence Berkley Laboratory, 1992.
[5] M.C. Wenji Wu, https://people.cs.clemson.edu/~westall/853/tcpperf2.pdf, 2006. Retrieved from https://people.cs.clemson.edu/~westall/853/tcpperf2.pdf.
[6] Linux Kernel Issues in End Host Systems from https://people.cs.clemson.edu/~westall/853/tcpperf2.pdf

13

INTERNET OF THINGS

Mark Kraeling*, Michael C. Brogioli[†]

*CTO Office, GE Transportation, Melbourne, FL, United States [†]Polymathic Consulting, Austin, TX, United States

CHAPTER OUTLINE

Software Engineering for Embedded Systems. https://doi.org/10.1016/B978-0-12-809448-8.00013-8

1 Introduction

The Internet of Things (IoT) is a vast topic that started to hit its stride in the 2010s. The phrase itself certainly has been thrown around a lot—and of course, when mentioned, it is something that gets a lot of attention and discussion. Simply put, it is any device that is connected to the Internet. It started with the concept that devices have useful information that could be offered to a larger cloud-based system. Before the IoT revolution truly set in this information was often relayed through a "smart" device that was connected and then sent to the cloud.

With the advent of both inexpensive and more accessible communications paths the focus shifted to the devices themselves, which send information to the cloud without having to use a relay node. The IoT still has the same communications paths but is now more focused on having devices able to make decisions locally as opposed to in the cloud. Often IoT devices use Internet communication paths to get the data they need to make smarter decisions or to send results of their calculations (as opposed to a giant stream of data) to the cloud or other devices.

This chapter discusses various IoT concepts, its history, and its progression. It then discusses factors associated with the technology and architecture for software when developing or using an IoT product.

1.1 Definition

The definition of an IoT device is one that has the intelligence to use a communication path connected to the Internet or a private network. An IoT device is sufficiently complex—a simple discrete device measuring light levels over a pulse width–modulated output does not qualify. IoT devices are required to have a communications and security stack to communicate.

Not only does an IoT device have the option of communicating with a centralized back office, it also has the capability to communicate with other IoT devices. Such devices can be similar or even the same in the case of a set of devices working together to produce a given result. Such devices can also interface with *humans* as opposed to just other *things*.

Any device that has an on-off switch and the capability of running software that enables connectivity can be considered an IoT device.

1.2 Examples

From fitness bands and watches to cellular phones and appliances any device that has something to say or something to listen to over the network becomes an IoT device. The power, however, is in having the IoT devices work together to drive an outcome that is useful. The following are a few examples of how IoT devices can be very effective.

A person sets an alarm for 6:00 a.m. The alarm is a connected IoT device that either communicates with a smart home server or is the smart home server. When the alarm is set it communicates with the coffee maker (another IoT device) to make sure water is in the reservoir and a filter with coffee is in the tray. If not, the person is informed when the alarm is set, otherwise no notification is required. At 5:50 a.m., 10 min before the wake-up alarm, the coffee pot is turned on automatically, so the coffee is ready. Over a period of time the coffee maker measures the amount of time that passes from when the coffee is done to when the coffee pot is lifted. It then makes the decision that, even if it is turned on through the alarm clock, it should wait a certain amount of time before brewing so that the coffee is fresh. Information such as who set the alarm clock can also be conveyed, just in case one person becomes ready for their coffee faster than another. This example shows there are many ways information can get from one IoT device to another to help make smarter and more useful decisions (Fig. 1).

In a separate example a person has a smart fitness device. Throughout the day the fitness device measures how active the person is and when they are the most active (steps, flights of stairs, calorie expenditure, etc.). Upon returning home the device communicates with other family members' devices to not only relay details on its performance data but to inquire about their data. The fitness device can then communicate with a cloud-based fitness server or decide locally to provide recommendations for family fitness.

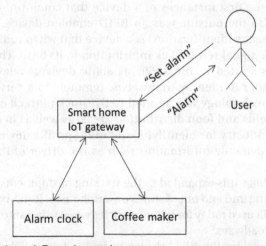

Fig. 1 Smart home IoT user interaction.

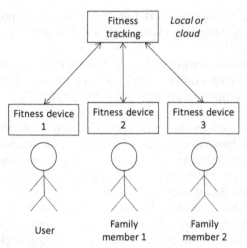

Fig. 2 IoT fitness trackers.

Perhaps a 1-mile walk will be recommended and sent to the smart home server so that if someone asks for a joint activity it has one already prepared (Fig. 2).

2 History and Device Progression

This section discusses the history of IoT and the progression of migrating cloud-based decisions to a smart device type of architecture.

2.1 History of Internet of Things and Cloud

One of the first instances of a device that communicated over a network to the outside was an RFID-enabled device. RFID (or radio frequency identification) is a device that when radiated with a source RF signal sends back information to its base. The devices themselves started in the 1960s as static devices, relaying back an ID number to identify themselves remotely to a server. In the 1980s this technology was applied to livestock to track movement between fields and food distribution areas, as well as in the transportation industry for identifying not only specific device ids, but also more dynamic information such as the driver of the vehicle (Fig. 3).

In the 1990s this expanded to the tracking of shipments and even goods coming into and out of stores or warehouses. RFID-based technology is still used today for billing vehicles when using automated toll booths on roadways.

Many consider the RFID device to be the first IoT device—as it communicates information over RF when requested. The usefulness

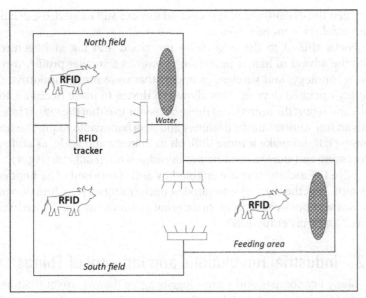

Fig. 3 IoT in agriculture.

of RFID has increased from simple identification of how assets have expanded to tracking and understanding product or assembly line flow. This led to thinking of devices as being smarter with the goal of getting devices to make more and more decisions.

Progression in the device space was overlaid with progression in the cloud and server space. Starting in the late 1980s there was a drive to push all the data that you could to a centralized server. Once there decisions could be made based on the data and then pushed back out as "decisions." These data could then be accessed from the Internet to see near real-time data from devices or reports depending on the need to be addressed. Changes in the algorithms or how to process the data could be made in one place along with the ability to scale to more servers or more disk space. The disadvantages of a cloud-centric architecture is the bandwidth cost of sending the data, handling situations where the path from device to cloud is broken (in and out of coverage), and latency and response time. Cloud-oriented architecture for IoT reached its peak in 2001.

Starting in 2001 more fog-oriented architectures began to appear. These configurations pushed the centralized server that held all the data to regional servers located closer to the data—typically on the same subnetwork. Data could then stay in country when looking at a global view or stay at a site or location instead of being pushed to a central server. Decisions could then be made at a more local level with decreased latency. Fog-oriented architectures brought about more data paths that had to be managed. However, problems with connectivity between the decentralized servers and devices still existed, especially with wireless or mobile devices.

Focus shifted to the edge from the cloud and fog architectures with the advent of higher processing power in a smaller profile, wireless technology, and wireless protocols that were more supportive of battery-operated devices. This allowed a device to make decisions locally and report the outcome of those decisions to either the fog or cloud. This architecture minimized latency and data bandwidth requirements. However, it did make it more difficult to manage and scale, as adding processing and storage resources at the edge is often difficult (Fig. 4).

Edge IoT architecture are enabled by and comprised of embedded systems. Whether battery-operated or performing a useful function on its own without needing a lot of external connectivity the device itself is well and truly embedded.

2.2 Industrial Revolutions and Industry of Things

Many economists and technologists agree that we are at the forefront of the fourth industrial revolution and IoT is a large factor in this (Fig. 5).

The first industrial revolution came about when we started harnessing energy and began large-scale deployment of mechanical devices in the late 1700s. The first scaled factories arrived and goods were produced to make lives easier—whether it was a mechanical loom or oil lamp.

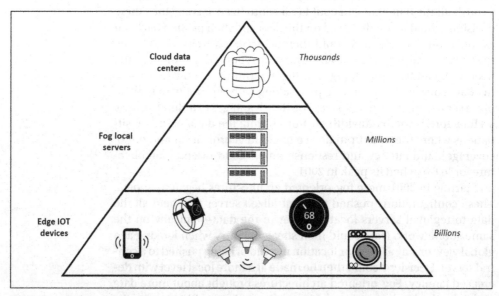

Fig. 4 Progression of data size.

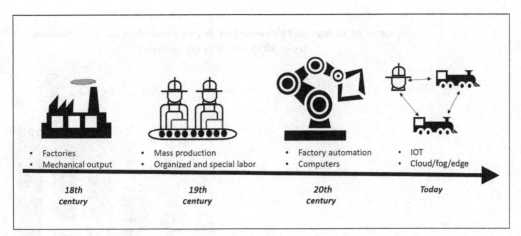

Fig. 5 Evolution to IoT.

The second industrial revolution started in the late 1800s when labor in factories became more specialized and organized. Electricity was also harnessed to create better products and provide more features in the home. Mass production also started in earnest—from locomotive factories to vehicle manufacture.

The third industrial revolution started in 1960. Throughout this period computers became increasingly important with automation in factories doing repetitive and defined work. Computers also became smaller and increasingly faster, reaching the point where cellular phones became more than just phones.

We are currently in the fourth industrial revolution. Most agree it started around 2010 with a renewed focus on cyber-physical systems. A great deal of investment has gone into making devices not only connected, but also smarter and able to do more complex tasks at the source. Machine learning also comes into play here—where devices get smarter and adapt to the tasks being performed to perform them better. All these traits are encompassed in IoT: devices being connected, devices being smarter, and devices performing more tasks locally. The fourth industrial revolution is often termed the "IoT Revolution."

2.3 Connected Devices

The premise underlying IoT is to have more and more connected devices. Fig. 6 shows the number of connected devices (in billions of devices) over time according to a statistical company.

As the number of devices increases, the importance of IoT expands. There are many applications for IoT in this connected device space and many enabling technologies are required.

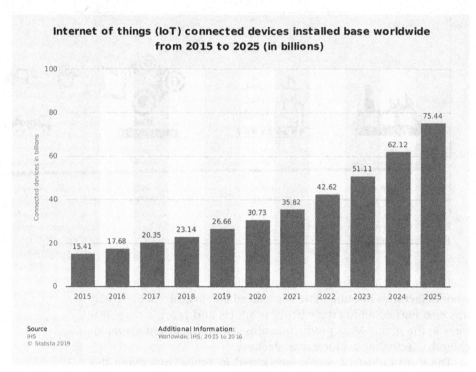

Fig. 6 Number of IoT-connected devices worldwide (https://www.statista.com/statistics/471264/iot-number-of-connected-devices-worldwide/).

3 Applications

Each market segment for IoT has its own unique requirements and needs. There are needs that are common across all market segments with respect to IoT such as data security and data latency. This section provides various IoT requirements for market segments and a few use cases where IoT gives value.

3.1 Factory Automation

One of the advances in the IoT space involves manufacturing and factory automation. Whether dealing with a small-scale facility that performs remelts on small batches of aluminum or a subassembly and vehicle production line, IoT is everywhere.

Typically, automation in the manufacturing environments of the past followed more of a cloud-like scenario when carrying out tasks. Data were collected from the entire operating facility and then tasks were run in the back office to alter manufacturing flow or reports were made on how the operation was running. Devices attached to the machine simply collected data and forwarded them to the back office—but did not communicate with each other.

This data architecture model has shifted. Now devices can talk to each other removing much of the data latency so decisions can be made quickly, a functionality the former server-based architecture could not support. For example, individual spray nozzles for coolant on an aluminum hot rolling mill can now communicate the volume of spray they are emitting to each other. If one isn't performing as well as it should the two adjacent spray nozzles can pick up the slack until proper maintenance is performed. This is just one simple example from an entire network of devices and sensors where IoT with its connected devices makes a big impact in manufacturing.

3.1.1 Use Cases

The following use cases provide examples of how IoT can be used in a factory or a manufacturing setting. They are meant to help illustrate how IoT could make a difference in this type of environment.

3.1.1.1 Overhead Crane in a Factory

This use case is for an overhead crane that runs along a steel track, so it can move back and forth in the X direction (the Y direction for the entire crane is negligible since it is on a track). The crane is used to pick up loads from one location and move them to another. The load can be shifted along the Y axis using motors so that items can be taken from one corner in the factory and placed in the opposite corner.

The IoT sensors included for this crane are:
- position sensors that determine the X and Y position of the entire crane
- position sensors that determine the X, Y, and Z position of the load being carried
- strain gauge that measures force in the cable between the crane and load
- Radar and LIDAR sensors for foreign object detection.

One use case involves using the strain gauge and the crane. Tension can be measured from the time the motor is activated until movement of the load being lifted is detected. Based on the weight of the load after it is free of the floor a measurement can be made by the IoT sensor over time to determine how much the cable has stretched. Maintenance flags can be set when outside a maintenance boundary.

Another use case for the crane could be measurement in the Y direction for the entire crane itself. There shouldn't be any—however, over time the wheels that keep the crane on the track could become worn. If more and more movement over time is detected the sensor can also flag that a maintenance inspection needs to occur.

A final use case for the crane involves sensors used to understand objects on the factory floor. Using object detection the crane can detect if the load will hit something on the floor. Even if the crane operator is commanding movement in a specific direction, because decision making can be made locally the sensors could prevent that movement from occurring or require a special override by the operator.

3.1.1.2 Aluminum Coils in a Plant

This use case involves the handling of aluminum coils produced in an aluminum rolling facility. For this use case the coils have been rolled in a hot rolling line and are waiting to be sent to the cold rolling line for further thickness reduction. There is a staging area, with an ideal "cooling" temperature range for the coil to be cold-rolled, based on alloy of the aluminum.

The IoT sensors that are included on a module placed on the spool are:
- temperature readings (ranging from outside of coil to inside)
- position in staging area (picks up positioning signals in local area).

The first use case involves monitoring the coil temperature and sending a wireless message every 5 min giving the predicted window of time that it should be cold-rolled. The alloy of the aluminum is sent to the IoT device as it is being hot-rolled onto the coil, which allows the device to determine the correct temperature for cold-rolling. The IoT device predicts the correct temperature based on its own cooling rate as proximity to other coils or its position in the staging area may be influential factors. This information is sent to the cold-rolling staging, so the coil can be pulled at the correct time and sent to the appropriate cold-rolling mill stand.

The second use case involves the positioning of the coil. The IoT device communicates its alloy and current temperature as soon as its coil is hot-rolled so that a decision can be made by the staging area controller on where to place the coil. Coils that will take much longer to cool for cold-rolling can be placed in an area where they are out of the way to minimize the number of coil movements required to "fetch" a coil for the next cold-rolling stage.

Another use case—not involving aluminum coil management—mentioned previously concerns using an IoT-based architecture on coolant spray used to cool aluminum as it is rolled. Individual coolant nozzles are used across the width of the aluminum sheet as it is rolled. The amount of spray from each nozzle can alter the thickness of the aluminum sheet as it is being rolled to adjust for variability across the sheet. If an IoT-based nozzle can measure the amount of coolant coming out vs. what is being commanded, it can try to compensate for this itself. If it cannot communicate with the other adjacent nozzle sensors immediately the spray volume can be increased to compensate (Fig. 7).

As the coil is being rolled at many feet per second quick decisions made locally are the only way this can be performed as sending data to a higher level centralized manufacturing facility computer would not suffice.

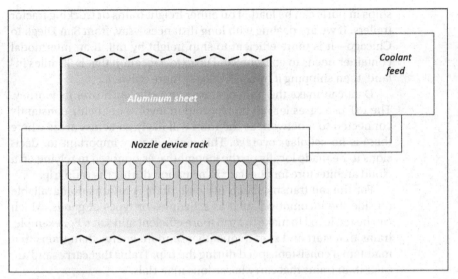

Fig. 7 Nozzle device rack.

3.1.2 Important Factors

In factory automation the important factors for IoT devices typically involve information that can positively contribute to safety, quality, and time efficiency. In the previously mentioned use cases device position is important. In enclosed factory environments constellation-based positioning, like GPS, isn't typically used so quality is enhanced as IoT devices bring in precision sensors like GPS. In addition to enhancing quality there are also many IoT device steps that can be put in place for safety. Even if the primary control path is triggered by human interaction, like movement of goods by an overhead crane, safety overlay technology can be put in place to assist the operator. This provides an opportunity to catch something that the operator hasn't. Finally, being able to shift to an IoT-based architecture enhances efficiency by enabling lower network latency and more local decision making. The example of dynamic adjustment of aluminum rolling nozzle cooling shows the importance of taking delay out of making decisions as an excellent way to improve product efficiency and quality.

3.2 Rail Transportation

Rail transportation, of course, involves using railway track to move goods or passengers from one location to another. In North America it is more efficient to move goods by rail vs. trucking over longer distances, especially when efficiency is measured using dollars spent per ton-mile. Intermodal containers that are offloaded from cargo ships in ports can be loaded on either freight trains or trucking tractor trailers. If we are dealing with long distances—say, from San Diego to Chicago—it is more efficient to ship freight by rail. If an intermodal container needs to go from San Diego to a location that is 80 miles inland, then shipping it by truck is likely more efficient.

Data can make the train operate more efficiently over its journey. The IoT use cases for rail transportation involve not being constantly connected to a network as the train will likely go across areas where there is no cellular coverage. This makes it very important for decisions to be made locally on the locomotive as opposed to relying on a cloud architecture for constant information during the train trip.

For the rail transportation use case there is information available outside the locomotive, such as schedules or types of goods, which can be retrieved to make the trip more efficient and safer. For example, trains that start and stop frequently use more energy than trains that maintain a consistent speed during the trip. Trains that carry sand are safer than trains that carry hazardous materials.

The importance of subnetworks is key to a locomotive because, even though cellular communications are not available, each of the subsystems can talk to each other. Centralized networking and wireless communications that can be shared by multiple applications can improve the uniformity of data used aboard a locomotive and hence help the data become more manageable.

3.2.1   Use Cases

The following use cases illustrate how IoT can be used on a railroad. They are meant to help illustrate how IoT could make a difference in this type of environment.

3.2.1.1   Rules-Based Decision Making

The use case for rules-based decision making involves using IoT devices to make decisions locally in the absence of cellular or other back office connectivity. For this use case there is an IoT edge device that can capture data from any of the locomotive subnets and use this information in a series of rule-based analytics. The analytics engine itself runs on board the locomotive (Fig. 8).

The IoT sensors that will be used in this use case are:
- sensors that measure the in-train forces of the railcars in a train
- throttle and braking positions used by the operator of the train.

The first use case for rules-based decision making involves collecting data on how an operator is driving the train. Measurement can be made for in-train forces because fast stretching or bunching of the train can break railcar knuckles and equipment. The throttle or braking commands could be evaluated on a locomotive locally so that

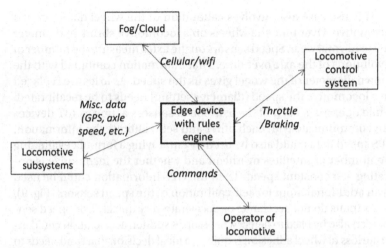

Fig. 8 Edge device IoT rules engine.

sudden changes made by the operator are flagged and information is passed back to the operator for better operation. The rules governing what constitutes inappropriate throttle or brake command progression could be created in the back office as rules and then as trains enter areas with good cellular or Wi-Fi communications the operator can grab those rules and implement them during the trip.

Other complex rules can be created on IoT edge devices on the locomotive to process data quickly and only flag areas of concern. Typically, the amount of information needed to operate a train is measured at around 1 TB per day. Instead of transferring all these data over a cellular or other network when the train enters a yard, smart rules-based decisions can be made by IoT devices on the train that only send data deemed important to be analyzed. This could be something like 10 min before and 10 min after a fault or error condition occurred, so that only the relevant data subset needs to be analyzed as opposed to sifting through a lot of data evaluated as "normal." Rules applied to a locomotive could provide immediate train operator feedback or be stored until offboard communications are restored and the data can be sent to the back office.

3.2.1.2 Smart Sensor Recalibration

The use case for smart sensor recalibration involves IoT sensors working with each other to understand the real speed of the train. Speed is measured based on wheel sensors mounted on the axles of the locomotive. The IoT sensors that will be used in this use case are:
- axle speed sensors
- GPS speed sensors.

This use case also involves calibration of the wheel diameter of a locomotive. Over time the wheels on a locomotive shrink in diameter due to normal wear. Speed sensors on the axles measure the number of revolutions of the axle over time. That information combined with the known diameter of the wheel gives us the speed. As axles are replaced on a locomotive the speed (diameter setting) needs to be recalibrated. While operating, each of the axle speed sensors acting as IoT devices can communicate with each other and self-calibrate the information. GPS speed data could also be used by employing a formula evaluating the number of satellites overhead and whether the locomotive is operating at a constant speed. Then the GPS information could be used as an additional input for self-calibration of the speed sensors (Fig. 9).

As trains do not accelerate or decelerate very quickly axle speed sensors can also be used to detect wheel slip. A sudden acceleration could indicate loss of wheel adhesion requiring a local decision that sand needs to be placed under the wheels to provide more traction. Communication between speed sensors can initiate automatic sanding, and at the same time modify the measured speed to only include axles that are not slipping.

3.2.2 Important Factors

Important factors relating to the rail transportation space using IoT devices as given in the examples include safety, local decision making, and operating efficiency. A locomotive not having consistent connectivity is a driver for all three of these factors. The operator of the train is responsible for its safe movement, but IoT devices can provide the operator better information (such as reliable speed data) and be able to flag when in-train forces as a result of throttling and braking need to be reduced. Local decision making needs to happen at the IoT edge as opposed to in a cloud-based IoT architecture because of the lack of constant connectivity. Finally, the operating efficiency of the entire train and train network is important so that goods, freight, and people

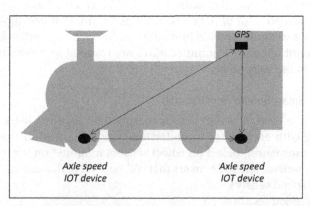

Fig. 9 Axle speed IoT device.

can be transported effectively. Using all the IoT information that is available can enable these smarter decisions.

4 Enabling Technologies

This section takes a look at the technology that is important for IoT devices. It provides a glimpse into the factors that have progressed to enable IoT technology to increase from wearable devices to IoT technology used in industry.

4.1 Processing

With the expansion of the IoT device space the processing device space has been more focused to larger volume more functionality-oriented processing. Microprocessors containing a CPU (and depending on scale a GPU) are typically used in computing but kept separate from microprocessor storage (RAM) and peripherals. Microcontrollers incorporate not only the CPU, but also the peripherals required for the application focus. This could include serial and communications interfaces, RAM, flash storage, and similar peripherals designed for the task, so incorporating several different chips for a device is not required (Fig. 10).

Using a microcontroller is often much simpler for the IoT device space than using a microprocessor. A full-featured operating system is often not required (depending on the device size), so simple programming is all that is needed to get it in correct functional operation. Having peripherals, such as discrete I/O and communications, integrated on the same die is also much simpler to get up and running as opposed to integrating separate components together in a microprocessor-oriented architecture.

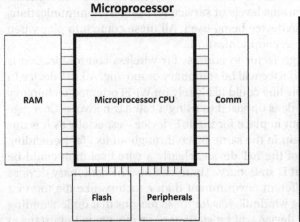

Fig. 10 Microprocessor vs. microcontroller architecture.

Developing a security feature set and plan is also easier on a microcontroller than on a microprocessor. All the buses and connections are internal to the silicon, so there is no need for specific protection to be given to them. The focus can turn to the security of the external interfaces of devices, which often come with specific recommendations that cannot be tampered with after shipping. Although there are still security concerns with the data to be sent on and off the device the hardware itself is often more difficult to hack than that of a microprocessor-oriented architecture.

The cost of a microcontroller for IoT is also typically much cheaper than designing a microprocessor architecture. Costs of RAM and flash or other peripherals are minimized as they are incorporated on the same die. There is no expense of designing a high-speed bus interface with peripherals because this has already been done and incorporated into the chip.

Processing for an IoT device needs to be scaled to the application. If the device is going to be battery operated, for instance, then running a higher level operating system, such as Linux or Windows, is not going to be appropriate. Having a more embedded-type system is going to be ideal where the device can power down to lower power consumption levels when it is not needed—or shut down parts of the system that are not being used at the time. Microcontroller architectures support this type of system much more easily and often come with drivers for lightweight operating systems that can turn on and off quickly or enable low power states.

4.2   Wireless Communications

When someone thinks of a typical IoT device it is often wireless. It has the capability of being wearable, such as a fitness device or watch, or accompanies its owner wherever they travel like a smartphone. There are also various levels of service for wireless communications depending on the device being used. All these come into play when addressing wireless communication.

The first design factor to address for wireless communication is whether the IoT device will be stationary or moving. An IoT device in a factory assembly line could likely rely on Wi-Fi oriented technology for a local network as opposed to using a cellular network. Coverage maps could be put in place for the IoT device—especially as it is understood to remain in the same place throughout its life. Depending on the location of the IoT device, clearly a wired solution could be evaluated since it is stationary. There can also be stationary devices in an entirely different environment than a factory, like the top of a power-generating windmill. It is for these reasons that understanding whether the IoT device will be stationary or moving is important to selecting the right wireless technology.

The second design factor is to understand how much data bandwidth is required for the IoT device. A cellular phone demands a large bandwidth. A 4G mobile communications standard known as the LTE cellular network is recommended if the user is downloading and watching movies or using Internet sites that transmit and receive large quantities of data. Releases of newer 3GPP and LTE standards focus on increasing bandwidth and reducing network latency, which are important for the cellular phone market. Advances in 5G will bring bandwidths measured in gigabytes per second. Having a bandwidth capable of 5 GB/s, which on the 5G roadmap is equivalent to a user watching 20 videos in 4K resolution at the same time, is not as important for smaller temperature IoT sensors or even battery-operated health wearable devices. These devices monitor physical values, and then using IoT device intelligence send periodic status or alert functions. In these types of cases a large bandwidth is not required and it is more important to be efficient with battery use than transferring large amounts of data.

Another newer technology for IoT devices involves the Cat-M1 and NB-IoT LTE standards. Cat-M1 and NB-IoT were designed to use existing LTE cellular networks but on different types of access methods that focus on devices that do not need to communicate very often. The protocols and modulation are optimized for longer distance communication and lower overhead communications in comparison with the typical 4G type of LTE. LTE Cat-M1 allows a device to transfer from one cellular tower to another, so it supports a nonstationary type of device. The effective bandwidth is measured in kilobytes per second, so if only small pieces of information are transferred it may be ideal for IoT devices. NB-IoT communications do not allow a device to seamlessly transfer between cellular towers, so it is meant for a stationary device. NB-IoT has an even lower bandwidth than Cat-M1; however, the device cost is expected to be much lower than Cat-M1 devices. The following table summarizes this functionality compared with today's LTE advanced technology available in cellphones and other advanced communications devices.

	LTE Advanced	LTE Cat-M1	NB-IoT
Bandwidth	800 Mbps	150 kbps	50 kbps
Latency	<10 ms	15 ms	5 s
Supports mobile applications	Yes	Yes	No
Device cost	$$$	$$	$
Cell plan cost	$$$	$	$

$$$ is expensive, $$ is moderately expensive, and $ is relatively inexpensive.

4.3 Wired Communications

Wired communications are often not considered for IoT devices because the more widely known use case is for a mobile-type application. However, there are plenty of IoT devices that are wired.

Often IoT devices need to communicate with devices that are not specifically "smart." This could include limit switches in a factory environment or an older RS-232 serial device that needs to be connected. The IoT edge device may have multiple communication paths to deal with when gathering information such as discrete inputs and outputs. In cases where the IoT device is stationary having a wired connection may make more sense.

Ethernet is the typical wired interface for today's stationary IoT devices. It has a given address and can transfer data to the fog or cloud based on its use. In that way it isn't much different than a wireless device, but again it will be stationary as far as its network is concerned.

IoT devices that require special timing for their communications can opt for a variety of technologies that are enabled over wired Ethernet. One such technology is time-sensitive networking (TSN) that enables the classification of Ethernet messages based on priority and queuing. The Ethernet nodes themselves that are TSN-enabled and connected to a TSN network can receive the queue and priority information for the TSN scheduler and then send its packets according to that classification and schedule (Fig. 11).

TSN provides the ability to have a packet sent and received at regular intervals, such as every second, while other Ethernet traffic that is not associated with this time slice will be delayed. This allows regular,

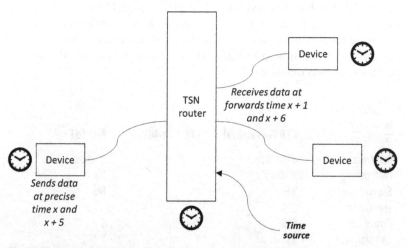

Fig. 11 Time-sensitive networking timing.

consistent messaging for devices regardless of other lower priority Ethernet traffic being processed in the system.

4.4   Power Storage

Power storage for an IoT device is another important technology consideration. An IoT device that is located near a power source and can be wired into that source clearly doesn't have to worry about the availability of power. IoT devices in factories, or automobiles, fall into that category.

Another category are devices that are not constantly connected to a power source, but instead are connected occasionally for charging. These devices call for an architecture and a storage system that meets a different set of needs of the IoT device and its user. A cellular phone or smart watch is clearly in this category—the user is expected to periodically charge both to keep them operating. It then becomes much more important to understand the specific power requirements of the use case for the IoT device. A cellular phone that only lasts an hour before it needs to be charged again would be returned to the store and classified as unusable!

The last primary category for power storage is an IoT device that will not have the opportunity of being recharged during its normal life. These IoT devices are ones that must be extremely power efficient, especially when still utilizing cellular and other communications options. They are certainly microcontroller based, with operating modes that allow them to "sleep" for long periods of time, then "wake up" to perform the required task, and then go back to a deep, low-power state. Every aspect of such an IoT device would be optimized for the power available. Current IoT devices that require this type of power storage typically perform some task every 24 h, provide a status report over wireless communications, and then go back to sleep. It is common to find devices perform this function over a 5-year time span before needing to be replaced or refurbished.

5   Internet of Things Architecture

The architecture of IoT applications can vary greatly depending on the use case. While a simple architecture may have a device at the sensing site that talks directly to a remote server or cloud node, increasingly more and more logic will be required to be intelligently dispersed across the network. This not only includes cloud computing, but also edge-of-network compute nodes in addition to the devices or "things" themselves. This section provides an overview and highlights the characteristics of various components that can comprise a modern IoT application deployment.

5.1 Cloud-Computing Nodes

As was mentioned earlier in this chapter, cloud computing can be thought of as the practice of using a network of remote servers that are hosted on the Internet. These servers are used to store, manage, and process data for a given application or service. This contrasts with the notion of having a local server or server farm or a locally based personal computer. To date, cloud-computing servers differ in several ways from various other compute nodes within a given IoT deployment. For example, at the time of writing a given general-purpose cloud-based server will generally be high-performance compute with RAID-based SSD storage. The server could have redundant 10-GB networking capabilities. In addition, disk I/O may be up to 35,000 4 K random read IOPS and 35,000 4 K random write IOPS. Such a cloud server may contain as much as 8 GB of RAM, 8 CPUs, and hundreds of gigabytes of SSD storage. It is also important to note that these types of servers would be capable of running robust operating systems, such as one of various Linux distributions or Microsoft Windows.

With such a resource-rich development environment, developers can often use very high-level programming languages to implement application and business logic. Modern frameworks for Web development can be used for the rapid development and prototyping of products, given the very resource-rich development and runtime environments that these devices can afford. In addition, as compute and network bandwidth demand increases, modern cloud infrastructures, such as Amazon, Microsoft, and Rackspace, allow system architects to rapidly deploy provision servers as needed to meet application needs, while taking into effect such things as load balancing, geographical proximity to users, and caching.

5.2 Fog/Edge-Computing Nodes

Fog or edge computing is another layer of computing that may be present in modern IoT applications and systems. Unlike cloud computing, whereby there is a network of remote servers hosted on the Internet, edge computing is a communications topology that includes edge-of-network compute devices that may be used to carry out significant amounts of computation, storage, and communication. These edge devices do not reside in the cloud but rather are located at the edge of the computer network in greater proximity to where collected data are sampled. As such, a fog compute node will often be connected to and have input/output from the physical world such as sensors, actuators, and mobile health care components. It is these edge nodes, as described below, that perform the physical input and output within the system.

The processing power of modern fog or edge compute nodes, such as those in high-performance networking, artificial intelligence computing, and autonomous vehicles, can be quite high power compared with traditional personal computers and modern mobile phones. Edge compute nodes in contrast to traditional cloud servers are enabled to reside at the edge of the network as opposed to traditional or virtual servers that reside in a data center. This allows a given fog or edge node to be in physical proximity to the end users and input/output devices within the system. In addition, due to the large amount of input and output as well as compute and storage capability increasing amounts of business logic may reside within the fog or edge compute node, thereby precluding the need for all traffic from the sensor site or extreme end of the network to travel all the way to the cloud compute nodes and vice versa.

An example fog or edge compute node would be the NXP QorIQ LS1043 reference design board, which is a quad-core, 64-bit ARM-based processor for embedded networking and industrial applications. The hardware that comprises such an edge compute node is the quad-core ARM Cortex A53 processor, up to 2 GB of RAM, and 10-GB Ethernet support with various UART ports. This particular edge compute device is also typical of many industrial IoT edge compute nodes in that it comes with embedded Linux and related software tools and development kits.

5.3 Device-Computing Nodes

Devices or "things" in the IoT also differ in marked ways from previously described cloud computing and fog/edge computing. Whereas cloud-computing nodes are robust servers with full operating systems and highly provisioned hardware resources and edge/fog nodes are multicore CPU-based devices with embedded operating systems, large amounts of RAM, and I/O capability, devices or things are typically at the opposite end of the computing resource spectrum.

Conceptually, a device will often comprise a programmable processor, a small amount of local memory, modest amounts of I/O, such as Bluetooth, Zigbee, Z-Wave, or similar, and possibly an embedded operating system or bare metal software development environment. These types of devices also often contain various sensing components, such as temperature, pressure, accelerometers, and the like. These devices are often not capable of connecting directly to the Internet, although in some cases that is possible. Rather, they are designed to perform a certain function to sense the real world, perhaps perform some lightweight computation on that sensed data and then transmit the data to either other devices via a device-to-device communications link, to edge compute nodes for additional processing or business logic steps, or in some cases ultimately up to a cloud-computing node for additional business analytics and computation and application logic layers.

Examples of devices can be rather sparse in technical features when compared with other parts of a given IoT deployment. A given device, for instance, may have a real-time operating system or some variant of embedded Linux; however, this is not always the case. Oftentimes, these devices have no operating system at all and application code must run on the bare metal itself. While some devices include a 32-bit CPU capable of running an embedded OS or Linux at higher clock rates and with larger memory footprints, some devices may only contain a single 8-bit MCU running at perhaps 5–10 MHz. Typically, these types of devices will come with supporting software development kits and libraries to aid the developer to bring a solution to market. In addition, even in the bare metal MCU case these devices can usually be programming in a low-level language such as C. Program space is often limited as well; for instance, some devices may be limited to as little as 32 KB of flash memory and perhaps as little as 2 KB of SRAM with possible scratchpad memory. I/O that is included is similar to that mentioned before, such as Zigbee, Z-Wave, or Bluetooth.

6 Communications Used in Internet of Things

IoT implementations can vary widely in terms of system architecture, communications technologies, and models used. As mentioned earlier in this chapter the notion of IoT is not a particularly new concept in computing, as networks to monitor and control remote devices have been in place for decades. Similarly, the use of IP (Internet Protocol) to connect devices other than traditional computers to the Internet has also been around for some time. At a high level, however, the current IoT is a conflux of multiple technologies and emerging market factors that is rapidly making it possible to connect orders-of-magnitude more devices in smaller form factors at lower cost and with increasing ease.

Some of the factors driving current trends for communication in the IoT are discussed in this section. For example, the large-scale adoption of IP-based networking is one such factor. IP-based networking is the primary communications protocol in the Internet Protocol Suite and is used for relaying datagrams or packets across network boundaries. As such, IP has become a global standard in computer networking. Accordingly, there are rich and robust platforms of software and tools that can be incorporated into networked devices of varying types, like low-cost, low-power, smaller form factors, that lead to its use within the IoT.

Similarly, the rise of cloud computing has been a factor in modern communications and architectural trends for the IoT. Cloud computing can be thought of as the use of remote servers hosted on the Internet to store, manage, and process data vs. a locally hosted and managed server. Cloud-computing devices offer relatively low-cost, highly scalable server solutions to which large networks of distributed devices may be connected to interact. In addition to this, since cloud-computing devices can be rapidly provisioned and configured, they provide an attractive solution for back end analytics.

As mentioned above, back end data analytics is another driving factor in the IoT that dovetails with cloud computing. As cloud-computing power continues to mature and cost of compute becomes more affordable this paves the way for new algorithms and computational complexity, data storage, virtualization, and related services. These services can support vast amounts of data, aggregation, and analysis with ever-increasing economy of scale. With the ability to effectively and efficiently handle such large and dynamic data sets new opportunities for deep insight and extraction of information have become available to IoT systems designers.

Keeping all of this in mind, there is still no widely accepted definition of what comprises the IoT across different architectures and deployments. For instance, some groups refer to smart objects like devices that have limited resources and are highly constrained. These can include devices with low power, low memory, limited compute power, or limited bandwidth. Others refer to IoT devices as devices that do not necessarily connect to the Internet, but rather are capable of communicating with a local gateway or with other machine-to-machine or device-to-device nodes. Still others refer to the IoT as devices and deployments that communicate with cloud services via the traditional Internet. Nevertheless, each case contemplates a deployment of objects, sensors, etc. that possess some level of network connectivity and local compute capability. The following sections break down the different types of communications used for various types of IoT deployments.

6.1   Device-to-Device Communications

While there are varying definitions of device-to-device communications depending on the infrastructure in which the devices are deployed, generally device-to-device communication is defined as representing two or more devices that communicate and connect directly with one another. This is in contrast to devices that might communicate through an intermediary application server or connection point. One subtle difference is device-to-device communication that occurs in certain cellular network technologies.

Device-to-device communication for the IoT can occur over a number of communication types like IP networks of the Internet. Unlike other types of computing, however, oftentimes these devices use protocols, such as Bluetooth, Zigbee, or Z-wave, to establish direct point-to-point communications. Fig. 12 shows a number of wireless device-to-device compute nodes talking to each other via bidirectional communication links. In this figure the channels may be Bluetooth, Zigbee, or Z-Wave. Note that there is no centralized hub through which each of the devices communicate, but rather communication is from device to device.

6.1.1 Device-to-Device Communications With Cellular Network

In the case of cellular communications, however, device-to-device communication can differ slightly in that communication may also require assistance from the cellular network. These are often referred to as device-to-device assisted networks. Fig. 13 encompasses the device-to-device communication that was described previously in this chapter (i.e., standalone device-to-device communication) in addition to network-assisted device-to-device communication.

As can be seen in Fig. 13 the difference between the two network structures is the presence of a cellular support infrastructure in (II). It is used to organize communications and resource utilization within a given cell of the cell network. Conversely, in (I) the devices organize the communications by themselves without the support of infrastructure help. The figure shows the difference in the fundamental network architecture of D2D communications across the two solutions.

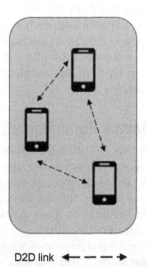

D2D link ◄ ─ ─ ─ ►

Fig. 12 Standalone device-to-device communication without network support.

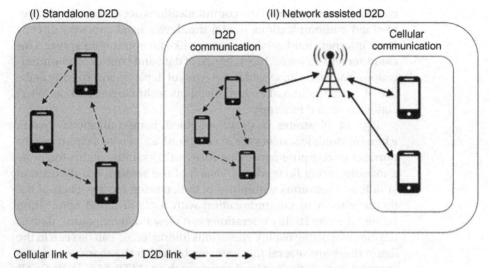

Fig. 13 (I) D2D communications without infrastructure (standalone D2D), and (II) D2D communications with infrastructure (network-assisted D2D).

D2D aggregators can be used to collect data from devices intending to connect to what is referred to as the cellular core network and send them to gateways that connect to the core network. The access network may be wired or may be wireless and the core network itself is what connects the devices with service providers to manage the different D2D services.

Nevertheless, device-to-device networks allow devices to adhere to the communications protocol of choice to exchange messages and information, as mentioned earlier. Exemplary applications are those with typically low data rate requirements that may only be required to transmit small packets of information between devices. These could be garage door opener systems, lighting systems, and home or commercial security systems.

One challenge with many device-to-device systems, however, is that systems oftentimes use device-specific data communications. This requires various vendors to implement the same functionality across product lines and requires capital investment in development and ongoing support for these specific data formats. This contrasts with the use of standardized data formats across vendors and products. As a result underlying communications protocols may not be compatible resulting in a silo effect across vendor offerings. An example of this would be a family of Zigbee devices within a given product offering that would not be compatible with Z-Wave based devices, and vice versa.

6.2 Device-to-Cloud Communications

In contrast to the device-to-device communications model discussed earlier device-to-cloud communications incorporate cloud compute nodes as part of the communications network. In the device-to-cloud communications model the device itself connects directly to an Internet cloud service or server like an application server. This cloud server facilitates the exchange of data and control of communications. As can be imagined these types of deployment typically make use of more traditional communications technologies such as Wi-Fi connections and Ethernet.

Fig. 14 illustrates a device-to-cloud communications system whereby multiple sensors are in communication with a cloud service provider or compute node. One real-world example of this might be a manufacturing floor whereby each of the sensors is a thermostat at different locations within the manufacturing facility. Each of the thermostats is in communication with a cloud-based application server whereby facility operations can view the temperature data of the manufacturing facility via various interfaces. As can be seen in the figure there are several mechanisms by which the thermostats may communicate with the cloud server such as HTTP, TCP, UDP, CoAP, and so forth.

Examples of a device-to-cloud communications model can be found in numerous consumer electronics devices such as modern smart televisions, smart thermostat devices, and certain smart speaker devices. One application of these device-to-cloud communication devices is to transmit data collected at the device to a cloud server, which in turn analyzes the data to determine certain metrics. These might include home temperature, lighting usage,

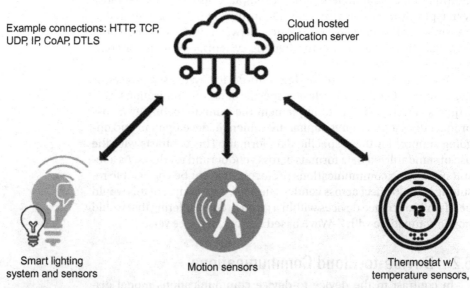

Example connections: HTTP, TCP, UDP, IP, CoAP, DTLS

Cloud hosted application server

Smart lighting system and sensors

Motion sensors

Thermostat w/ temperature sensors,

Fig. 14 Device-to-cloud communications.

home access patterns, and energy consumption. Since intelligence is performed on a cloud-based application server this also allows the user a number of different ways to observe these data such as via a traditional Web interface, a mobile phone app, mobile browser, or tablet computer.

Like the device-to-device communication model above, however, interoperability can also be an issue when trying to connect device-to-cloud services and devices from different manufacturers. Oftentimes, the device and cloud service are from the same manufacturer such as a smart thermostat manufacturer. If a manufacturer uses proprietary data protocols or web interfaces, then a consumer is often locked into a given manufacturer's product line. By creating a barrier to exit for the consumer or barrier to integration of other manufacturers' devices a given product manufacturer can increase the certainty of locking in a customer.

6.3 Device-to-Gateway (Fog) Computing Communications

Up to now we have considered point-to-point communications models whereby either a given device communicates directly with other devices as a standalone configuration or with network assistance, or a given device communicates directly with a cloud server itself. Device to gateway adds another layer to the communications and network architecture. In this model of communications the device itself does not connect directly to a cloud application service, but rather talks to a gateway or fog compute node as an intermediary. In turn this gateway node will then itself communicate directly with a cloud-based application server as a proxy for the devices themselves. This provides several benefits to system developers, as will be seen below.

Since the gateway itself acts as a form of application software there will hence be application software operating locally on the gateway device. This allows the gateway to act as an intermediary between cloud-computing services and device services. The gateway node can perform several different services such as security, data translation, or protocol translation. In addition, it is increasingly more common for the gateway itself to contain business logic operations thereby removing the requirement of the devices themselves to communicate all the way to a remote cloud service provider. Fig. 15 illustrates an example of a device-to-gateway communication path.

As can be seen in the figure the sensors themselves talk to a local gateway node. They may communicate by any number of means, such as HTTP, TCP, UDP, and so forth (as described in Fig. 15). The local gateway node in turn communicates with the cloud service provider,

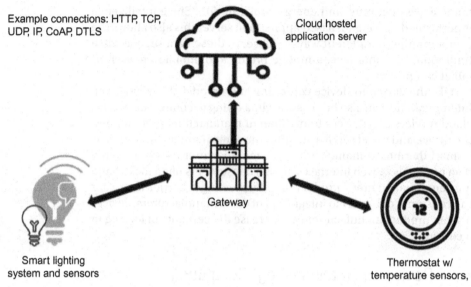

Example connections: HTTP, TCP, UDP, IP, CoAP, DTLS

Cloud hosted application server

Gateway

Smart lighting system and sensors

Thermostat w/ temperature sensors,

Fig. 15 Device-to-gateway (fog) to cloud example.

cloud node, or application server using standard IPv4 or IPv6. The wireless connectivity used to pair a given sensor to the gateway can be one via Wi-Fi, Bluetooth, or other means. The local gateway node itself may perform a number of different functions such as data formatting and conversion, various application or business logic, and other functionality.

The model illustrated in Fig. 15 is currently common to a number of consumer electronics devices. For example, with some products the local gateway device is actually a modern smartphone capable of Bluetooth or other technology. An app would run on the smartphone that may pair with other consumer devices like a smart watch. In many cases the smart watch itself does not have the built-in capability to connect to a local Wi-Fi connection or cellular network, as this would commonly increase the cost of the device's bill of materials, increase price, or have an adverse effect on power consumption. Instead, the smart watch would pair to the local gateway or smartphone and the smartphone in turn would operate as the intermediary local gateway facilitating the connectivity of the smart watch to the cloud server. Other examples of this communication topology are certain home security systems that connect to Internet services that may use Z-Wave or Zigbee technology for connectivity between devices and local gateways. As will be discussed in greater detail later in this chapter the addition of a local gateway device brings enhanced system flexibility, but at the cost of increased development resources, system complexity, and cost.

6.4 Back End Data-Sharing Model

While this chapter has largely focused on models where sensor data are collected and processed throughout the network for a given IoT application, there are also data models in which data can be shared across IoT applications as well as entities and management systems. One term for this type of data sharing is back end data sharing. In summary, back end data sharing refers to a data model whereby the system architecture supports the ability for the export of data for consumption via other systems. That is to say, application data for a given IoT application or infrastructure can be shared outside a given use case or organization. For example, data could be collected at the sensor level, potentially preprocessed or postprocessed, and ultimately made available as data objects at the cloud or application server level. The data can then be consumed by other users or management systems to combine with yet other third-party data sets to perform various analyses. In summary, the system architecture has been set up with the goal to grant access to uploaded cloud-based data to third parties.

One use case might be a large hospital system, perhaps spread across multiple geographic locations, comprised of HVAC systems, various lighting systems, automated control, and perhaps asset-tracking systems comprised of RFID or other enabling technology. In the traditional device-to-cloud architecture or even device-to-fog-to-cloud architecture the data collected within the system will sit on a given server of a series of servers that support the underlying architecture. In this case the data are often walled off from consumption by third-party applications either due to connectivity to the server holding the data, or due to the fact that data aren't shared in a normalized or meaningful manner (data format, UI, etc.). By designing a data-sharing model such data can be readily accessed and analyzed by the organization at the cloud level using an assortment of modern graphical and analytical tools, as well as parsing and consuming data across the enterprise and across the various types of sensing devices and infrastructure deployed within the enterprise. In addition, the data can be packaged or made accessible via various mechanisms and APIs for consumption by third-party organizations or services. This allows the breakdown of what are commonly referred to as silos within IoT. It is important to note, however, that these are data silos rather than *development* silos, which are addressed later in this chapter.

Fig. 16 illustrates what a data-sharing diagram for the facility-based use case described above might look like. Various HVAC sensor data as well as potential RFID tracking data can be collected and aggregated for a given facility. The data collected across the various sensors and solutions can be hosted at application service provider A. At the same time such data may also be made accessible to application service

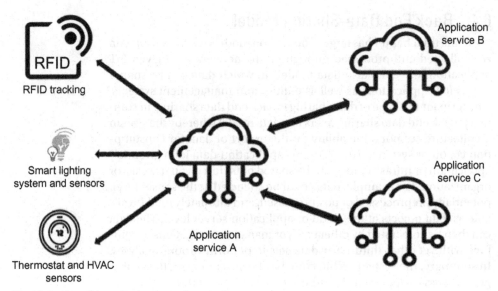

Fig. 16 Back end data-sharing example.

providers B and C for further consumption and postprocessing. This use case will be discussed further in the next section on data analytics.

7 Data Analytics

As many readers are probably aware the term "big data" has frequently been used in the literature. Big data can simply be thought of as large data sets of information that can be collected and analyzed for varying purposes. The types of data collected can often be categorized and characterized by the volume of data that is being collected, the variety of data being collected, and the velocity at which the data are collected. That said, there is no shortage of opportunities to analyze and use the vast amounts of data collected in modern, and especially future-facing, IoT applications. Simply envisioning use cases, such as deployments for intelligent transportation, power grid, energy and smart metering, health care and smart cities, immediately brings to mind the scale and speed at which data will be generated via the various types of systems described in this chapter.

In general terms IoT and analytics can be thought of as the various steps that are taken via a system, either in real time or offline, in conjunction with analysts and automation whereby a variety of IoT data are examined to reveal trends in the data. These may be as simple as analysis of the raw data at a sensor collection point to the revealing of underlying trends, unseen patterns, correlations, and other new secondary information that can be created with the goal of businesses, data miners, and data scientists to make efficient and well-informed decisions.

As most readers will be aware many techniques for data analysis have been developed for both application-specific domains and more generalized use cases. IoT, however, is quite a bit different. Rather than having normalized and regular data sets to work with IoT data characteristics vary in a number of important ways. Data collected within a given system may vary due to the sensors used and various solutions used within the infrastructure. This can result in highly heterogeneous data, noise in the sampled data, variety, and unforeseen and rapid growth in data consumption and analysis requirements.

7.1   IoT and Analytics/Big Data

The volume and rate at which data are generated by various sensors, devices, health care applications, temperature systems, and myriad other applications and services is ever increasing. This will only continue with the rollout of additional IoT applications and systems. At the same time the data continuously generated are oftentimes unstructured or semi-structured. As such, traditional database systems are not able to meet or are prohibitively inefficient when storing, processing, and analyzing rapidly growing amounts of data.

For data miners and scientists to be able to analyze these types of data at large volumes and with reasonable processing times, analysts require tools and technologies that enable them rather than hinder them. These tools must be able to transform a vast amount of structured, unstructured, and semi-structured data into more easily compressible data and metainformation. It is not enough for these tools to simply analyze and format data, however, they must also be able to generate visualized findings into tables, graphics, and spatial charts for proper decision making within the organization. The integration of disparate data systems for comparison and analysis, similar to the back end data-sharing model described previously, must also be made feasible.

7.2   Analytical Systems for Internet of Things

When analyzing the data collected via a given IoT system oftentimes different types of systems must be employed according to the requirements and characteristics of the application. These different types of analysis can be characterized by their footprint, timing requirements, and end goals to name a few examples. These can be characterized roughly as real-time systems, postprocessing systems, in-memory systems, business intelligence, and large scale.

Type of Analysis	Characteristics
Real time	This is often performed on data that are collected from sensors themselves. The data have the potential to change over time, often quite rapidly. To this end rapid analytics are required to process the data in real time. One advantage of these types of solutions is that they can benefit from parallel processing implementations.
Memory level	These are analytics solutions that are capable of processing data sets wherein the size of the data is smaller than that of the memory available on the cluster or compute node. At the time of writing compute clusters can often be at the terabyte scale. These types of solutions may or may not be capable of real-time processing and analysis. Similarly, these solutions can also provide the addition of real-time processing capability.
Postprocessing	Postprocessing or offline analytics solutions are attractive when a real-time response is not required. Systems such as Hadoop are capable of performing these offline analyses. One advantage of these types of solutions is that they can provide efficient data acquisition and reduce the cost of data format conversions for subsequent analysis.
Business intelligence	Business intelligence–style analytics can be adopted when the size of data sets is beyond that of the memory level of the cluster itself. In this case data may be imported into the system itself for processing. These systems currently support terabyte-level data sets and can be used to discover business opportunities from the vast sets of data. Oftentimes these solutions can be used in both offline and online modes of operation.
Large scale	These types of analysis solutions are employed when the size of the data to be analyzed exceeds the maximum capacity of business intelligence products and/or traditional database solutions. They often use distributed file systems for data storage and map reduce-type technologies for analysis. Large-scale solutions, like those described above, are most often only available in nonreal-time and offline modes of operation.

8 Internet of Things Development Challenges

IoT development can often be thought of as occurring in silos. This is true in a number of different capacities including not only data collection and storage, but also the technology development process itself. It is important to note that unlike other areas of computing, such as general-purpose application development or mobile application development, IoT development is much more heterogeneous in nature. Rather than writing a mobile app UI or perhaps a business logic application, end-to-end IoT application and system development crosses a number of heterogeneous areas of computing. These can include but are not limited to security, general-purpose or high-performance computing, Linux-based application development perhaps on real-time operating systems, and even true embedded systems development that requires a deep understanding of the underlying processor and peripheral architecture.

This creates a siloed development environment in which engineers and practitioners who work in a given part of the system may not have the sufficient understanding or skill sets to operate on other parts of the system. For example, a mobile app developer or UI person likely does not have the skill set to do edge-of-network layer development in the C or C++ programming languages. Similarly, it is widely known that there is an ever-increasing shortage of embedded engineers available for edge-of-network and embedded device development. The development processes for embedded firmware and cloud applications have taken largely divergent paths. Embedded development has stayed close to the metal, focusing on coping with extremely limited computing resources. Cloud development has raced toward frameworks and abstractions that eliminate the individual hardware nodes as a developer concern and try to enable transparent access to computing resources limited only by budget.

This section characterizes some key traits of system development that managers and system designers should be mindful of not only when ramping teams, but also in terms of the skill sets required for ongoing development and maintenance of legacy systems.

8.1 Cloud-Computing Development

Cloud computing and application development on high-performance, high-resource computers is alive and well. The availability of industrial-grade cloud service providers over the last decade has evidenced this with Microsoft Azure, Amazon Web Services, and the related solutions such vendors offer. Cloud-computing developers are used to rich development environments, highly powerful integrated development environments, and high-level languages that may be interpreted or scripting-based. Similarly, myriad development frameworks are available to accelerate and alleviate the burden faced by application developers. Rich libraries may be used that are often very large in memory and compute requirements without generally affecting overall system performance. As mentioned earlier the price of cloud computing is quite low with various commercial vendors, and hence the provisioning of additional hardware, memory, and operating systems is not usually of huge concern when resources do become limited or underperforming.

The developers of these applications often rely on multigigahertz CPU speeds, gigabytes of RAM, and terabytes of disk space. Networking is reliable, fast, and relatively cheap. In addition, more compute nodes can usually be provisioned in a matter of minutes with command line tools or the push of a button on a Web UI. In fact, it is not unrealistic to assume that the application developers are often largely unaware of the underlying hardware architecture that they are developing for. As bare metal machines gave way to virtual machines, which in turn are giving way to containers, software components are built to communicate via lightweight APIs and message buses that eliminate developer considerations of where or how much of a component is deployed. In this space the concept of fixed storage, finite compute cycles, hardware failures, and all the concerns of deploying to physical computers are abstracted away. Developers are allowed to think purely in terms of data flows through the application, communication over invisibly fast network meshes, and storage in limitless reliable data pools.

8.2 Embedded Device Development

Embedded development and design, especially when considering embedded device nodes acting as agents for powerful cloud software, is a markedly different animal from the above characterizations of cloud computing. Embedded developers often develop in far more resource-constrained environments. Chapter 2, titled "Development Process," gives a good overview of the application development cycle for embedded systems..

Here developers often do not have the robust integrated development tools that are afforded to other areas of computing. The system being developed may be as little as a bare metal MCU with only a few kilobytes of program memory. There may be hardware assists for things like computational acceleration or direct memory transfer, but these are accessed at a very low level often using complex data structures and memory-mapped registers. A far cry from the runtime of a cloud server, application developers must possess an understanding of the minute details of the underlying architecture.

The system may have a variant of embedded Linux or a real-time operating system, but it is entirely possible the device will not. While there are an increasing number of development tool chains available for the embedded developer, assembly language is still used in certain cases. C and C++ may be the more desirable programming language when available, either due to system resources, vendor tool chain, memory, or performance considerations. Even in this case, however, memory resources and compute resources may severely mandate how the development works. If the device has no underlying floating-point hardware, for example, computation may need to be done with fixed-point or saturating arithmetic. This often requires developers to understand proprietary intrinsic functions for the device. Memory alignment, real-time compute deadlines, and other aspects further burden the developer.

8.3 Integration of Development Silos

With the future of embedded computing and cloud software working together in tandem, program managers and developers alike must be mindful of the human resources and capital required for building and deploying these solutions. High-level cloud developers likely do not possess the intricate knowledge of embedded hardware and oftentimes do not possess the programming skills required for these systems. Similarly, embedded developers more than likely are not aware of the high-level development frameworks and tools rapidly emerging for cloud development. Care must be taken to align the many moving parts for unified and heterogeneous development of these systems if they are to successfully deploy these applications in the future.

The integration of silos goes far beyond just the development for a given target node, however. Various communications channels between devices must be accounted for as well as the maintenance of infrastructure against which a given IoT application is deployed. In addition, the deployment of security configuration information and security layers themselves must be continuously maintained throughout the life cycle of the application's deployment. When the application code, security layers, or configurations for one compute node within an IoT application are updated they need not only be tested and validated against the other software and systems enabling the application, but also deployed. This iterative deployment cycle, often comprised of multiple interconnected software modules and layers, some proprietary and some third-party or open source, must be accounted for across development teams and enterprises.

Exercises

1. **Q**: What are the three types of architectural configurations that have evolved over time?
   A: Cloud-centralized server, fog-regional servers, and edge on the device.
2. **Q**: Why is decision making on the edge better than in the cloud or fog?
   A: Because it is faster and enables more real-time decision making.
3. **Q**: What KPIs do IoT devices typically contribute positively to in a factory setting?
   A: Quality, efficiency, and safety.

14

SECURITY AND CRYPTOGRAPHY

Ruchika Gupta*, Pankaj Gupta†, Jaswinder Singh‡

*Software Architect, AMP & Digital Networking, NXP Semiconductor Pvt. Ltd., Noida, India †Senior Software Staff, AMP & Digital Networking, NXP Semiconductor Pvt. Ltd., Noida, India ‡Software Director, AMP & Digital Networking, NXP Semiconductor Pvt. Ltd., Noida, India

CHAPTER OUTLINE

1 What Is Security?

The word "security" can have different meanings to different readers. A few definitions, which may come to mind when you hear the word "security," are provided here:

Software Engineering for Embedded Systems. https://doi.org/10.1016/B978-0-12-809448-8.00014-X

Security means safety.

Security means taking measures to be safe or protected.

Security is the extent of resistance to, or protection from, harm.

Security is the state of being free from danger. The state of being free from danger is called a secure state.

Security includes the measures taken to be in a secure state.

Security involves the degree of measures taken by an entity to attain a secure state.

A secure state is quantified as the extent of resistance to, or protection from, harm.

Security is a kind of protection which creates a separation between assets and the threat to assets.

Clubbing all these definitions of security into one gives:

Security is the degree of measures/steps taken by an entity, to offer the degree of resistance to, or protection from harming resources, for which the entity bears responsibility.

1.1 Embedded Security

1.1.1 What Is an Embedded System?

An embedded system is an electronic product that comprises a microprocessor or multiple microprocessors executing software instructions stored on a memory module to perform an essential function or functions within a larger entity.

1.1.2 What Is Embedded Security?

Considering securing entity as an embedded system that takes degree of measure to offer degree of resistance to, or protection from harming the resources of the overall system, to which either the embedded system is:

- connected, or
- in which it is subsumed.

1.2 Embedded Security Trends

Embedded security trends clearly show increasing system complexity.

1.2.1 Embedded Systems Complexity

With ever-growing demands for enhanced capability, increased digitization of new manual and mechanical functions, and turning dumb embedded devices into smart ones via interconnectivity, the complexity of the embedded system is increasing. Although these electronic complexities bring betterment to mankind they also bring security vulnerabilities.

The vulnerability of security cannot necessarily be attributed to electronic complexity if such complexity can be effectively managed.

Complexity gives birth to flaws, which are later misused to circumvent system security.

Complexity cannot only be measured by code size or transistor count. Let's take few examples to understand it better:

1. Software complexities
   a. Deciding to use Linux in systems which require higher security levels is debatable.
      i. The open-source nature of the Linux code is considered its strength as the code gets greater exposure and is reviewed by the worldwide community.
      ii. Linux has its disadvantages too, as the code undergoes continuous modification which can lead to the introduction of vulnerabilities.
2. Hardware complexities
   a. Network connectivity to embedded systems.
      i. Traditionally, by not being connected, embedded systems were immune to the risks associated with the Internet.
      ii. Nowadays, there is a growing need for remote management of devices and device life cycle management.
   b. Embedded system consolidations.
      There is a growing trend to have single, powerful embedded systems performing multiple tasks or components, in order:
      • To have better interworking between them.
      • To save costs in some cases.
      Doing so, typically leads to mixing of high-security tasks with tasks that are low security of not security critical.

1.3 Security Policy

To evaluate an electronic product for its security strength, it is important to understand its security policies first so that the robustness of the security of a product can be evaluated against its adherence to its own set security policies.

Security policies are a set of defined steps/measures to achieve a defined level of protection for specific resources. Policies are simply created to counter threats.

Hence, to define a security policy, a prerequisite is to identify the resource requiring protection and granularize the expected protection offered by the entity, considering:

a. The type of attacks an entity is guaranteed to safeguard against.
b. The limitations of the entity.

1.3.1 CIA Triad and Isolation Execution

Each of the defined security policies can be mapped to one, two, or all three of the CIA Triad:

1. Confidentiality: restrict access to authorized and authentic users only.

2. Integrity: resources should be protected against any modification done:
   a. intentionally by unauthorized users.
   b. unintentionally by authorized/authentic users.
3. Availability: resources should be always accessible for intended usage.

To minimize the impact of a violation of security policy, another aspect is added to the CIA Triad, called isolation (Fig. 1).

If the CIA Triad is broken on one isolated execution plane it will not hamper the CIA Triad on the other isolated execution planes.

1.3.2 Policies for Information Flow Between Isolation Execution

There can be separate security policies governing the communication between two isolated execution planes. Security policies depend on the following questions:

- Is communication allowed?
- What information can be exchanged or accessed?

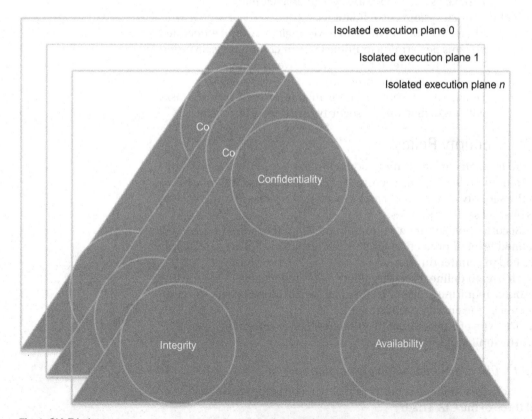

Fig. 1 CIA Triad.

1.3.3 Physical Security Policies

Physical security policies are security policies detailing counter-measures to the physical threat to embedded systems.

2 Cryptology

Cryptology is the science of secure data communication and data storage in a secret form. It comprises cryptography and cryptanalysis (Fig. 2).

2.1 What Is Cryptography?

Let's consider the problem of two legitimate people, Alice and Bob, who want to communicate data secretly over a communication channel. This channel is deemed unsecure as any illegitimate user, say Eve

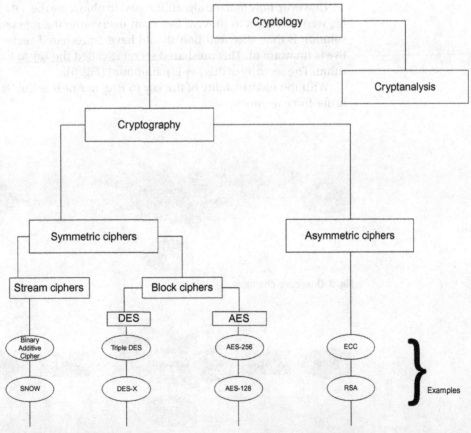

Fig. 2 What is Cryptology?

(an eavesdropper), has access to the channel and can easily hamper confidentiality and data integrity (Fig. 3).

2.1.1 How to Solve This Problem?

Alice and Bob can encode/encrypt the data while sending and decode/decrypt the data upon receiving. This would block the illegitimate user Eve from decoding the data sent over the unsecure channel. This technique is called cryptography. Cryptography refers to communication techniques derived from mathematical concepts and a set of rule-based calculations, called algorithms, to transform messages in ways that are hard to decipher, for secure communication (Fig. 4).

In an ideal world, Alice and Bob should keep secret the algorithm/technique used to encrypt and decrypt the data, so that Eve can not decode it. Keeping the algorithm secret is neither sensible nor practical. Moreover, making the algorithm public hardens it by allowing cryptoanalysts to evaluate and challenge the algorithm. Using an algorithm which is publicly unannounced is never recommended (Fig. 5).

However, now that the algorithm used to obfuscate the data is public, we need ways to prevent Eve from decrypting the message. The solution is that Alice and Bob should have a preshared secret which Eve is unaware of. This preshared secret is called the *key* to the algorithm. The security of this key is paramount (Fig. 6).

With the unavailability of the key to Eve, her next action is to use brute-force techniques.

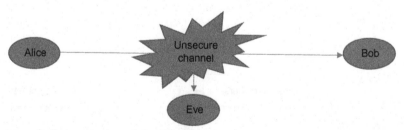

Fig. 3 Unsecure channels.

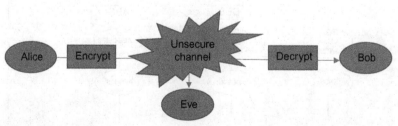

Fig. 4 Basic cryptography.

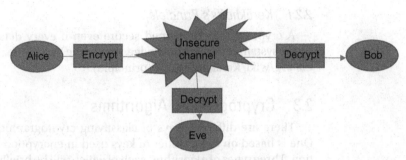

Fig. 5 Weakness of basic cryptography without keys.

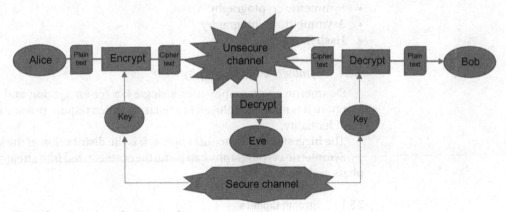

Fig. 6 Complete view of cryptography.

2.2 What Is a Brute-Force Attack?

A brute-force attack is a trial-and-error technique where the attacker tries all the permutations and combinations of keys to decipher the plain text, with meaningful results. This attack is both time and resource consuming. Since it is a trial-and-error technique, the time taken using this attack method depends on the key space.

What is key space? Key space is the finite number of keys which can be applied to an algorithm to decipher meaningful content.

The robustness of any algorithm is governed by the following three things:

1. Time taken to locate the key.
2. Required computation/resource power.
3. Size of the key space:
    a. The larger the key space, the more time and resource consumption is needed for deciphering. Hence, the larger the key space, the greater the algorithm strength.
    b. To increase the key space, the key size needs to be increased. Increasing the key size increases the effort required to share it between authentic users.

2.2.1 Kerckhoffs's Principle

A cryptosystem is deemed secure even if every detail about the cryptosystem is public knowledge, except the keys (https://en.wikipedia.org/wiki/Kerckhoffs%27s_principle).

2.3 Cryptographic Algorithms

There are different ways of classifying cryptographic algorithms. One is based on the number of keys used in encryption and decryption. Three types of algorithm, each of which will be briefly elaborated, are given below:

- Symmetric cryptography
- Asymmetric cryptography
- Hash functions

2.3.1 Symmetric Cryptography

Symmetric cryptography uses a single key for encryption and decryption. It is mainly deployed in scenarios which require privacy and confidentiality.

The biggest difficulty to this approach is the distribution of the key.

Symmetric cryptography can be further categorized into stream ciphers and block ciphers.

2.3.1.1 Stream Ciphers

Stream ciphers operate on a single bit at a time and have a feedback mechanism such that the key is constantly changing.

A category of symmetric key cipher is where digitalized plain data digits are combined with a pseudo random digit stream (also called a keystream) sequentially. Each plain data digit is ciphered by XORing it with its corresponding bit in the keystream, one at a time. These ciphers are also called state-ciphers since encrypting each plain data bit depends on the current state of the cipher.

The simplest example of a stream cipher is the binary additive stream cipher.

2.3.1.2 Block Ciphers

Block ciphers represent an encryption method where blocks of data are encrypted with a deterministic algorithm using a symmetric key that has been securely exchanged. DES and AES are two popular examples of block ciphers.

2.3.1.2.1 Data Encryption Scheme (DES) Key size: 56 (+ 8 parity) bits.

Block size: 64 bits.

Successors of DES ciphers are: Triple DES, G-DES, DES-X, LOKI89, ICE.

2.3.1.2.2 Advanced Encryption Scheme (AES) Key size: 128, 192, or 256 bits.

Block size: 128 bits.

Successors of AES are: AES-GCM, AES-CCS.

2.3.2 Asymmetric Cryptography

Asymmetric cryptography uses one key for encryption and another for decryption. It is usually used for cases which require authentication, key exchange, and nonrepudiation. This cryptography scheme is also referred to as public key cryptography.

In this type of cipher, a pair of keys is used to encrypt and decrypt the data. The key pair consists of two unidentical large numbers, where one of the keys can be shared with everyone, called the public key, and the other part, which is never shared, is called the private key.

A digital signature scheme can also be implemented using public key cryptography. This scheme has two algorithms, one for signing and the other for verification. Signing requires the use of a secret/private key while verification is done using a matching public key. RSA and DSA are two popular digital signature schemes.

These algorithms are computationally expensive. The hardness of RSA comes from its integer factorization while the hardness of DSA is related to a discrete logarithmic problem. More recently, elliptic curve cryptography (ECC) has been developed and is gaining traction over the RSA algorithm for the following two reasons:

- with the same key sizes, ECC provides a higher cryptographic strength compared with RSA.
- for the same cryptographic strength, the ECC key size is smaller than that of RSA.

2.3.3 Hash Functions

Hash functions use mathematical transformations to irreversibly encrypt information without the usage of any key. These are usually used for cases where message integrity is required, for example, the digital fingerprint of a file's content and encryption of passwords. These functions are also called message digests or one-way encryptions. Some popular examples of hash functions are:

- SHA (Secure Hash algorithm), SHA1., SHA256.
- MD (Message Digest Algorithm), MD2, MD4, MD5.

2.4 Random Number Generator (RNG)

A good random number generator is fundamental to all secure systems. Lots of security protocols require random bits to remain secure. You may often find the word "unpredictable" is used instead of "random." Either way, the idea is to make it difficult for the attacker to guess a value. Random numbers may be required for applications that use one of the following:

- Keys and initialization values (IVs) for encryption
- Keys for keyed MAC algorithms
- Private keys for digital signature algorithms
- Values to be used in entity authentication mechanisms
- Values to be used in key establishment protocols
- PIN and password generation
- Nonces

There are three types of RNGs:

1. True RNG (TRNG).

   Random numbers that are generated from random physical sources, such as thermal noise, flipping a coin, and mouse movement. The strength of TRNG is that it is not able to regenerate a number once generated.

2. Pseudo RNG (PRNG).

   Random numbers that are computed and deterministic. Every PRNG computational functional needs an initial value called a seed (typically generated from TRNG and shared between the sender and receiver), which is used to compute the PRN. PRNG is nothing but a computational function which always generates the same number, if the seed is same.

   It can be explained as follows. If s_0 denotes the seed, then the number generated by the PRNG can be denoted as:

   $s_1 = f(s_0)$

   Generalizing the above equation gives:

   $s_{i+1} = f(s_i)$.

3. Cryptographically secure PRNG or cryptographic PRNG (CSPRNG or CPRNG).

   Compared with PRNG, CSPRNG has two additional properties:

a. It should pass statistical randomness tests. In other words, it should satisfy the next-bit test, such that given the first n bits of a random sequence, no polynomial-time algorithm exists that can predict $n+1$ bits with a success rate better than 50%.

b. It should withstand a situation where even if the state is compromised, it is not possible to reconstruct the state or reach the state even with reverse engineering.

2.5 Implementation of Cryptographic Algorithms in Embedded Systems

Cryptographic algorithms can be implemented:

- using a hardware accelerator
- in software

OpenSSL layered architecture within the Linux kernel (CryptoAPI) is one example that can be used to understand how cryptographic algorithms are realized (Fig. 7).

a. Using software

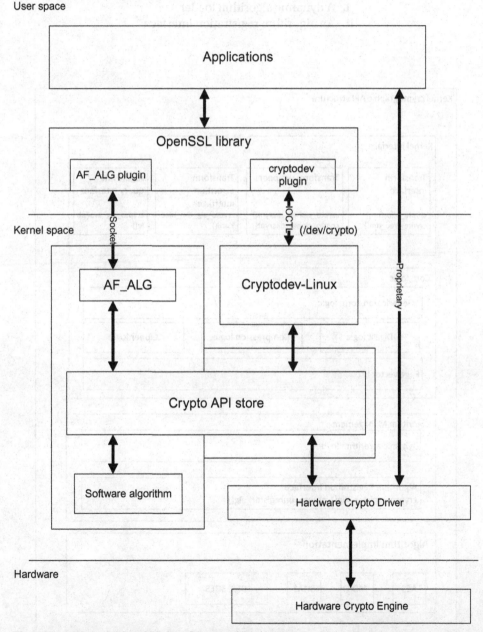

Fig. 7 View of Linux crypto APIs in software and hardware.

All cryptographic algorithms are implemented in the software as part of the Software Algorithm Block in the Linux kernel, as shown in Fig 8.

In Fig. 8, the block above the Algorithm Implementation Block, the algorithm management block, provides:

i. A dynamic algorithm loader

ii. An algorithm registration interface

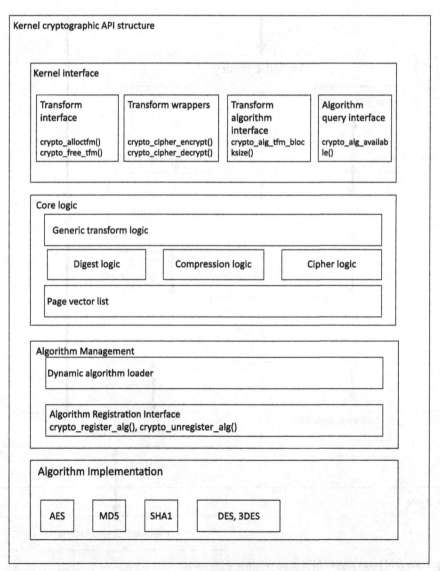

Fig. 8 Kernel-level structural details of crypto API implementation.

This is the block which is responsible for off-loading requested cryptographic operations to:

- The SW algorithm block of the kernel
- The HW accelerator

There are multiple reasons why a hardware accelerator is preferred to software implementations:

- Saving CPU cycles
- CPUs are not optimized for this work

The strength of software implementation is:

- Portability.
- Support of arbitrary algorithms.
- No requirement for hardware accelerators, that have an additional expense.

The abovementioned strengths become less effective with:

- Increasing standardization across different product domains.
- Increasing trends in on-chip supporting crypto accelerators, to support the increasing demands of crypto hunger applications.

b. Using a hardware accelerator

Let's look at the example of an NXP crypto accelerator. The flow mentioned below shows that an engine "cryptodev" is register with openSSL. The cryptodev engine uses CryptoAPI to realize its cryptographic algorithms. As mentioned in the previous section, the algorithm management functional block in Fig. 8, is configured to select the SEC driver to off-load all of the cryptographic algorithm to the NXP SEC H/W (Fig. 9).

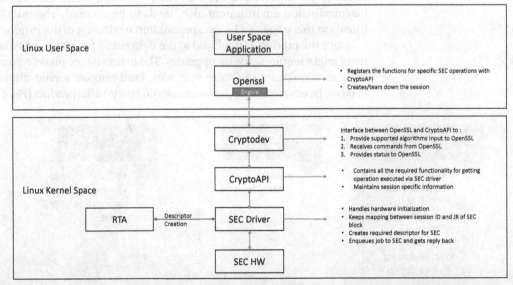

Fig. 9 On-chip SEC block h/w accelerator on the NXP SoC (http://cache.freescale.com/files/training/doc/ftf/2014/FTF-NET-F0352.pdf).

3 Life Cycle of a Secure Embedded System

Security has for a long time been focused upon in the world of enterprise. With the advent of the IoT, security in embedded devices is becoming a hot topic. OEMs are usually in a hurry to get products to market and security is usually an afterthought. However, security negligence right from the very start of product development makes devices vulnerable, exposing them to a vast number of attacks.

To ensure the security of embedded systems against a wide range of attacks the entire ecosystem must be protected. Typically, the life cycle of an embedded device involves phases, such as design and development, product deployment, maintenance, and decommissioning. Security needs to be embedded at each of these phases.

The design and development phase starts with conceptualization of the product. This phase further incorporates stages such as requirement gathering, design, development, validation, and integration. Security requirements need to be added at the requirement gathering stages and mitigations need to be propagated through all the different phases. Threat analysis is an important step which needs to be added at this stage. This ensures the correct security requirements are identified and propagated throughout the development process. Another important aspect at this stage is to ensure the secure design of the device. Based on threat analysis a list of security requirements is made available to the product developer. Devices need to be made secure at this stage by utilizing both hardware and software solutions.

Once a product is developed, it is produced on a mass scale and needs to be deployed to the customer's premises. The manufacturing/production environment also needs to be secured. The possible threats at this stage can be overproduction or cloning of the product.

Once the product is deployed at the customer's premises, the equipment might require software upgrades. The maintenance phase typically involves taking care of secure upgrades. Furthermore, secure disposal needs to be ensured at the decommissioning stage of the product (Fig. 10).

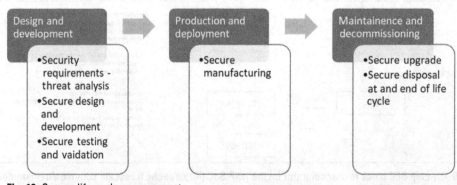

Fig. 10 Secure life cycle management.

3.1 Security During the Software Development Life Cycle

Secure design practices need to be incorporated at various stages throughout the SDLC.

Given below are various secure design practices that apply at different levels of the development cycle (Fig. 11).

3.1.1 Design

One of the important aspects during the design phase of an embedded product is to get the requirements right. "Threat modeling" helps define the security requirements for an embedded product. We will look in detail at how to do a risk assessment, followed by threat modeling, in the next section.

It is important to keep the design as simple and small as possible. Complex designs increase the likelihood that errors will be made during implementation, configuration, and use. Additionally, the effort required to achieve an appropriate level of assurance increases dramatically as security mechanisms become more complex.

3.1.2 Development

3.1.2.1 Secure Coding Guidelines

It is essential to decide and follow certain coding standards that govern how developers write code. The coding standard helps in increasing reliability by advocating good coding practices.

Given below are some practices for secure coding that can be added to the coding guidelines.

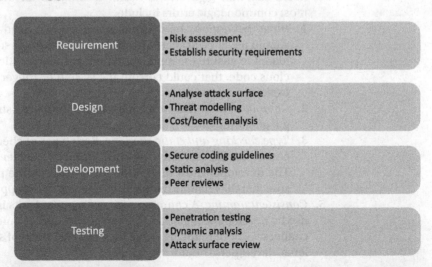

Fig. 11 Security during SDLC.

1. *Minimize size and complexity and increase traceability.* Secure code should implement its functions in the smallest number of lines of code possible, while maintaining readability and analyzability. Using multiple small, simple, single-function modules instead of one large, complex module that performs multiple functions will make the system easier to understand and document thus making it easier to verify the security and correctness of individual components and of the system.

2. *Code for reuse and maintainability.* The features of code that make it reusable—simplicity, comprehensibility, traceability—are the same features that help make it secure. Developers should never assume that their source code will be self-explanatory. All source code should be extensively commented upon and documented, reviewed, and tested to ensure that other developers and maintainers can gain a complete and accurate understanding of the code, which will enable them to reuse or modify it without introducing exploitable faults or weaknesses.

3. *Use a consistent coding style.* A consistent coding style should be maintained throughout the system's code base, regardless of how many developers are involved in writing the code. Coding style includes the physical appearance of the code listing, that is, indentation and line spacing. The physical appearance of the code should make it easy for other developers and code reviewers to read and comprehend the code. The entire development team should follow the same coding style guide. Coding style should be considered as an evaluation criterion for open-source software, particularly for software that will implement trusted and high-consequence functions.

4. *Avoid common logic errors.* Useful techniques for avoiding the most common logic errors include:

    1. *Input validation.* Input from users or untrusted processes should never be accepted by the system without first being validated to ensure the input contains no characteristics, or malicious code, that could corrupt the system or trigger a security exploit or compromise.

    2. *Compiler checks.* Take advantage of the compiler's strictest language settings.

    3. *Type checking and static checking.* Both types of checks expose consequential (security-relevant) and inconsequential faults. The developer must then distinguish between the two, to ensure that they handle the security-relevant faults appropriately.

5. *Consistent naming.* A common cause of security faults is incorrect use by developers of aliases, pointers, links, caches, and dynamic changes without relinking. To reduce the likelihood of such problems, developers should:

1. Treat aliases symmetrically. Every alias should be unique and should point to only one resource.
2. Be cautious when using dynamic linking to avoid unpredictable behaviors that result from runtime introduction of components.
3. Minimize the use of global variables. When such variables are necessary give the variables globally unique names.
4. Clear caches frequently.
5. Limit variables to the smallest scope possible. If a variable is used only within a single function or block then that variable should be declared, allocated, and deallocated only within that function or block.
6. Deallocate objects as soon as they are no longer needed.

3.1.2.2 Static Analysis

Static code analyzers help to find code sequences that could result in buffer overflows, memory leaks, and other security and reliability problems. These are designed to analyze an application's source, bytecode, or binary code to find security vulnerabilities. These tools find security flaws in the source code automatically.

Listed below are the most common errors static code analyzers can detect:

- Use of uninitialized data.
- Memory leaks.
- Potential NULL pointer dereferences.
- Buffer overflows.
- Write to read only memory.
- Out of scope memory usage.
- Use of memory that has been deallocated.

In order to identify all software flaws via static analysis, organizations should use multiple tools from different vendors.

3.1.2.3 Peer Reviews

Peer reviews are an important part of the development phase and help in catching issues at a much earlier stage. Most code reviews are aimed at finding coding bugs, design flaws, and coding standard violations. Though these ensure reliable software, emphasis should also be placed on security analysis.

The reviewer should also consider security-relevant characteristics in the code like:

- Attack surface. The reviewer should think like an attacker and try to find weaknesses and entry points in the system.
- Least privilege. The reviewer should give suggestions if code can be refactored such that the least privileged component gets the least amount of access to resources.

3.1.3 Secure Testing and Verification

A comprehensive test suite that includes functional, performance, regression, and coverage testing is well known to be one of the best mechanisms to assure software is reliable and secure.

Before final penetration tests are performed by specialized teams, a formal security evaluation stage is needed. This is accomplished through dynamic runtime testing methods [1, 2], for example, fault injection systems can also be used to check for the presence of flaws.

4 Threat Analysis

The term "threat analysis" refers to the organized, systematic process of examining an embedded system's points of target and sources from which it might be attacked. A thorough threat analysis is required before any design decisions can be made regarding what methods of protection to use against any attack on the system.

A sound security policy needs to be established as the foundation for embedded system design. To establish this security policy, all threats to the system need to be identified and possible mitigations should be integrated within the product's life cycle.

The following section provides steps that can be taken to complete threat analysis when designing an embedded system.

4.1 Steps to Complete a Threat Analysis

1. Identify what needs to be protected

   It is easy to say that a system needs protection, but we need to specifically identify the assets in the system that need protection. Answers to the questions that follow will enable the identification of assets needing protection.

   a. Who are the possible attackers?

      To help identify what needs protection, we first need to identify who the attackers are? This may include anyone starting from the developer, device manufacturer, distributor, and end user. Some of these attackers may be clever outsiders who have a limited knowledge of the system. They may try to exploit the weaknesses of the system. These attackers may be knowledgeable insiders who have in-depth knowledge of the system. They may have access to sophisticated tools. Another class of attackers may be funded organizations who are specialists. In this case, time and money might not be a constraint.

    b. What could be the possible entry paths for attacks into the system? The entry paths would depend whether an attacker has physical access or remote access to a device.

    - If an attacker has physical access to a device, he can exploit entry interfaces like flash interfaces, UART, or USB. Furthermore, if they have administrative access to a device, they can exploit its services.
    - There can be scenarios where attackers do not have physical access to a device but can work within its proximity, , for example, devices with wireless, Bluetooth connectivity.
    - For devices that have internet connectivity, access to a device becomes very easy for a potential attacker.

2. Identify the threats and create a threat matrix

Once the assets in the system that need protection have been identified the next step is to identify the threats that could compromise them.

You can consult the list of "common threat vectors" to build a risk matrix. Details on possible attack vectors are described later in this chapter.

3. Develop a mitigation strategy by doing a risk assessment

Do a risk assessment to determine possible threat mitigations. This helps to determine the security requirements of the system. Based on the risks associated with the threat, threats can be considered acceptable or requiring mitigation.

4. Identify whether these mitigations can be circumvented or whether they introduce some additional threat to the system—representing new threats to the system.

5. Repeat Steps 3 and 4 until all the threats have either been mitigated or are considered acceptable.

4.1.1 Modeling Threat Analysis

Hunter et al. [3] demonstrate an iterative threat modeling flow for an embedded system. Initially, the first point of weakness and attack needs to be identified. This is followed by a detailed review of security requirements and objectives. Next, modeling must occur for each security requirement, considering the three points of view of an attack scenario discussed above, that is, asset, attacker, and mitigation/defense built in to address the threat. Both the modeling of the specific security requirements and system objects are then iteratively evaluated until a high level of certainty is reached that the model developed provides adequate security against the identified threats and that no new vulnerabilities have been introduced.

4.2 Common Threat/Attack Vectors in Embedded Systems

Attack vectors are ways in which an attacker attempts to gain access to a system and exploit its vulnerabilities to achieve an objective. To understand how to attack a system it is important to understand the objectives of such attacks. Ravi et al. [4] broadly classify attacks into three categories, based on their functional objective:
- Privacy attacks where the objective is to extract secret information stored on an embedded system.
- Integrity attacks where the objective is to change how the embedded system behaves by changing the data or the applications executing on it.
- Availability attacks where the objective is to make the system unavailable to its users. These are typically denial-of-service attacks.

These attacks can be launched on either the hardware or the software of an embedded system. Let's first look into some of the most common attacks. These attacks can be:
- Side-channel attacks.
- Timing attacks.
- Fault injection attacks.
- Physical tampering.

4.2.1 Physical Tampering

There are several ways an attacker, who has physical access to a system, may tamper with it. This tampering can be used to extract secret information or destroy the device. Methods used to achieve this include removing or adding material to the IC to access information. Etching or FIB can be used to remove such materials. Optical inspection can be carried out to read internal signals, probe bus, or memory to extract secret information.

The goal of achieving tamper resistance is to prevent any attempt by an attacker to perform an unauthorized physical or electronic action against the device. Tamper mechanisms are divided into four groups: resistance, evidence, detection, and response. Tamper mechanisms are most effectively used in layers to prevent access to any critical components. They are the primary facet of physical security for embedded systems and must be properly implemented to be successful. From the design perspective, the costs of a successful attack should outweigh the potential rewards.

Specialized materials are used in tamper resistance to make access to the physical components of a device difficult. These include features such as locks, encapsulation, hardened-steel enclosures, or sealing.

Tamper evidence helps ensure that visible evidence is left after tampering has occurred. This can include special seals and tapes which area easily broken, making it obvious that the system has been physically tampered with.

Tamper detection enables hardware devices to be aware they are being tampered with. Sensors, switches, or circuitry can be used for this purpose.

Tamper responses are the countermeasures enacted upon detection of tampering. Measures that can be taken by hardware devices include deletion of secret information and shutting down to prevent an attacker from accessing information.

4.2.2 Side-Channel Attacks

Side-channel attacks are typically noninvasive attacks where things like timing information, power consumption, or electromagnetic radiation from the system can be used to extract secret information. As the name suggests, an attacker does not tamper with the device under attack in any way but uses side channels, observations to mount a successful attack. The observation can be made either remotely or physically, using the right tools. The most common side-channel attacks are architectural/cache, timing, power dissipation, and electromagnetic-emission attacks. Let's try and understand these attacks in a little detail to appreciate how secrets can be extracted using these side channels. The underlying idea of SCAs is to look at the way cryptographic algorithms are implemented, rather than looking at the algorithm itself.

SCAs can be instigated because it is possible to find a correlation between the physical measurements taken during computations and the internal state of an embedded device, which itself is related to a secret key. It is this correlation—with a secret key—that the SCA tries to find.

The power consumption of a device can provide information about the operations that take place and the parameters involved. An attack of this type is applicable only to the hardware implementation of cryptographic algorithms. Such attacks can be divided into two categories—Simple Power Analysis and Differential Power Analysis (SPA and DPA) attacks. In SPA attacks, the attacker wants to essentially guess from the power trace which instruction is being executed at a certain time and what values the input and output have. Therefore attackers need precise knowledge about implementation to mount such an attack. The DPA attack does not need knowledge about implementation details and alternatively exploits statistical methods in the analysis process. DPA is one of the most powerful SCAs that can be mounted, using very few resources. These attacks were introduced by Kocher et al. in 1999 [5]. In this particular case DES implementation in the hardware was under attack.

Popular countermeasures against SCAs include masking and hiding. With masking the line is broken between the processed and algorithmic intermediate value. This means that an intermediate value can be masked by a random number. To remove the mask, changes must be tracked through operations. Such a mask is not predictable and is not known to an attacker. Masking can be done at the architectural or chip level.

"Hiding" helps break the link between processed intermediate values and emitted side-channel information. It tries to make the power consumption of a device independent of the processed intermediates. There are two different strategies adopted for hiding:

- Random power consumption in each clock cycle (SW and HW).
- Equal power consumption in each clock cycle (mainly in HW).

No perfect solution has been found so far for hiding. A list of countermeasures proposed by various authors is available for the further study of this topic [6].

4.2.3 Timing Attacks

Timing attacks are popular and occur when the time taken by cryptographic operations can be used to derive a secret. Cryptographic implementation can be done via hardware or software libraries—both implementation types are vulnerable to this kind of attack.

OpenSSL is a popular crypto library which is often used in Linux web servers to provide SSL functions. Brumley and Boneh [7] demonstrated that timing attacks can reveal RSA private keys from an OpenSSL-based web server over a local network.

In recent times there have been cross-VM timing attacks on OpenSSL AES implementations [8, 9]. Here cache Flush + Reload measurements were used. This type of attack takes advantage of the fact that the executable section of the code is shared between processes. When a process is run for the first time, the operating system loads the process into physical memory. If another use launches the same process for a second time, the operating system will set the page tables for the second process to use the copy that was loaded into memory for the first process. Here, by calculating the time it takes to access any data in shared memory, it is possible to determine if another process accessed it.

In the Flush + Reload attack, both the attacker and victim have some shared memory mapped into their own virtual space. The attacker flushes the lines it is interested in, waits for some clock cycles, and then calculates the time it takes to read those lines again. If the read is fast, it means that the victim's process accessed these lines. These lines can be either in the code area of the victim or some other data the attacker is interested in.

One possible countermeasure for these attacks is to adopt timing-invariant implementation of the algorithms in the hardware as well as the software.

In recent times, these cache attacks have been used in the popular Meltdown and Specter attacks. We will discuss in more detail the Meltdown and Specter attacks in a case study later in this chapter.

4.2.4 Fault Injection Attacks

Fault attacks are active attacks against cryptographic algorithms where the hardware is exposed to random and rare hardware and software computation faults. These faults result in errors which in turn can be used to expose secrets on the chip.

The most common fault injection techniques include underpowering, temperature variation, voltage bursts, clock glitches, optical attacks, and electromagnetic fault injection. All these fault injection methods manipulate the physical layer of the device causing the transistors to switch abnormally.

Clock glitches and voltage bursts/spikes are the most popular form of noninvasive fault attacks where no damage is done to the equipment. Clock glitches can be introduced by supplying a deviated clock signal to chips while voltage spikes are introduced by varying the power supply to the chip. Both these techniques can affect the program as well as dataflow. These glitches can cause the processor to skip the execution of instructions, can change the program counter, or tamper with loops and conditional statements. Effects on the dataflow include the possibility of invalid data in memory reads, computation errors, and corrupt memory pointers. Both these fault attacks are easy to implement and are very inexpensive. Some other examples of noninvasive attacks include exposing the chip to very high or low temperatures and underpowering the device.

Another type of fault attack is the optical attack. These are semi-invasive/invasive in nature. In such attacks, a decapsulated chip is exposed to a strong light source. These attacks require expensive equipment and a complex setup. With a focused laser beam it is possible to set or unset a bit in memory.

An external electromagnetic field can also be used to change memory content. These EM field changes induce eddy currents on the surface of the chip and can cause single bit failures.

Fault attacks can be used to change program flow by attacking critical jumps. For example, say we have an authentication code sequence where a decision needs to be made to pass control to the next image if authentication passes or to stop in the case of failure. An attack on the authentication check can be critical to the execution flow (Fig. 12).

We have just considered an example where attacks on program flow can lead to the skipping of security branches or bypassing of security settings.

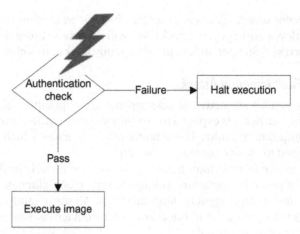

Fig. 12 Fault injection attack.

There are attacks which are launched on I/O loops in the code. Typically, all programs have I/O loops where copies are happening from buffers. Attacks can be launched on the copy loops to:

- Copy either more or less than expected.
- Copy from a different source.
- Copy to a different destination.

These attacks can lead to the wrong initialization of data or keys.

One type of countermeasure against fault attack can be applied at the hardware level to prevent fault injections. Examples include active and passive shields. An active shield can consist of a wire mesh over the chip to detect any interruptions on the wire. Passive shields are metal layers that either cover the chip completely or partially to prevent optical injection or probe attacks. Light sensors and frequency detectors can be added to the chip to detect clock and voltage glitches. However, these countermeasures are costly, and attackers are always on the lookout for ways to bypass these and come up with novel fault injection methods.

Other countermeasures include protecting the software and hardware, so that faults can be detected. These employ redundancy checks to check if a computation has been tampered with and incorporate fault checks to detect and report faults. A detailed discussion of these countermeasures can be found in [10]. In embedded systems, these faults can be detected in different parts of a processor.

1. Input part

If inputs are supplied to an algorithm or implementation externally, any miss on checking these input parameters can result in a fault. For example, typically in a Chain of Trust, for authenticating an image, signatures, public keys, and their lengths are provided externally. If the software does not do bound checks on these externally supplied parameters, attacks on copy loops may result. Say the user supplies a public key length which is greater than the buffer length allocated internally in the software. If no bound checks are performed on key length this would result in the copying of a key greater than the buffer allocated to it. An attacker can intelligently use this buffer overflow attack and modify some decision-making data lying in the periphery of this buffer. Thus proper checks on any input parameters are essential to prevent these kinds of attacks. Validity checking of input parameters is essential before doing any computation.

2. Processing part (memory and data path), program flow

Attacks on processing parts usually attempt to change the program flow by skipping instructions or modifying memory content. Some of the countermeasures against such attacks include:

- Adding parallel and redundant computations. This would mean to compute some operations more than once and comparing results.
- Adding checksum over memory content.
- Adding redundant forms in the coding of flags.
- Checking the specific properties of an algorithm.

4.3 Case Study: Meltdown and Spectre Attacks

4.3.1 Meltdown Attack (CVE-2017-5754)

Meltdown [11] breaks the most fundamental isolation between user applications and the operating system. This attack allows a program to access memory and thus all the secrets of other programs and the operating system. One way an attacker can trigger a Meltdown attack is by making use of an exception. An exception can occur when a user tries to access something from kernel memory. This exception is handled by the kernel. Architecturally, the isolation mechanisms will not allow the user to access kernel memory, but in the short window between when the exception has been handled and control returns to client memory, some user space instructions might get executed out of order. These instructions in the user space can be used to deliberately access kernel memory. Due to execution being out of order, the content, though not visible to the user process, will be in the processor's caches. After the exception is handled, before returning to the user process, the processor would do a cleanup. However, caches do not get cleaned up as part of this process. From the user space, an attacker can run the Flush + Reload attack, as described earlier, to extract this information.

You might be wondering about the practical aspects of this attack. In the Linux kernel, keys are usually stored and can be used for various purposes. One purpose being disk encryption. A user may be able to access this key by using the attack outlined above.

Specter [11] is like Meltdown in the sense that data being accessed speculatively will end up in cache, which is vulnerable to cache attacks looking to extract data. While Meltdown breaks the isolation between the kernel and user process, Specter attacks the isolation between different applications. It allows an attacker to trick error-free programs into leaking their secrets. Examples have been given in a white paper [11], showing how Specter attacks can be used to violate browser sandboxing. A java script code can be written to read the data from the address space of a browser process running it. There are two variants of Specter attack, one which exploits conditional branch misprediction and another which leverages the misprediction of the targets of indirect branches. We will discuss these variants in some detail in the following text.

4.3.2 Specter Variant 1—Bound Check Bypass (CVE-2017-5753)

Given below is the example stated in the Specter white paper [12].

```
if (x < array1_size)
    y = array2[array1[x] * 256];
```

Let's try and define what we are going to steal using this attack. Let us assume that the target program has some secret data stored right after array [13]. This is what the attacker wants to get his hands on.

In the code snipper above, input to the program is x. Bound checking is done on "x" as expected so that extra data beyond the array [13] index in array2 does not get accessed. When the code executes, if the array1_size variable is not in cache, the processor will speculatively fetch and execute the next set of instructions. In this case, if the value of x is greater than array1_size, due to speculative fetch, bound check would be by-passed and the processor would fetch the data in array2 at the location denoted by array1[x]. However, his doesn't solve our problem, does it? The attacker wants to find out the value of array1[x], with x being his invalid input pointing to some secret data in the target program. How does he find that value? For this he would utilize cache timing attacks as discussed in the previous sections.

Assuming array1 is a uint8_t-type variable, the possible values of array1[x] range from 0 to 255. So, this means access would be happening from array [3] at locations 0 *256,255 * 256. The attack process can use cache attacks (Prime + Probe) to fill the entire cache with values. If the value at array1[x] is 0×20, then the location pointed out by array2[0×20] will be evicted out of cache. The attacker can then measure the timing they need to fetch their data and predict the value. This is a very simplified example to help users understand how the attack can be implemented. For further details refer to the Specter white paper [12].

4.3.3 Specter Variant 2—Branch Target Injection (CVE-2017-5715)

Systems with microprocessors utilizing speculative execution and indirect branch prediction may allow unauthorized disclosure of information to an attacker with local user access via side-channel analysis. Here the user tricks the branch predictor and influences the code which will be speculatively executed. Usually processors have a branch target buffer (BTB) to store the branch target prediction.

5 Components of Secure Embedded Systems

5.1 Building a Trusted Execution Environment

Embedded devices, specifically smart phones, provide an extensible operating system giving the user the ability to install applications and do variety of things. With this flexibility, comes a wide range of security threats. This highly extensible and flexible environment is usually referred to as the rich execution environment (REE). To protect the assets of the system and assure integrity of the code being executed along with the REE, we need a trusted execution environment (TEE). In simple words a TEE can be defined as the hardware support that is required for platform security.

Global Platform defines a TEE as "a secure area that resides in the main processor and ensures that sensitive data is stored, processed and protected in a trusted environment" (Fig. 13).

A TEE needs to ensure that:

1. The code executing inside it can be trusted.

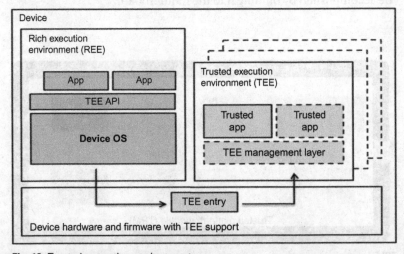

Fig. 13 Trusted execution environment.

2. The code that runs on it executes in an isolated environment in order to protect its assets from the REE.
3. It provides secure storage that the REE doesn't have access to.

To realize a TEE an external security coprocessor, like TPM, can be connected to the SoC. Since this is a separate chip it provides complete isolation.

Another way of realizing a TEE is to have an on-chip security subsystem which can fulfill its requirements.

Another architecture can be such that the processor and the other peripherals provide a secure environment without the need for a different entity. ARM TrustZone is an example of such a TEE. This will be considered in detail in the following sections.

5.1.1 TPM

The Trusted Computing Group (TCG) is a not-for-profit organization formed to develop, define, and promote open, vendor- neutral, global industry specifications and standards, supportive of a hardware-based Root of Trust, for interoperable trusted computing platforms. They define standards for what is called TPM. TPM is a dedicated secure crypto-processor are designed to secure hardware or software by integrating cryptographic keys into a device. TPM chips are passive and execute commands from the CPU.

The main objectives of TPM include:

1. Protecting encryption and public keys from external stealing or misuse by untrusted components in the system.
2. Preventing malicious code from accessing secrets from inside it.
   Given below is a high-level block diagram of a TPM chip (Fig. 14).
   TPM must be physically protected from tampering. In PCs this can be accomplished by binding it to the motherboard.

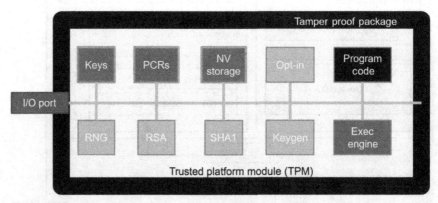

Fig. 14 TPM.

There is an I/O port which connects the TPM chip to the main processor. The data transmitted over this I/O port follows standards specified by the TCG. The I/O block is responsible for the flow of information between the components inside TPM, and between TPM and the external bus.

TPM consists of a lot of cryptographic blocks which help provide cryptographic isolation. The RNG block is the true random bit stream generator. Random numbers produced by the RNG can be used to construct keys, provide nonce, etc. It has a SHA1 engine to calculate hashes which can be used for PCR extension, integrity, and authorization. There is an RSA engine to execute the RSA algorithm. The RSA algorithm can be used for signing, encryption, and decryption.

TPM also has some nonvolatile storage to store long-term keys. Two long-term keys are the Endorsement Key and the Storage Root Key. These form the basis of a key hierarchy designed to manage secure storage. NV storage (nonvolatile) is also used to store authorization data like owner passwords. Such passwords are set during the process of taking ownership of TPM. Some persistent flags related to access control and Op-In mechanisms are also stored here.

PCRs are used to store integrity metrics to store measurements. These are reset every time the system loses power or restarts.

Keys form part of nonvolatile memory that is used to store them for crypto operations. For a key to be used it needs to be loaded into TPM.

TPM acts as the Hardware Root of Trust, providing:
- Root of Trust for measurement.
  - TPM uses PCR (Platform Configuration Registers) to save the state of the system.
- Root of Trust for reporting.
  - TPM acts as an entity to report information accurately and correctly.
  - PCR and RSA signatures are used for this purpose.
- Root of Trust for storage.
  - TPM uses PCR and RSA encryption to ensure that data can be accessed only a when platform is in a known good state.

5.1.2 Secure Element

A secure element (SE) is a tamper-resistant platform (typically a one-chip secure microcontroller) capable of securely hosting applications and their confidential and cryptographic data (e.g., key management) in accordance with the rules and security requirements set forth by a set of well-identified, trusted authorities.

The main features of a SE are:
- Hardware-supported cryptographic operations.
- Execution isolation.
- Data protection against unauthorized access.

5.1.3 ARM TrustZone

ARM TrustZone technology provides protective measures in the ARM processor, bus fabric, and system peripheral IPs to provide system-wide security. TrustZone technology is implemented in most ARM modern processors including Arm Cortex-A cores and the latest Cortex-M23- and Cortex-M33-based systems.

ARM TrustZone starts at the hardware level by creating two worlds that can run simultaneously on a single core: a secure world and a nonsecure world. Software either resides in the secure world or the nonsecure world. A switch between the two worlds can be done via a monitor call in Cortex-A processors or using core-logic in Cortex-M processors. This partitioning extends beyond the ARM core to memory, bus, interrupts, and peripherals within an SoC.

The ARM core or processor can run in two modes: secure and nonsecure mode. This state of the processor is depicted by a flag called NS. This flag is propagated to the peripherals though the bus (AMBA3 AXI system bus). Between the bus and the various peripherals, such as external memory and I/O peripherals, sits a gatekeeper. This gatekeeper allows/restricts access to these external resources based on policies set. Examples of these gatekeepers are:

1. TZASC (TrustZone Address space controller) for external memory.
2. TZPC for the I/O peripherals.

TrustZone technology within Cortex-A-based application processors is commonly used to run a trusted boot and a trusted OS to create a TEE. Typical use cases include the protection of authentication mechanisms, cryptography, key material, and digital rights management (DRM). Applications that run in the secure world are called Trusted Apps [14].

TrustZone for Cortex-M is used to protect firmware, peripherals, and I/O, as well as provide isolation for Secure Boot, trusted update, and Root of Trust implementations while providing the deterministic real-time response expected for embedded solutions [14].

5.2 Hardware Root of Trust

What is the definition of a Trustworthy Embedded System? A trustworthy system is a system which does what its stakeholders expect it to do, resisting attackers with both remote and physical access, else it fails safe.

Such a system should have features that allow its stakeholders to prevent or strongly mitigate an attacker's ability to achieve the following attacks:

• Theft of functionality.
• Theft of user or third-party data.
• Theft of uniqueness.

A trusted system can be built by rooting the trust in hardware and continuing the Chain of Trust to ensure only authentic software runs on the system (Fig. 15).

Root of Trust begins with a piece of immutable code which can not be changed during the life cycle of an embedded product. This code lies in the ROM (read only memory) of embedded systems and is the first to execute after boot. It is the responsibility of this code to ensure the authenticity of the next-level code before passing control to it. This next-level image is responsible for authenticating the next image, this is how Chain of Trust continues. Typically, ROM code authenticates the boot loader and the boot loader authenticates the operating system which further authenticates user space applications (Figs. 16 and 17).

This CoT is also referred to as a Secure Boot Chain of Trust. Let's try and understand the significance of authenticating the images. Authentication ensures that the image is from a genuine stakeholder who has the required private key.

Images are typically signed offline using a private key from an asymmetric key pair (e.g., RSA). This signature is then verified using the corresponding public key on the Silicon before the image is executed. It is essential to tie this public key, used in the authentication process, with the underlying hardware Root of Trust. This can be done via a comparison of the hash of this public key with the hash stored in some secure immutable memory on the SoC. This memory can be in the form of one-time programmable fuses which are programmed as part of production when a device is manufactured.

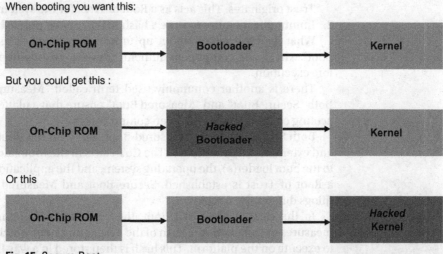

Fig. 15 Secure Boot.

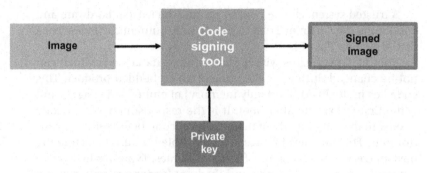

Fig. 16 Signing an image using a cryptographic key.

Verifying the signed image insures that the right image is

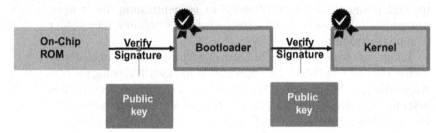

Fig. 17 Chain of Trust.

To build a Root of Trust in embedded systems, two things are essential:

1. Immutable code which authenticates the next level of code (ROM generally), called the trusted building block (TBB), where Chain of Trust originates. This acts as a Root of Trust for the system.
2. Immutable memory to store a hash of the trusted public key.

What we have talked about up to now is referred to as "Secure Boot" where one component authenticates the next component before execution.

There is another commonly used term called "Measured Boot." Both "Secure Boot" and "Measured Boot" ensure that a platform is executing code which has not been compromised.

Both Secure Boot and Measured Boot start with the Root of Trust and extend a "Chain of Trust." The CoT starts in the root, and spreads to the boot loader(s), the operating system, and the applications. Once a Root of Trust is established, Secure Boot and Measured Boot do things differently.

In the case of Measured Boot, the current running component measures or calculates the hash of the next component which is going to execute on the platform. This hash is then stored in a way that it can be retrieved later to find out which components ran on the system.

The Measured Boot does not make any decision in terms of good or bad, neither does it stop the platform from running. These measurements are used for attestation with remote servers to ensure that the required software is running on the SoC.

One of the main requirements for Measured Boot is that these hashes (measurements) need to be stored in a location which can be trusted and not easily manipulated. This location would serve as Root of Trust for Storage. The TPM is typically used to store these measurements.

The TPM is a small self-contained security processor that can be attached to a system bus as a simple peripheral. More details on the TPM are available in the next section. Here we will discuss in brief the function provided by the TPM which helps in the Measured Boot. One of the functions a TPM provides is called PCRs, used for storing hashes.

These registers in the TPM are cleared only at hardware reset and cannot be written to directly. The value in these PCRs can be "extended," that is, the existing value of the PCR is taken along with the new value, and they are concatenated, producing a 40-byte value. Then, the hash of that value is taken and stored in the PCR. Thus as the platform boots, each measurement is stored in the PCRs in a way that unambiguously shows which modules were loaded.

TPM can report these values, signed by a key that only the TPM can access. The resulting data structure, called a Quote, gives the PCR values and a signature, allowing them to be sent to a Remote Attestation server via an untrusted channel. The server can examine the PCRs and associated logs to determine if the platform is running an acceptable image.

Secure Boot with CoT and Measured Boot together help to create an architecture which is resistant to any malware in the boot software, generally referred to as rootkits.

5.3 Operating System Security Considerations

We have come a long way from the time when embedded systems were meant to run a single application, to where present-day embedded systems behave like mini computers. The smartphone market is an obvious example of this. Gone are the days when devices used to operate in isolated environments. With the advent of the IoT (Internet of Things), we have devices which are always connected to a variety of public networks or proprietary networks. Connection to the Internet, increases the possibility of the exposure of embedded devices to cyberattacks. Such attacks are not limited to the Internet, even proprietary networks are vulnerable—a good example of this being the Stuxnet attack. In the Stuxnet attack, traditional malware techniques were used to take over a proprietary network in a locked down facility in Iran.

Over time operating systems have become more and more complex. This increase in size and complexity means that it is not possible to examine all the OS software for security vulnerabilities and issues. These cyberattacks increase the need for built-in security in OSs as these attacks can easily work their way into devices through OS vulnerabilities.

The OS needs to have built-in security features to thwart these attacks regardless of how they enter the system. These features need to focus on:

1. Providing application isolation.
2. Ensuring the integrity and authenticity of applications.
3. Ensuring confidentiality of data, that is, protecting data at rest. This includes applications as well as user data.
4. Protection from network attacks.
5. Protection when data is in transit or motion.

These can be achieved if the operating system can enforce fine-grained separation between the user and access to resources. This can be done by defining security policies. Furthermore, it is important to ensure that this separation is effective by making sure that execution is completed through a trusted execution path. This path should be free from any flaws and vulnerabilities.

The following sections describe some key security features that can be built in to operating systems for application and data security.

5.4   Application Level Security

5.4.1   Access Control

Access control is required to ensure that only authorized users and processes can access resources they are entitled to access. These resources include not only data files but memory, I/O peripherals, and other critical resources of the system.

At a high level, access control policies of a system can be divided into following categories:

1. Discretionary Access Control (DAC).
2. Mandatory Access Control (MAC).

DAC, as the name suggests, is discretionary, that is, at the discretion of the user. The user decides the policies on its objects. For example, when a user creates a file, they decide who can has certain permissions on that file. These permissions are stored in the inode associated with the file. Thus each object on a DAC-based system has what is called an ACL (access control List). The ACL has the complete list of users and groups which the creator of an object has granted access to. Here the user who owns the resource gets total control whether they want it or not. One important point to note is that a user can set/change permissions for resources they own. There is another category called the super user, where the DAC policy for managing the system can be bypassed.

In MAC, it is not the user but the system administrator or a central user that controls what resources each user gets access to. It is stricter than DAC and helps in containing bugs in the user space software. In MAC, the object is associated with a security label instead of an ACL. The label contains two pieces of information:

- The security classification of an object—if the object is secret, top secret, or confidential.
- The category of the object which indicates the level of user that is allowed access to that object, for example, management level or the project to which the object/resource is available.

Each user account also has a classification and category associated with them. A user is allowed access to an object only if both the category and classification of the object match that of the user. SELinux, AppArmour, and SMACK are some of the widely used implementations of MAC in the Linux world [15].

5.4.2 Application Sandboxing

Application sandboxing helps to isolate applications from critical system resources thus adding a security layer to prevent malware from affecting systems. Sandboxing is also sometimes referred to as "jailing." It provides a safe environment which is monitored and controlled, such that the unknown software cannot do any harm to the host computer system [16]. It ensures that a fault in one application doesn't bring down the complete system.

Virtualization is one of the ways of achieving application sandboxing. Virtualization is the use of hypervisors or virtual machine monitors to create and manage individual partitions that contain guest OSs on a single real machine. The hypervisor allocates system hardware resources, such as memory, I/O, and processor cores, to each partition while maintaining the necessary separation between operating environments. A hypervisor enables hosting multiple virtual environments over a single physical environment. A critical function of the hypervisor from a security stand point is to maintain isolation between partitions and continue running even if another OS crashes [17]. The ability to maintain isolation is highly dependent on the robustness of the underlying hypervisor. There are two types of hypervisors:

1. Type 1 hypervisors.
2. Type 2 hypervisors.

Type 1 hypervisors run on bare metal while Type 2 hypervisors have an underlying operating system. Since the security of a Type 2 hypervisor depends on the underlying host operating system, these hypervisors are not used in mission-critical deployments.

CPU hardware assists are generally used for implementation of hypervisors in embedded systems. Popular CPU architectures like PowerPC and ARM have hypervisor extensions defined in them.

Apart from CPU extensions, ARM architecture also provides another capability called ARM TrustZone which provides ways to partition systems into two zones: secure and nonsecure. Trusted software which uses secrets, like keys, completes cryptographic operations, and operates digital rights management software, etc., can be run in the secure world which is isolated from the nonsecure world. Further details about ARM TrustZone are provided in the Section 5.1.

Linux containers are also used for providing application isolation, apart from virtualization. While hypervisors are used to provide virtualization, containers use the functionality of underlying OSs, like namespaces, to restrict applications from accessing certain system resources, files, etc. This effectively means that applications share the same operating system but have separate user spaces. With containers, the operating system, not the physical hardware, is virtualized (Fig. 18).

Let's discuss the security aspects of the two approaches. If an application running in a container has some vulnerability and affects the operating system, all other applications running in the container would be affected. In a similar situation in the case of a virtual machine, only the OS running that application is affected, leaving the host OS and other VMs unaffected. While a container uses software mechanisms to

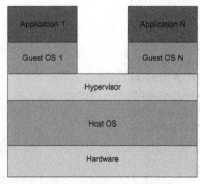

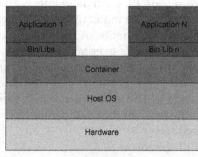

Fig. 18 Virtualization and containers.

achieve isolation, virtualization is tied up with hardware and provides more security. However, this added security comes at a price—performance is lower in the case of VMs since many context switches are involved. Containers have lower overheads and are less resource-heavy. So, you need to choose the right sandboxing methodology based on your use case and requirements. For a constrained embedded device, not capable of running virtual machines, containers seem to be the first practical virtualization technology.

Container adoption is on the rise in IoT devices that have limited storage space, bandwidth, and computing power. Docker is a popular container technology which is built on LXC and has an "easy button" to enable developers to build, package, and deploy applications without requiring a hypervisor.

5.4.3 Application Authenticity

Attackers usually try to modify existing code or inject malicious code into a system, tricking the user to run it. This can be prevented if applications are authenticated before execution. Authentication helps ensure that application code is from a trusted source and has not been modified. A typical way of doing this is by using certificates and signatures. The hash of an application can be compared with the hash present in the certificate that comes along with the application. The certificate always comes from a trusted authority. For example, Apple iOS implements this by enforcing all applications through the App Store.

Normally asymmetric cryptography, using public and private keys, is used for authenticity, but this same effect can be achieved using symmetric key hashes too, like HMAC. Both methodologies have their pros and cons. When using a signature, the confidentiality of the private key needs to be ensured by a single authority and the system just needs to protect the integrity of the public key used for verification. However, since the system doesn't have the private key, it cannot resign the application or file in case any changes are made to it. This schema can be used for read-only files and applications which don't change during the lifetime of a system. However, if there are security- critical files which change during the lifetime of a system, they need to be protected by a local symmetric key. The caveat being that this local symmetric key needs to be carefully protected to prevent attackers from using it to sign a malicious application.

Linux IMA (Integrity Measurement Architecture) and EVM (Extended Verification Module) are example frameworks which have been built in Linux for application integrity and authenticity. These frameworks have been available in the Linux kernel since 2.6.30. Linux integrity frameworks provide the capability to detect whether files have been accidentally or maliciously altered, either remotely or locally, appraise a file's measurement against a "good" value stored as an extended attribute, and enforce local file integrity. These goals are complementary to Mandatory Access Control (MAC) protections provided by LSM modules, such as SElinux and Smack, which, depending on the policy, attempt to protect file integrity [17].

At a very high level, the IMA and EVM provide the following functionality:

- *Measurement* (hashing) of file content as it is accessed and tracking of this information in an audit log.
- *Appraisal* of files, preventing access when a measurement (hash) or digital signature does not match the expected value.

The IMA maintains a runtime measurement list which can be anchored in hardware (e.g., TPM) to maintain an aggregate integrity over the list. Hardware anchoring in TPM, or in any other way, helps to ensure that the measurement lists can't be compromised by a software attack. Further details about this infrastructure can be found in the Linux kernel documentation [17].

5.4.4 Case Study: Chain of Trust Along With Application Authenticity Using IMA EVM on Layerscape Trust Architecture–Based SoCs Without Using TPM

The Secure Boot mechanism (as shown below) is provided by NXP on its Layerscape trust architecture SoCs. This mechanism establishes a Chain of Trust with every image being validated before execution.

Root of Trust is established in boot ROM execution phase, by validating the boot loader image before passing control for its execution. Each firmware image is appended with a header. This header contains security information related to the image, such as an image signature or public key. The Chain of Trust ends after validation of the fit image or kernel image (Fig. 19).

Rootfs is another important entity that needs to be authenticated before passing control for its execution. Rootfs can be used in following ways:

1. Rootfs is the part of the fit image (combined kernel/device tree/initramfs image) that is validated using the standard Secure Boot mechanism in the Chain of Trust. In this case, rootfs will always be expanded in RAM and hence no new application or image can be added at runtime.
2. Rootfs is placed on some persistent storage device, such as an SD, SATA, or USB device. In this case, rootfs can be expanded at runtime but cannot be validated using a standard Secure Boot mechanism. On each boot to validate rootfs content we need to leverage the mechanism provided by the Linux kernel.

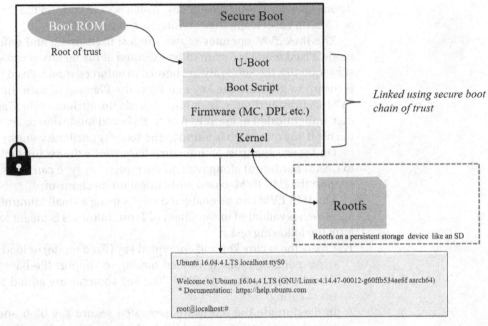

Fig. 19 Chain of Trust.

One such mechanism provided by the Linux kernel for validating rootfs content is the IMA EVM feature. This provides file-level authentication as discussed in the previous section.

The IMA EVM uses an encrypted-type key. An encrypted key blob for a user is derived by the kernel using the master key. The master key can be a:

1. User key type. This is independent of any hardware and is dependent on the user mechanism to protect its content.
2. Trusted key type. This is dependent on TPM hardware and is tied to it. The user is only able to access the blobs signed by the TPM hardware.
3. Secure key type. This is dependent on the NXP CAAM (Cryptographic Accelerator and Assurance Module).

A secure key is generated using the CAAM security engine which constitutes random bytes. The key contents are stored in kernel space and are not visible to the user. The user space will only be able to see the key blob.

Blobs are special data structures for holding encrypted data, along with an encrypted version of the key used to decrypt the data. Typically, blobs are used to hold data to be stored in external storage (such as flash memory or in an external file system), making the contents of a blob a semipersistent secret. The secrecy of a blob depends on a device-specific 256-bit master key, which is derived from the OTMPK or ZMK on Layerscape trust architecture–based SoCs.

The IMA EVM operates in two modes: fixed mode and enforced mode. Fixed mode is meant to be executed at the factory setup stage, subsequently, the SoC is always booted in enforced mode. Fixed mode is meant to generate the key and label the file system with the IMA EVM security attributes. In enforced mode the attribute values are either authenticated or appraised only. Enforced mode denies access to the file if any mismatch is found in the security attribute values.

On Layerscape trust architecture–based SoCs the secure key (tied to CAAM hardware) along with an encrypted key type can be used to support the IMA EVM–based authentication mechanism.

The IMA EVM can be enabled on SoCs using a small initramfs image which is validated in the Chain of Trust. Initramfs is meant to perform the following tasks:

1. Create the secure key and encrypted key (fixed mode) or load their corresponding blobs (enforced mode), to support file-based authentication by the IMA EVM. The key contents are added to the user keyring.
2. In fixed mode the hardware-generated secure key blob and the kernel-generated encrypted key blob are saved on the main rootfs mounted over some storage device.
3. Enable the EVM by setting an enable flag in the securityfs/evm file.
4. Switch control to main rootfs for its execution.
The Chain of Trust with the IMA EVM is shown in Fig. 20.

5.4.5 Application Execution

Attackers have in the past exploited (and will probably continue to exploit) applications through user supplied input. One of the most common and oldest forms of attack is the "buffer overrun," where user-supplied input goes unchecked and ends up writing directly to the operating system and application memory that is normally used to store the application execution code and temporary and global data. Instead, an attacker supplies sufficient data to take control of application execution (by manipulating the stack pointer) and executes, within the application context, the data and code they have written to memory rather than continue to execute the application. To mitigate this attack, several platforms and operating systems, such as Windows XP onwards, Apple iOS, Android, and SELinux, all mark application data as nonexecutable so that even if the attacker manages to write data to memory they will struggle to execute that data.

To mitigate the nonexecution of the overwritten data or where space available is too small to contain all the malicious instructions, attackers attempt to use another technique, "return to-lib-c/return oriented programming." In this case, they attempt to use already

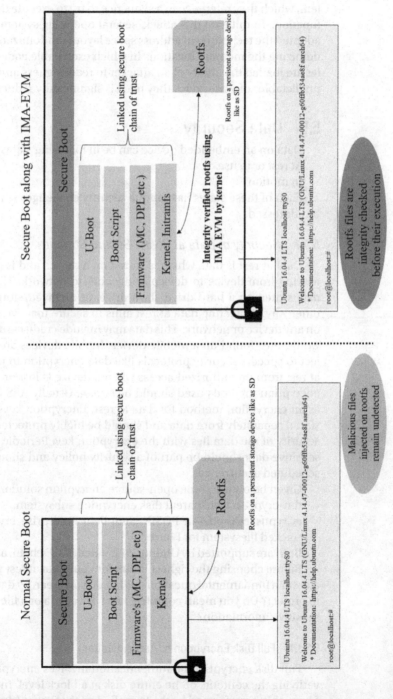

Fig. 20 CoT with the IMA/EVM.

preloaded and existing libraries and the code of the operating system, which they reference in a sequence to try to execute their desired function. To mitigate this attack, several operating systems have also adopted the technique of address space layout randomization. By randomizing the memory locations in which executable code and libraries are loaded the ability of an attacker to readily guess and access the predictable software codes they need is significantly reduced.

5.5   Data Security

Data on an embedded device can be in following states:
- At rest or in use
- In motion

Each of these states has unique security challenges which need to be addressed.

5.5.1   Security of Data at Rest—Secure Storage

Data at rest is data which is stored on a device and is not actively moving from device to device or network to network. This includes data stored on a hard drive, flash drive, or archived/stored in some other way. Protecting data at rest aims to secure inactive data stored on any device or network. This data may include credit card pins, passwords, etc., found on a mobile device. Mobile devices are often subject to specific security protocols like data encryption to protect data at rest from unauthorized access when a device is lost or stolen. The encryption methods used should be strong. Usually AES is preferred as an encryption method for data at rest. Encryption keys should be stored separately from data and should be highly protected. The total security of the data lies with the encryption key. Periodic auditing of sensitive data should be part of a security policy and should occur at scheduled occurrences.

Given below are some open-source encryption solutions in Linux:
- dm-crypt—a transparent disk encryption subsystem.
- eCryptfs-eCryptfs—a POSIX-compliant enterprise cryptographic stacked filesystem for Linux.

Both are supported by Ubuntu, SLES, RedHat, Debian, and CentOS.

Before choosing the right solution for your data at rest you need to answer a fundamental question—What do you mean by data you want to protect? Do you mean complete hard drive data or a file containing sensitive information?

5.5.1.1   Full Disk Encryption or Authentication

Full disk encryption or authentication involves encrypting and/or verifying the contents of the entire disk at a block level. In the case of Linux, this is performed by the kernel's device mapper (dm) modules.

This method can be used with block devices only (e.g., EMMC and NAND). This software is called dm-crypt and works by encrypting data and writing it onto a storage device (by way of the device driver) using a storage format called LUKS.

Linux Unified Key Setup (LUKS) is a specification for block device encryption. It establishes an on-disk format for the data, as well as a passphrase/key management policy. LUKS uses the kernel device mapper subsystem via the dm-crypt module. This arrangement provides low-level mapping that handles encryption and decryption of device data. User-level operations, such as creating and accessing encrypted devices, are accomplished using the cryptsetup utility (Fig. 21).

5.5.1.2 Directory/File Level Encryption/Authentication

Data can be protected at the directory or file level too. Some mechanisms available in Linux that offer this protection are:
- ubifs.
- ecryptfs.

5.5.2 *Protecting the Key Used for Encryption*

The key which is used for encryption needs to be protected. Mechanisms that can be used to protect this key include:
- Storing the key in external crypto or a security chip–like SE.
- Using TPM.
- Encrypting the key using an SoC mechanism (details are given below).

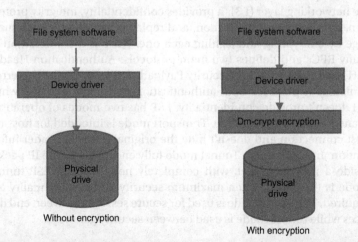

Fig. 21 Encryption using dm-crypt.

5.5.3 Security of Data in Motion—Secure Communication

To protect data in motion, an encrypted channel needs to be created for moving data. This encrypted channel can exist at the application layer or at the transport layer. We often select transport layer protections given our desire for code reuse and the wealth of battle-hardened encryption technologies available at the transport layer. The two most widely used mechanisms for transport layer encryption are Transport Layer Security (TLS) or IPsec.

IPsec, TLS, and SSH share a common goal, that is, to provide a secure connection between two peers/devices/endpoints. The difference between them is the layer at which they execute. There is no preferred protocol in that they all offer certain benefits. To decide which one to use, you really need to understand what you are trying to secure. Once you understand that, the choice of which network security protocol to use becomes easy! The security services provided by these protocols include:

- Message integrity—ensuring that a message has not been altered to prevent theft of service. This is achieved by message signing using digital signatures, HMAC, etc.
- Authentication to provide mitigation against theft or masquerading attacks. This is provided via user authentication using X 509 certificates.
- Confidentiality to mitigate against eavesdropping using symmetric algorithms.
- Nonrepudiation to ensure accountability (i.e., the person cannot later deny sending the message).

5.5.3.1 IPSec

IPsec (RFCs 2401, 2406, 2409, 2411) is a protocol suite that runs at the networking layer (L3). It provides confidentiality, integrity protection, data origin authentication, and replay protection for each message by encrypting and signing each one. IPsec is a combination of many RFCs and defines two main protocols: Authentication Header (AH) and Encapsulating Security Payload (ESP). ESP is the preferred choice as it provides both authentication and confidentiality while AH doesn't provide confidentiality. ESP has two modes of operation: Transport and Tunnel Mode. Transport mode is intended for host-to-host connection and doesn't hide the original packet's header information. In comparison Tunnel mode fully encapsulates the IP packet inside a new IP packet, with completely new headers. ESP tunnel mode is the choice when maximum security and confidentiality are required. Transport mode is used for secure sessions between end devices while tunnel mode is used between security gateways.

In order to establish an IPsec Security Association (SA) between two endpoints, the SAs need to be dynamically established via a key management protocol. This is normally done via IKEv1/IKEv2 in the Internet world. The peer who wants to establish a secure connection with a remote host sends the host its identification information in the form of a certificate. It also sends random data to check that messages are still alive and not being replayed. The data is signed by the initiator to assert their origin. The receiving peer verifies the signature to authenticate the sender and then signs the same data and sends it back to the initiator for the converse operation. Each peer computes session keys, based on the exchanged data and agreed algorithm, typically a variant of the Diffie Helman algorithm. These keys are used during session communication.

New generation embedded processors have security engines/ crypto accelerators that can to help accelerate the performance of IPSec dramatically. For example, NXP QorIQ, Layerscape family processors have IPSec off-load engines. These engines have flow-through capability which means that they can handle the bulk of the IPSec procession without intervention by the core.

5.5.3.2   SSL/TLS

Transport Layer Security (TLS—RFC 2246, 4346, and 5246) is based on SSLv3. It is a Layer 4 protocol as it runs directly on top of TCP ONLY. It uses PKI to provide user authentication as well as symmetric keying for confidentiality protection. It is designed to prevent eavesdropping, tampering, and message forgery. It establishes a secure connection between two peers using origin authentication and session establishment. TLS authentication can be mutual authentication or only server-side authentication:

- Mutual authentication. Both parties (server and client) exchange certificates when establishing a session. Each party must validate the other's certificate. This adds to security but is computationally expensive as it is based on public key cryptography.
- Server-side authentication. Only the server provides a certificate when establishing a session. This is the common solution we see today when using secure web sessions between a user and a web server (HTTPS request). This approach avoids the extra computational overhead of a PKI operation on the User Equipment. It provides medium-level security, with lower CPU requirements being placed on the UE. Also, the user is not mandated to own a valid certificate.

OpenSSL is a popular open-source SSL/TLS stack. It consists of two major components: libssl, implementation of the SSL/TLS protocol and libcrypto, which is a cryptographic library. GNUTLS is another open-source SSL/TLS library.

Questions

1. What is embedded security?
2. At what stages should security be considered during the development life cycle of embedded systems? Give examples.
3. How do you build Chain of Trust in an embedded device?
4. What is "data in motion"? What kind of security measures can be applied to protect this data?
5. What is "data at rest"? How can you protect data at rest?
6. What are the common threat vectors to an embedded device?
7. List some security features which need to be built-in in an operating system.
8. What is application-level security? How can you achieve it?
9. What is cryptology?
10. Describe the CIA Triad.

References

[1] J.R. Larus, T. Ball, M. Das, R. DeLine, M. Fh-ndrich, J. Pincus, S.K. Rajamani, R. Venkat-apathy, Righting software, IEEE Softw. 21 (3) (2004) 92–100.
[2] C. Wang, J. Hill, J. Knight, J. Davidson, Software tamper resistance: Obstructing static analysis of programs, in: Technical Report, Univ. of Virginia, 2000.
[3] https://etd.auburn.edu/bitstream/handle/10415/4889/Secure%20Design%20Considerations%20for%20Embedded%20Systems_ETD_fixes.pdf?sequence=2.
[4] S. Ravi, A. Raghunathan, S. Chakradhar, Tamper resistance mechanisms for secure embedded systems. in: Proceedings of the 17th Internatioanal Conference of VLSI Design, (VLSID' 04), IEEE Xplore Press, 2004, pp. 605–611, https://doi.org/10.1109/ICVD.2004.1260985.
[5] P. Kocher, J. Jaffe, B. Jun, Differential power analysis. CRYPTO'99, LNCS 1666 (1999) 388–397.
[6] https://csrc.nist.gov/csrc/media/events/physical-security-testing-workshop/documents/papers/physecpaper19.pdf.
[7] D. Brumley, D. Boneh, Remote Timing Attacks Are Practical, in: Proc of 12th Usenix Security Symposium, 2003.
[8] https://eprint.iacr.org/2014/248.
[9] https://eprint.iacr.org/2014/435.
[10] https://lirias.kuleuven.be/bitstream/123456789/395845/3/article-2204.pdf.
[11] https://meltdownattack.com/meltdown.pdf.
[12] https://spectreattack.com/spectre.pdf.
[13] https://etd.auburn.edu/bitstream/handle/10415/4889/Secure%20Design%20Considerations%20for%20Embedded%20Systems_ETD_fixes.pdf?sequence=2.
[14] https://www.arm.com/products/security-on-arm/trustzone.
[15] https://www.linux.com/learn/overview-linux-kernel-security-features.
[16] https://www.uni-obuda.hu/journal/Vokorokos_Balaz_Mados_57.pdf.
[17] http://linux-ima.sourceforge.net/.

Further Reading

[18] https://www.theseus.fi/bitstream/handle/10024/135754/Nayani_Srinivas.pdf?sequence=1&isAllowed=y.

[19]   http://www.grandideastudio.com/wp-content/uploads/secure_embed_paper.pdf.

[20]   https://www.itu.dk/~/media/d602e06412af44b69e3c86924fca9820.ashx.

[21]   https://www.arm.com/products/security-on-arm/trustzone.

[22]   https://asokan.org/asokan/Padova2014/tutorial-mobileplatsec.pdf.

[23]   http://www.cloudauditcontrols.com/2014/09/mac-vs-dac-vs-rbac.html.

[24]   Embedded Systems Security, Kliedermacher and Kliedermacher; Chapter 2; Feb, 2013 http://www.edn.com/design/systems-design/4406387/1/Embedded-Systems-Security.

[25]   https://www.linux.com/learn/how-encrypt-linux-file-system-dm-crypt.

15

MACHINE LEARNING AT THE EDGE

Markus Levy*, Filip Naiser†

NXP Semiconductors, Eindhoven, The Netherlands, †Center for Machine Perception, Czech Technical University in Prague, Czech Republic

CHAPTER OUTLINE

Software Engineering for Embedded Systems. https://doi.org/10.1016/B978-0-12-809448-8.00015-1

1   Introduction

In this chapter we start by introducing machine learning (ML). We explain the basic terminology such as supervised and unsupervised ML. We explain the basic ML tasks called classification and regression. Then we introduce basic algorithms such as nearest neighbor or support vector machine (SVM), and speak about decision trees and in reference to an example of decision trees we explain ensemble techniques as well as boosting and bagging techniques.

In the second part of this chapter we focus on neural nets (NNs) and explain the basic concept of a neuron and how neurons are arranged into networks and how the basic mechanics of NNs work. Then we examine the learning process in these networks and the associated backpropagation algorithm.

This is an appropriate place to insert a disclaimer. Machine learning (mainly the domain of deep learning) is changing so rapidly that what you read might not be 100% valid. The best approach to getting the most out of this chapter is to take it as a starting point, learn the principles, follow the suggested literature references, and see if they are cited in fresh papers—the optimum way to keep up to date with current state-of-the-art information.

We assume that a reader has basic knowledge of linear algebra, probability and statistics, calculus, and optimization.

1.1   Coding Examples

In this chapter you will find code examples. However, unlike the other chapters in this book these examples will be written in Python, primarily because most of the ML libraries and supporting tools are based on this programming language. We are aware that Python is slower than C++ and similar languages. On the other hand, most of the libraries, like NumPy [1], are written in C++ and well optimized and Python is just an interface enabling much easier and faster development, and the computational overhead is negligible. Python code is also more readable and shorter allowing us to provide more examples. Furthermore, Python is more suitable for ML when there is a need for a fast way to test hypotheses. Once a proper way is found Python be rewritten and optimized into C++ format.

To be able to run the code examples the reader needs to have access to Python 3.7.* and the following libraries: NumPy, Matplotlib,

Scikit-Learn, PyTorch, torchvision. These are widely used on most platforms, and installation issues can be solved with solutions found using Google Search.

1.2   The Machine Learning Revolution

Lately the primary focus on machine learning has been on neural networks (NNs) and so-called deep learning. The main reasons for this focus are:

- There are huge data sets available and the information continues to expand.
- Processing hardware is more powerful than ever before.
- Industry experts have discovered that deep NNs are able to outperform the current state-of-the-art techniques that have been handcrafted and tuned for decades. Previously, no one was able to sufficiently train these deep architectures because there just weren't enough data or processing power to do it.

In this chapter we will discuss deep learning and the methods used before its widespread adoption. These classical methods are usually lightweight in size and might also be faster than NNs. For some applications they might provide comparable accuracy, but with much smaller costs and it would be a huge mistake to ignore them. Sometimes they are also used in combination with neural nets.

2   What Is Artificial Intelligence

Artificial intelligence (AI) has many possible definitions. One is that it is an ability of a computer or computer-controlled device to perform tasks commonly associated with intelligent beings. Herein lies the first problem—it is not easy to define an intelligent being—but it should probably include an ability to reason, discover meaning, generalize, and learn from experience.

There is a famous Allan Turing test [2] in which artificial intelligence is accepted when a human observer is not able to distinguish the results of a given program from results provided by a human performer. Today there are chatbots able to converse on various topics that might be indistinguishable from a human. In some specific domains machines have even surpassed human performance, but only on a well-trained, given task (e.g., image classification, translation). If we want to speak about general AI (i.e., a program able to perform any task a human can do), we are still far away from achieving such capabilities. Current complex solutions are usually an assembly of various subsystems.

AI is an interesting domain because it represents a combination of mathematics, computer science, engineering, neuroscience, philosophy, and other studies—thereby making it difficult or impossible for any one person to be a complete AI expert. As you'll see later in this chapter the good news is that there are a growing number of open-source technologies and tools that are helping to bring AI to the masses.

3 What Is Machine Learning?

Machine learning, a field of computer science and a subset of artificial intelligence (AI), is comprised of many algorithms. These algorithms have in common an ability to learn (or be trained) to perform a given task based on the data provided. It can be viewed as a prediction tool that can deal with visual, textual, vibrational, and many other types of input data. For example, given an image of a face ML can predict whether the image is that of a man or woman. Given newsfeeds that include a previous price and current price of gold ML can predict the gold price in the next hour. Given sensory data, like vibrations, ML can detect or predict a motor failure.

This raises another important aspect to consider since predictions are guesses. Algorithms will never be 100% correct and it is difficult to predict when they will fail (however, the better the training, the better chance the prediction is closer to 100%). Predictions can easily fail when presented with kinds of data the algorithm has never previously "seen," especially if the training was not generalized properly. The main issue is that most of the algorithms are not aware of "not being sure" (e.g., when we train a system to label images either "pigs" or "cows" and we don't introduce a category like "others," it cannot do better than label an image of car with the high probability of being a "pig" or a "cow" (in the best case it will be irresolute—giving 50% to both).

In machine learning no code is written except for the implementation of training and the inference of an algorithm. There are no handcrafted rules. If you consider face detection, for example, we as humans don't really know how to do it, it just happens somewhat automatically. It would be difficult (if not impossible) to provide instructions showing a person how go about face recognition. If a person had the ability to learn we would try to teach them by sharing images, and on the images they would point to positions where they think the face is present. Then we would provide them with an answer (e.g., in this image there is no face or the face is more to the right). ML is also called pattern recognition since that is what you are seeking—patterns.

In this chapter we won't discuss a concept called general artificial intelligence. Algorithms described in this chapter can perform one exact task—even if the machine-learning algorithms are an assembly of many components to perform complex tasks, like driving, you won't expect the same machine to be able to clean your table and put dishes into your dishwasher.

This approach is called supervised learning, which represents most ML algorithms. When we want to categorize ML one criterion is based on how much we know about provided data:

- *Supervised*—Learning with a teacher. In supervised learning in addition to providing data (e.g., images) we also obtain the desired results (e.g., x, y, width, height—parameters of a rectangle defining the position of a face in the image). These results are usually called *ground truth*. Such images are called *labeled* and/or *annotated*.
- *Unsupervised*—Learning without a teacher. In this case we are given unlabeled data only. Based on assumptions there are different approaches to choose such as clustering, anomaly detection, autoencoders, etc.
- *Semisupervised*—As the name suggests, this lies between supervised and unsupervised. Typically, we get a small set of labeled data and a large unlabeled data set usually because of the high cost of acquiring labeled data. Semisupervised learning attempts to use this combined information + assumptions (e.g., continuity—the probability of having the same label is higher for data points closer to each other). It also outperforms unsupervised algorithms in situations when labels are completely removed and outperforms supervised algorithms running only on a small subset of data.

Intuitively, we can think of the semisupervised learning problem as an exam and labeled data as the few example problems that the teacher solved in class. The teacher also provides a set of unsolved problems. In this setting these unsolved problems are a take-home exam and you want to do particularly well on them. In the inductive setting these are practical problems of the sort you will encounter in the in-class exam. The supervised analogy would be almost the same, but without the unsolved, take-home examples.

The following sections refer to a supervised machine that can be summarized as: given data X and corresponding results Y find model parameters θ such that they will minimize the given loss function L (loss function explained later, but it basically measures the difference between prediction and ground truth). Herein the two most common tasks are *classification* and *regression*.

In *classification*, for a given data X, predict the correct label/category Y. The classification task is further divided into concrete categories:

- In binary classification inputs are classified into two groups (classes). Binary classification is considered a better understood problem, while multiclass classification is more complex. Some algorithms are built based on this relaxed property (only two classes) and are not able to solve tasks with three or more classes. As an example of binary classification—based on weight and height predict whether we observe a basketball player or a jockey. Another example might be given some email messages and their headers—decide if it is or is not SPAM.

- Multiclass classification makes binary classification more general, and the number of classes is bigger. For example, given a face image classify which friend it is (e.g., John, Bruno, Markus, Philip, or Rob). It is not rare for binary classification algorithms to be used in multi-class classification tasks in "one vs. all" (OvA) or "one vs. one" (OvO) mode [3]. OvA is when a single binary classifier is trained per class with the samples of that class being positive samples and all the others being negative. OvO is when a single binary classifier is trained for all class pairs, then during the prediction phase all classifiers predict and the result is based on "voting." This allows using binary classifiers, like SVM, in multiclass tasks as well.

- Multilabel classification assigns to each data sample a set of labels. For example, given categories (classes) {blues, rock, funk, jazz, hip-hop, classical, country, R&B, soul} and given the audio file return all genres that it can be categorized into. A valid result might be {blues, rock, soul}, for example (Fig. 1).

For *regression* tasks the goal is to predict one or multiple continuous values. For example, if the algorithm is detecting a face in an image we want to predict values $(x1, y1)$, $(x2, y2)$, the corner points of a bounding box defining the face's position. Algorithms used for regression are usually different than those used for classification tasks, but some can be used for both (e.g., decision trees, neural nets).

One interesting thing to note when talking about classification tasks is object/region detection vs. classification. In many cases a key step is finding the region(s) (which might be a regression) of an image (camera or stored picture) where something exists that is worth classifying. This is because an HD image (say 1920×1080 or bigger) is a lot of pixels—doing a detection to locate possible things of interest is an important consideration (especially as you can

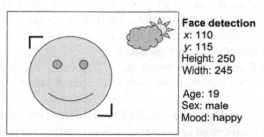

Fig. 1 The difference between regression and classification task. Regression is when we detect face position in an image or predict age. Binary classification is when deciding if a given face image is a man or woman. Multiclass classification would be when we want to decide whether the face is happy, neutral, or sad.

downsample to do this). Likewise, this comes up in voice recognition (waiting for voice band and sound that is not continuous or background), control systems (waiting for plant to change), sensing classifications (e.g., metal fatigue, etc.). Detection as a form of reduction to a small set is really important for embedded systems as it makes the problems practical as well as saving power (in many cases).

One popular way of carrying out a vision task is via background/foreground segmentation where in the simplest case a background model is computed as an average of previous frames, and the current frame is pixel-wise compared with this model. When the difference is bigger than some threshold a pixel is considered to be a foreground. This is not an ML approach as it is an exact algorithm.

3.1 Bias vs. Variance Trade-off

There are two sources of error that prevent supervised learning algorithms from generalizing beyond their training set:

- *Bias* is an erroneous assumption in the learning algorithm. It can cause relevant relations between features (input data) and outputs (e.g., fitting polynomials of degree 1 (linear) into data with quadratic relations) to be missed. This issue is also called an underfitting problem.
- *Variance* is an error that results from sensitivity to small fluctuations in the training set. It can cause an algorithm to model random noise in the training data and somehow mimic memorizing the dataset. It is also called an *overfitting problem*.

Bias can be prevented by using a more complex model, but as the number of parameters increases the model will be more prone to overfitting. Overfitting can be prevented with more data. There is always some trade-off between bias and variance when we are choosing the proper model. To support intuition see Figs. 2 and 3.

Regularization techniques are another way to prevent overfitting. These techniques add an associated cost into the loss function for each parameter used. This allows the model to have the flexibility to choose the proper complexity. There is no silver bullet for choosing a proper technique, and it usually requires lots of experience and intuition to set up everything properly.

To support the intuition behind overfitting go back to the example of learning to detect faces in images—what happens if we only show 10 examples to a lazy student with a good memory (big number of model parameters). The student will probably find it easier to memorize the answer instead of learning to generalize/use reasoning. Then we can determine whether the student was able to generalize if we show him previously unseen examples. If he just memorized the question and answers, then he will fail. The same approach is

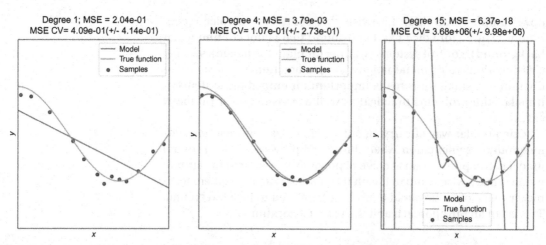

Fig. 2 The problem of bias vs. variance tradeoff explained on modeling x/y dependency with a polynomials of various degrees. Data distribution follows a cosine curve with noise added on top for correctness. MSE stands for mean squared error. MSE CV is an average MSE when 10-fold cross validation is done. (Left) Example of bias (underfitting) issue. Polynomial of degree 1 doesn't have enough expression power to fit well into given data. (Middle) This seems to be an optimal model—a polynomial of degree 4. (Right) Even though this model has the lowest mean square error (MSE) on training set it is far from capturing the trend in data and will probably totally fail during prediction phase. Here cross validation reveals that the mode overfits (MSE $6.37e{-}18$ vs $3.68e{+}06$).

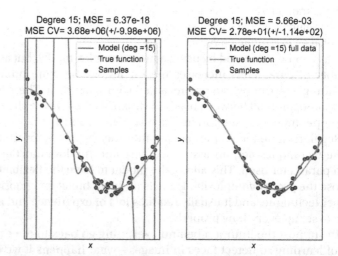

Fig. 3 (Left) one can observe an effect on overfitting when real data are present. Such model completely fails to predict them. (Right) the same polynomial complexity as in the left but trained on bigger dataset, one can see it better fits into the data (effect of overfitting can be reduced/prevented with more training data).

used in machine learning, whereby the previously unseen data are called test data, and with this approach we can estimate how the algorithm will perform in the future (additional details are provided in the next section).

4 Feeding Your Brain—Data

In machine learning data are crucial and can have various forms including audio/sound, images, videos, aggregated data/statistics, and text strings. What is important is that in the end the data should be preprocessed and combed into a numerical matrix X where each row represents a single data sample. The annotations should be preprocessed into a vector/matrix y where each row bears the required result(s) for the data sample in the matrix X.

4.1 Data Are Crucial

- If you don't have enough data, then state-of-the-art algorithms won't help you.
- If you don't have enough data, then complex models (with a high number of parameters) will tend to overfit (see Section 3.1) and fail dramatically when deployed.

The natural question arises: How many data are enough? The higher the model's complexity, the higher the demand for data. There are some theoretical boundaries, like the Vapnik-Chervonenkis theory for VC dimension [4], providing rough estimates. Some basic rules are that we need at least k independent examples for each class; there must be more independent examples than the number of input features and more independent examples for each parameter in the model. If we take, for example, an input to be an image of size $500 \times 500 \times 3$, we at least know that we should have more than 0.75 M images to train a given model. Another approach to estimate the required amount of data is to look at similar problems and published papers.

Another thing we should be aware of is that a lot of ML algorithms assume that data are independent and identically distributed (i.i.d.). This is a terminology from probability theory and statistics. I.i.d. means that each random variable has the same probability distribution as the others and all are mutually independent. But what happens when we violate that? What happens when we have fewer training data? Usually, the model still learns something, but might be more prone to overfitting. This is task dependent as well as algorithm dependent.

4.2 Data Preprocessing, Grooming, and Preparation

The goal of data preprocessing is to prepare data into a "tabular" numerical arrangement. This means that ultimately all data must be transformed into numbers. For example, this might mean encoding categorical variables using one-hot encoding. Tabular means that each data sample is in a row, whereas columns represent features and a few of the last columns have the right answers (e.g., class number).

The entire process can be reduced to:

- Data cleaning—a process in which wrongly completed questionnaires, invalid experimental data, or data not passing through logical control are removed.
- Data labeling and tagging—for cases where some of the data are unlabeled we must "get our hands dirty" and create annotations. This step is usually also educative because we can learn a lot about the nature of the given task. This task is usually outsourced.
- Augmentation—allows us to automatically enlarge our data set (explained later in more detail).
- Selection of training, validation, and testing sets (described later).
- Normalization—this step normalizes data. The reason is that most algorithms work better after this step. The only algorithm class (of those covered in this chapter) that doesn't need this step are decision trees (see end of Section 6 for an explanation of why the algorithm might fail without normalization).
- Feature extraction (covered later).

4.3 Training/Test and Validation Data Split

To prevent overfitting, data are usually divided into three categories. The first and biggest category is the training data set; it is usually 80%–90% of the size of the regular data set. To have a balanced data set the training data set is typically pseudorandomly chosen from the whole data set. For example, 80% are randomly chosen from each class to ensure all classes are represented.

The remaining 10%–20% of the data are typically divided into two equal-sized data sets—a test set and a validation set. The test set will only be used in the final process to evaluate and predict how well the model generalizes for new data (i.e., not used during the training phase!). The validation set is used to tune hyperparameters for a given ML algorithm.

A slightly more complex but commonly used technique is called k-fold cross-validation [5]. This technique divides the data set into k balanced folds (whereby the user determines the number of folds k and balanced means in which the percentage distribution of classes stays approximately the same proportion as in the full data set). Then the training process is done for all except the kth fold (this is used as the test and validation set). The entire process is done k times. By measuring the accuracy average and standard deviation we get a much better estimate on how the model generalizes. Commonly used values of k are between 5 and 10. Typically, higher values are better, but this increases the training time. This k-fold technique is not commonly used with neural networks because the training time is usually more demanding than a suitable non-NN algorithm.

4.4 Semantic Gap

Humans have a very advanced perception of the world around us. For example, when we see something we don't think about each photon or atom. Instead, we perceive objects, faces, textures, etc. The same applies to audio; we don't think about frequencies in time, instead we perceive words from which we build sentences and on top of that we reason.

On the other hand, observe how a machine sees the world (Fig. 4). The bottom section of the images are 10×10 matrices representing the highlighted *red blocks*. Notice how different the matrices look even after rotating the block by only 2°, although humans have no problem recognizing the rotated image. This demonstrates the huge semantic gap within data and how humans interpret them. For obvious reasons

Fig. 4 An example of how the computer sees the world— when the left image is rotated slightly by 2°. The 10×10 matrices represent the red channel of an image selection highlighted in *red* (gray in print version).

Fig. 5 An easy way how to increase dataset size without spending your entire salary on Amazon Mechanical Turk using augmentation (various transformations (e.g., translation, scale, rotation, brightness shift, …) are applied on top of original image). Each image should be classified as a dog, but from computer's point of view, each example looks very different.

we want to design our systems to be resilient to such changes (e.g., slight rotations, translations, noises, intensity changes) (Fig. 5).

4.5   Data Augmentation

We've learned that data are crucial and that usually it is difficult to obtain enough training data to achieve close to 100% accuracy on predictions. Data augmentation is a great way to increase the number of training examples. Using an example of image classification we want to classify an image of a dog as a dog even if the image is horizontally flipped, contains a little random noise, or is scaled, translated, or rotated. In computer vision these data modifications are straightforward and applicable beyond traditional image classification (e.g., it can be used for regression like object detection). The only rule is that the augmented image still maintains the same label or at least we know how to change the label (e.g., in the case of object detection, if we rotate an image slightly, we must change the object's digital position as well). It is possible to do the same for audio (e.g., adding background noise, increasing the tempo, phase shifting, etc.). Similar techniques can be applied to other domains.

Why does data augmentation work? Looking at Fig. 4 we don't think the image changed significantly, but from the computer's perspective the data matrix is completely different even when small changes are applied. This process is artificially accomplished by adding more examples of the same class, and more examples should help to prevent overfitting. Furthermore, compared with manual labeling the computational cost of augmentation is negligible.

4.6 Introducing an Image Classification Problem

The best way to learn something is to apply what you've already learned to solve a problem. In this section we will apply ML algorithms to an image classification task (we'll train our own convolutional neural net or CNN). We will use a task from the computer vision domain because it is more intuitive to visualize what is happening (Fig. 6).

Fig. 6 A preview of CIFAR-10 dataset [6]. Small 32×32 images are in 10 classes.

The CIFAR-10 data set is a relatively small, publicly available data set with 32×32 color images sorted into 10 different classes (e.g., airplane, automobile, dog). CIFAR-10 is considered a sandbox data set that is useful for testing but not real-world inferencing, replacing the previously popular MNIST example (28×28 images of handwritten digits). The simple MNIST model has become well understood (many recent algorithms have less than 0.25% accuracy error whereas state-of-the-art algorithms on CIFAR-10 achieve rates of around 2%).

4.7 Feature Extraction

For any type of machine learning the input data should be preprocessed. Take audio classification, for example—the incoming audio signal must be filtered to remove ambient noise. For vision applications various types of filtering or color conversion might be utilized to enhance specific colors or lines.

Here we list a few commonly used feature extraction techniques (we can also view feature extraction as a dimensionality reduction): in computer vision a search for key points (e.g., corners, dark/light blobs, etc.) is done and then each key point is described using algorithms like SIFT, SURF, ORB, Tf-idf for text analysis, or MFCCs and MPEG-7 for audio, etc. Handcrafted features can also be utilized.

For image classification, for example, an entire $32 \times 32 \times 3$ image can be unraveled into a 3072×1 vector. This vector might be our feature space, the dimensions where the variables live. However, most ML algorithms will suffer from such a big feature space, and this is often referred to as "the curse of dimensionality"—a situation in which data are organized in high-dimensional spaces. The main problem is that as the number of dimensions increases the data points become sparse, statistical significance drops, and this in turn damages the methods that rely on it. A big feature space will also harm the speed of inference as well as training. Finally, another problem of a big feature space in the case of an image is that there is information hidden in the pixel distribution (spatial relations are very important and if you do random permutation of pixels you will no longer be able to recognize the image). If we compare images at the pixel level, there is a huge semantic gap; shifting an image a small amount causes completely different values of the given pixels (refer to Section 4.4). This highlights the fact that we need a better way to represent each image.

An easy and straightforward way to represent images is to make us of a color histogram. A histogram divides a given range into bins

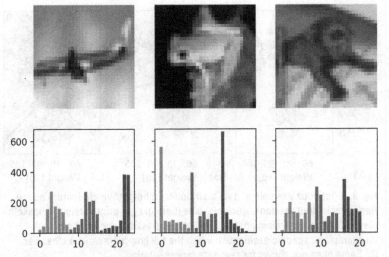

Fig. 7 An example of a color histogram for three small $32\times32\times3$ images. Each color channel is divided into eight bins. The x-axis represents bins, y-axis frequency. Notice, that color histogram won't change when we flip the image, it won't change dramatically when we rotate it.

and then counts the number of occurrences in each bin. For example, we can divide each color channel into k bins and compute how many pixels of a given intensity in that given channel are present in an image. Using this approach we can reduce the feature space from 3072×1 into a much more manageable $3k\times1$ vector. See examples of histograms in Fig. 7.

4.8 A Baseline

In the following sections we will introduce several famous ML algorithms, but before starting let's think about a small experiment. What about having a naive classification algorithm returning a class randomly? This algorithm will have an accuracy of $1/K$, where K stands for the number of classes. It is good to consider this as a baseline and sanity check. When our algorithms perform below this number it is an indication of something being broken. Therefore, it is good practice to have these types of sanity checks included during development.

5 Support Vector Machine

Here's a task—given weight and height, classify whether an observed man is a jockey or basketball player. We are also provided with training examples (see Fig. 8A). Let's consider red to be jockeys and blue to be basketball players. A simple way to solve this task

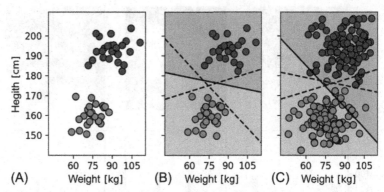

Fig. 8 A fictitious example of data distribution in height/weight feature space. *Red* (gray in print version)—jockeys, *blue* (dark gray in print version)—basketball players. (A) shows training dataset, (B) shows possible dividing lines and an example of space classification when the full line represents a classifier. (C) Same lines are shown for real data representation.

algorithmically would be to define a line and everything to the left is considered class 1.

Mathematically, for the linearly separable case any point x lying on the separating line satisfies $x^T w + b = 0$, where w is the vector normal to the line, and b is a shifting constant from the origin. The distance of the line from the origin is $\dfrac{b}{\|w\|}$. If we change equality to inequality we have a decision maker: $x^T w + b \geq 0$ is a jockey and < 0 is a basketball player. If we look at Fig. 8B there is an infinite number of possible lines. Fig. 8B shows that there is an infinite number of possible lines. From Fig. 8C we can see that when we use these three classifiers on real data (not seen during training) some lines are better than the others.

The support vector machine (SVM) algorithm is built on the following idea—choose a line that has the biggest distance to all data points, because such a line will better generalize on the inference. We can enforce this policy by slightly modifying our equation: $x^T w + b \geq m$, where m is a margin we want to have for all data. If we define our training labels y_i as 1 for jockeys and -1 for basketball players, we can then define the training criteria as $y_i(x^T w + b) \geq m$. Using the formulation of distance from the origin of three lines we can show the margin $M = \dfrac{2m}{\|w\|}$. Without any loss of generality we can set $m = 1$, since it only sets the scale of w and b. Now it is clear that if fraction $\dfrac{1}{\|w\|}$ is maximized the margin M is also maximized (Fig. 9).

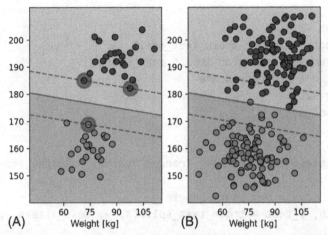

Fig. 9 SVM classifier visualization. The *dashed line* shows margin. Data points in *grey* are support vectors. The left image shows performance on train dataset and the right one shows performance on possible real data.

Finding the parameters for the best dividing line can be done by solving for the following quadratic programming problem:

$$maximize \frac{1}{\|w\|}$$

$$subject\ to\ y_i\left(x^T w + b\right) \geq \frac{1}{\|w\|},\ \forall i = 1...N$$

As this is not a convex function t is reformulated in practice as a dual $minimize \frac{1}{2}w^2 ...$, which is convex and hence easier to optimize. Luckily, as an ML engineer you can usually stand on the shoulders of giants and use ML libraries. Usually, we just construct an SVM classifier with given parameters, train it with data, and then use it for class prediction.

Given this problem definition there is still one issue to resolve. In a case when even one point is positioned in such a way that linear separation is not possible the entire optimization problem won't work because it won't be possible to satisfy all the constraints. This is solved by adding a penalty for each data point not satisfying the equation. All these penalties are weighted by parameter C which controls the trade-off between a smooth decision boundary and classifying the training points correctly. It is a hyperparameter of SVM and it is a good idea to try multiple choices. A reasonable starting value is 1.0:

```
1   import numpy as np
2   import matplotlib.pyplot as plt
3   from sklearn.model_selection import train_test_split
4   from sklearn.datasets.samples_generator import make_blobs
5   from sklearn.metrics import accuracy_score
6   from sklearn.svm import SVC  # "Support Vector Classifier"
7   from my_utils import plot_svc
8
9   # generate example data, 2 blobs
10  X, y = make_blobs(n_samples=300, centers=2, random_state=3, cluster_std=2.0)
11
12  # split data into random train and test subsets
13  X_train, X_test, y_train, y_test = train_test_split(X, y, test_size=0.2,
    random_state=42)
14
15  model = SVC(kernel='linear', C=1.0) # SVM classifier
16  model.fit(X_train, y_train)
17  y_train_pred = model.predict(X_train)
18
19  # visualize data
20  fig, axs = plt.subplots(1, 2)
21  plot_svc(X_train, y_train, model, axs[0])
22  plot_svc(X_test, y_test, model, axs[1])
23  plt.show()
24
25  y_test_pred = model.predict(X_test)
26  print("model accuracy on train set: {:.2%}".format(accuracy_score(y_train,
    y_train_pred)))
27  print("model accuracy on test set: {:.2%}".format(accuracy_score(y_test,
    y_test_pred)))
```

This code sample describes how to train our own SVM classifier. The same code template might be used for any other classifier. The interface is usually the same—construct, fit data, predict.

All these approaches can be easily scaled into higher dimensions, where the line becomes a plane in 3D and a hyperplane in higher dimensions; the math stays the same.

So now we have at least a rough idea how to solve a linearly separable task. But what about a linearly nonseparable task? To see how our approach performs on such a task look at Fig. 10. Linear SVM cannot solve this task, but one way to deal with this issue is to transform the feature space into another format (see example

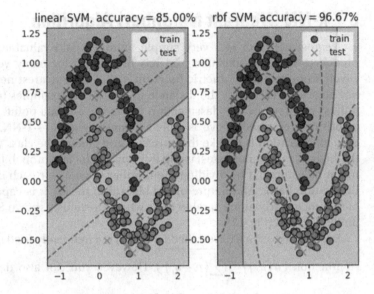

linear SVM, accuracy = 85.00%   rbf SVM, accuracy = 96.67%

Fig. 10 Performance of linear SVM and SVM with Radial basis function (rbf) kernel on linearly-nonseparable data. Gamma = 1.0 for rbf SVM. Train examples represented by circles, test examples used for evaluation are marked by crosses.

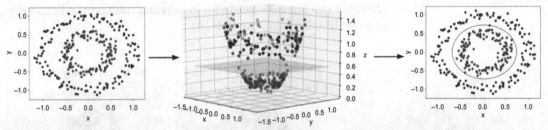

Fig. 11 The intuition behind feature space transformation. Data that are linearly nonseparable are lifted into higher space (in this example $[x, y] \rightarrow [x, y, x^2 + y^2]$) where it is possible to divide them using a hyperplane. This is commonly used in SVM with the so-called kernel trick for efficient computation. Famous kernels are radial basis function (rbf), polynomial, or sigmoid. Your own kernels can also be implemented.

in Fig. 11). We can change the previous code example to use the kernel just by modifying the previous code example around line 15 in the following way:

```
model = SVC(kernel='linear', C=1.0) # SVM Classifier
model = SVC(kernel='rbf', C=1.0, gamma=1.0)
```

Further information with detailed proofs of the SVM algorithm can be found in any ML textbook (e.g., [7]), and more information on kernel methods can be found in [8].

6 *k*-NN (Nearest Neighbor) Algorithm

Nearest neighbor is a very simple, intuitive, and valuable approach to data classification and has been used for many years. Given a query Q and a metric function $d(Q, X_i)$, the nearest neighbor algorithm measures the distance to all training examples (data points) and returns the classification of the closest data point. The algorithm described above is called nearest neighbor—1-NN. An easy modification is to take the closest k and choose the class with the highest number of votes; using this approach class probabilities might be estimated by dividing the number of votes for each class by the number k. The nature of the algorithm means it is capable of solving linearly nonseparable tasks where previously shown SVM methods failed (Fig. 12).

Euclidean distance (also called L2 norm) is a frequently used metric function ($d(p,q) = \sqrt{\sum_i^n (p_i - q_i)}$). However, you can also define your own metric function suitable for the given task—there are no limits.

We can use NN Classifier on the previous task from SVM just by changing two lines. The code for generating similar visualizations is inspired by scikit-learn tutorials [9]:

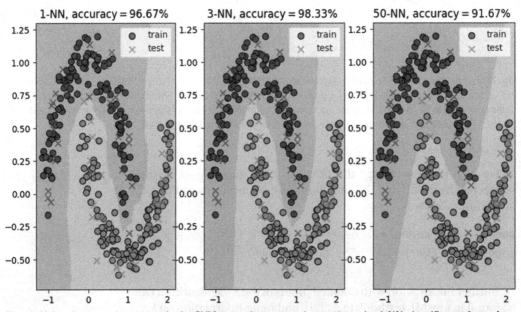

Fig. 12 Using the same data set as in the SVM example we can observe how the *k*-NN classifier performs for different *k* values.

```
1   from sklearn.neighbors import KNeighborsClassifier
2   from sklearn.datasets import make_moons
3   # use moon-like shaped dataset instead of blobs
4   X, y = make_moons(n_samples=300, random_state=3, noise=.10)
5   clf = KNeighborsClassifier(k) # k = number of nearest neighbors
```

The main advantages of the k-NN algorithm are that it's an easy implementation, it works on nonlinear cases, and has constant training time $O(1)$ (this stands for big O notation, an upper bound complexity estimation that in this case is constant—not dependent on input size as no actual training is performed). The main disadvantage is that when the number of dimensions increase a lot of data are needed because in high dimensionalities data entries become rare and sparse. Next all training data must be stored and the inference time is $O(N)$, where N is a number of training examples (which means it is slow). These limitations can be overcome by using more advanced data structures like k-dTrees [10], where inference time is on average $O(\log N)$ with the worst time is still $O(N)$.

Even though the nearest neighbor method is very simple it can still find use cases where the data dimensionality is low and space is well covered. New attempts have been made to improve this concept (e.g., [11]).

It is important to note that, as described in Section 4.2, most ML benefits derive from data normalization. Let's return to our example from the SVM section (jockey/basketball player classification). What happens when we have weight measured in kilograms but height in millimeters? The Euclidean distance will change by the same amount whether we add 10 kg or 10 mm, thus information about weight will be mostly ignored.

7   Decision Trees

A decision tree, another important and frequently used concept in ML, is a mathematical structure from graph theory (specifically, it is a directed acyclic graph or DAG). Each tree starts with a root node that branches into descendant nodes. When a node has no descendants it is called a leaf. Binary trees are very common, so-called because each node has up to two descendants. A decision tree implies that a decision function resides in each node, and based on the output flow continues to branch into descendant nodes until a leaf is reached. Decision trees are most commonly used for classification tasks, implying that each leaf holds a class label (or class label distribution) (Fig. 13).

A benefit of decision trees is that once trained the testing phase of a decision tree is relatively fast ($O(\log N)$). Additionally, decision trees

Fig. 13 Decision tree example. Given information about an observed fruit, decide whether it is a lemon, banana, apple, or we are not able to decide. In each node (e.g., "Is it red?") a decision is made. These decisions are based on features. In our case there are two features, color and convexity (shape). Nodes not having successors are called leaves and they contain class labels.

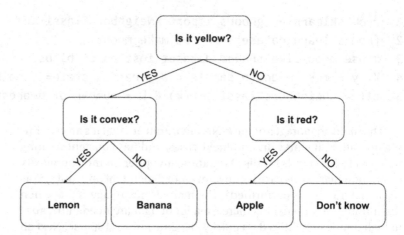

can work on nonlinear tasks. Another great value of a decision tree is in the way it can interpret a model (a path to the leaf can be followed and each decision can be interpreted). Interpretability is an important aspect of an algorithm. For example, imagine a task where ML is used to help a doctor with disease diagnosis. There is minimal value in saying: "I think it is cancer"; it's better if an algorithm can infer there's cancer with a probability of 98%, allowing a decision to be made based on these criteria.

One disadvantage of decision trees is that they can badly overfit, thus they might not generalize very well (ensemble learning, introduced in the next section, can be applied to overcome this issue).

7.1 Ensemble Learning

We will explain ensemble learning here as it relates to decision trees, but these approaches are general and can/should be used with other algorithms. Ensemble learning usually comes with higher computational and memory cost, but it leads to higher accuracy. In accuracy-critical applications these techniques are also used for computationally intense algorithms like NNs. Ensemble learning is a technique that can make decisions based on multiple models which can be homogeneous (e.g., 10 decision trees) or heterogeneous (e.g., decision tree + SVM + neural network). A heterogeneous example is based on the idea of combining the results of different algorithms, assuming that each has its own strengths and weaknesses. Why does it make sense to use it on homogeneous models? Aren't the models the same? First, since we can use different hyperparameters for each model they might be slightly different. There is also a technique in which the hyperparameters are the same, and yet it still makes sense (one of them is called a random forest classifier/regressor; see Section 7.3).

7.2 Bagging

Using a technique called bagging (bootstrap aggregating) we combat overfitting by taking k-same models. With this technique we randomly sample a subset of data (e.g., 85%) from the training data set for each model. Then we train the model, perform k predictions, and in the case of a classification task take the argmax. We can also recast this information as a probability distribution over classes. This approach works because each trained model saw just a portion of the data set, thus it cannot fully overfit. In combination with other models it generalizes better on a given task.

7.3 Random Forest

The decision trees concept has many versions, and one of the most frequently used is the random forest approach [12]. This uses the ensemble-learning concept called bagging, but this is not the only randomization performed. During training a vanilla decision tree uses all features to choose which one best divides the data. The random forest approach randomly samples a subset of features in each node (a frequently used subset size is $\sqrt{F}$, where F is the number of features). This helps prevent overfitting and speeds up the training process. In the end the final decision is based on voting (as in any other ensemble method):

```
1   from sklearn.ensemble import RandomForestClassifier
2   clf = RandomForestClassifier(n_estimators=10)
```

As mentioned in Section 6 it is easy to reuse the code for classification and use a different algorithm. We can try to do the classification using a random forest with 10 trees. Classifier has many more parameters like max depth. A great way to build intuition about this is to play around with parameters and observe what Classifier does (Fig. 14).

7.4 Boosting

Boosting is another ML technique that produces a prediction model as an ensemble of so-called weaker models. Boosting is done in an incremental way where each new model emphasizes the training data misclassified by the previous model. Sometimes boosting might have better accuracy than bagging, but it might be more prone to overfitting. AdaBoost [13] is an example of a boosting algorithm. Because of its efficiency and flexibility XGBoost [14] is a commonly used implementation of gradient boosting. It is usually good practice to compare multiple approaches (e.g., gradient boosting vs. random forests) on a given problem and choose the better one (using trial and error).

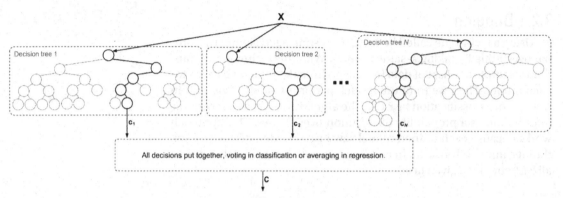

Fig. 14 A scheme of a Random Forest, there is N trees. After training, each might end up with a different structure. For a given input, each tree produces an output, results are ensembled and the final decision is made.

8 Neural Nets

Neural networks are another category of algorithms that are getting lots of attention, but remember you should avoid using them unless your application requires complex classifications (etc.). Classical machine-learning algorithms are easier to train and typically require fewer computations. Anyway, NNs are brain-inspired algorithms known since the 1950s (Warren McCulloch and Walter Pitts created a computational model for NNs based on threshold logic), and the general methodology has not changed significantly over the decades. However, what has changed is the usage model—NNs were previously only useful for PhD mathematicians—now, through a variety of proprietary or open-source frameworks and tools, NNs are in the hands of the masses.

8.1 Motivation

A distinct historical landmark is the 2012 success of AlexNet [6] in the ILSVRC image classification challenge utilizing the ImageNet data set [15]. The task was to classify ImageNet's images into one of 1000 classes including animals, dog breeds, cello, cradle, car wheel, volcano, seashore. These algorithms were compared using top-5 metrics, which means the top-5 predictions are returned and if the correct class is among them it is accepted as a correct prediction. As seen in Fig. 15 the last successful classic computer vision approach was done in 2011 where there was a 26% top-5 error. The following year AlexNet [6] reduced this error by almost 10%; this ignited the spread and domination of convolutional neural nets (CNNs). In 2014 AlexNet was followed by the larger VGG network [16] as well as the more complex GoogLeNet [17]. Finally, in 2015 the even deeper CNN architecture ResNet [18] outperformed human performance in the top-5 classification tasks (Figs. 16 and 17).

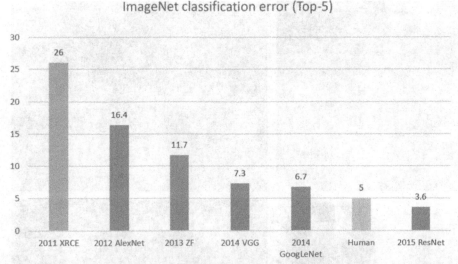

Fig. 15 Winning results of the ImageNet large scale visual recognition challenge (LSVRC) on the top-5 classification tasks. The *green bar* (gray in print version) indicates the best computer vision approach, whereas the *blue bars* (dark gray in print version) are all deep neural network architectures. The human score is represented as the *orange bar* (light gray in print version).

Fig. 16 Neural style transfer. A style of a reference image (B) is applied on top of an input image (A). The result looks very realistic. Image Credit: Ref. [19] where other interesting examples can be found. It is obvious that NN needs to be able to understand concepts like the sky, building, etc.

8.2   What Is a Neural Network?

A neural network is a brain-inspired algorithm in which the basic building block is called a neuron—so-called because it is a simplified, mathematical model of a biological neuron. Each neuron has one or more inputs and a single output. Similarly, inputs are called synapses and there are typically other synapses connected to the output—sending information to other neurons. Each synapse is represented by three properties: a starting neuron, a weight, and an ending neuron. The synapse weights are the primary NN parameters. Where does the "magic" occur? Each neuron takes a signal from each input, multiplies

Fig. 17 Example of object detection, classification and segmentation. Results shown and image credit are from Mask R-CNN paper [20].

it by a synapse weight, and sums these products together (i.e., a multiply-accumulate function).

On top of that, an *activation function* is applied. The main purpose of an activation function is to bring nonlinearity into the whole system. This is a very important aspect. Compared with a vanilla SVM (which is linear), NN can solve nonlinear tasks. Recall from Fig. 11 that when we want to separate data points it is not possible to do it with a line (this is called linear nonseparability). Thanks to the nonlinearity added in an activation function the NN is projecting (warping/upscaling) its input feature space into a new one where linear separation might be possible in the end.

Another goal of an activation function might be to scale the output into a given range (e.g., <0, 1> for a sigmoid function). For more examples of activation functions see Fig. 18. The output of the activation function is then sent to all other connected neurons, multiplied by the respective weights, and so on through each layer of the network.

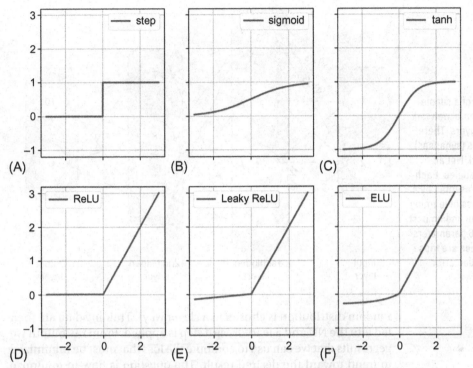

Fig. 18 An overview of the most common activation functions. Starting with step, sigmoid and hyperbolic tangent which were replaced with modern activations like Rectified Linear Unit (ReLU), Leaky ReLU or Exponential LU (ELU).

In general, neurons are connected and form a graph structure called a directed acyclic graph (DAG). DAGs can have almost any imaginable topology. When referring to currently used deep-learning architectures, DAGs are best organized into layers (Fig. 19). Layers are groups of neurons. As we'll see in the following sections, layers can be more complex than a straightforward fully connected one (which means that each neuron has a connection with all neurons from previous layers). One example of a more complex layer is a convolution layer, which is described in the Section 8.4. The convolution layer is one of the most important concepts in NNs for vision and other tasks.

8.3 How Training Works

8.3.1 Backpropagation—Key Algorithm for Learning

Section 4.3 describes how a training data set can be divided into three subsets. In the first stage of training we'll start with an untrained NN architecture initialized with pseudorandom weights (actually this

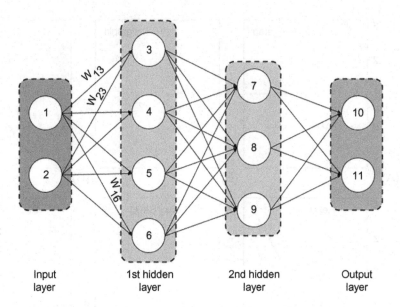

Fig. 19 A schema of a simple fully connected neural network with two hidden layers. There are 26 connections (synapses) (e.g., w_{11}, w_{21}, etc.). Not all connections are labeled. Each connection is represented by a weight. The process of learning is nothing else than finding best values for these 26 parameters/weights. Best values are those minimizing prediction error (loss function).

Input layer   1st hidden layer   2nd hidden layer   Output layer

random distribution is chosen in a clever way). Training data are then fed into the NN and a feedforward step is applied. From the output we get results that we can use to compute the loss that must be minimized to trend toward the desired result. The question is how to minimize the loss function in an efficient way—this is where backpropagation comes into play (Fig. 20).

Backpropagation is a process in which each parameter/weight of a network is systematically updated to minimize the loss function. An example of a loss function might be a mean squared error (MSE) defined as $L = \dfrac{1}{N}\sum_{i=1..N}(y_i - \hat{y}_i)^2$, where y is predicted and $\hat{y}$ is the ground truth value. In a simplified example where we have an extremely small model with only two parameters to tune (albeit unrealistic), imagine a loss function output as a landscape (see Fig. 21) where each position represents a configuration of the two parameters, and altitude represents the loss function's value. Once the initialization is done we are at the exact position, but are blindfolded. Our goal is to find the lowest valley in the landscape, but it is obvious that it won't be tractable for a full space search because the weights are continuous (infinite number of options). Furthermore, we usually have millions of weights and our measurement of altitude is a costly operation (feedforward step).

Therefore, we must establish a reasonable way to find the lowest (or at least a low enough) point in the landscape. We can try to step in four directions and estimate which direction yields the sharpest descent. Once we have an estimate of direction and the magnitude of the

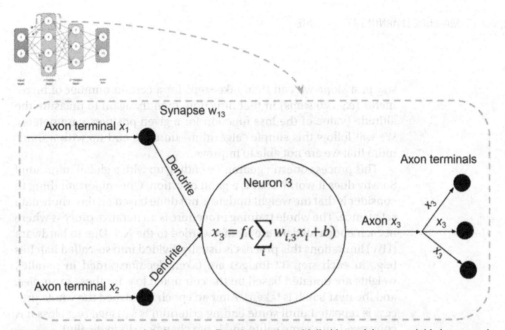

Fig. 20 A mathematical model of a neuron. Each input x_i is multiplied by weight $w_{i,3}$ and this is summed up and then an activation function is applied on top of this sum (see activation functions examples in Fig. 15). An output is an input of another neuron or the output of the whole network.

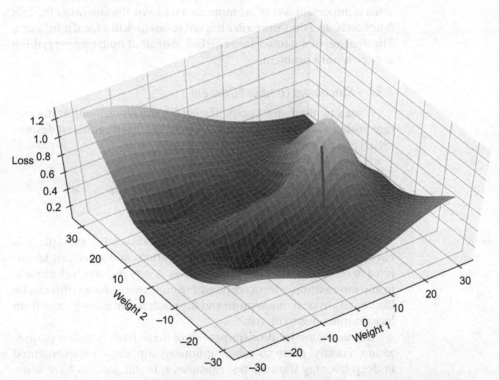

Fig. 21 We can the imagine loss function of two weights as a landscape. The goal of training NN is to tune parameters in such a way which provides smallest loss value. Computation of all possible combinations of thousands of weights is intractable in practice. Instead, after each feedforward phase loss gradient is computed which provides us with a clue in which direction each weight should be tweaked in order to minimize loss. Each weight is updated. And whole process repeats.

sharpest slope we can then take steps for a certain number of increments (e.g., 20 steps) in that direction, then try again to measure the
altitude (value of the loss function for a given position/parameters).
We will follow this simple "algorithm" until we end up with a minimum that we are not able to improve.

This process doesn't guarantee ending up with a global minimum!
So why does it work? That is a good question. One important thing to
consider is that the weight update is not done based on the whole data
set at once. The whole training procedure is an iterative process where
in each epoch all data are feedforwarded to the NN. Due to hardware
(HW) limitations this process is usually divided into so-called batches
(e.g., in each step 32 images are taken, feedforwarded in parallel,
weights are updated based on the computed loss for these examples,
and the next batch is taken). After an epoch is finished the whole process is repeated until some ending criterium is satisfied (e.g., loss has
converged into some value and is not changing dramatically).

There are many tactics and techniques to get a global minimum.
Usually, data are randomly divided into batches in each epoch. The
output of the loss function (landscape) looks different for each batch.
What is important is that the more data we have, the smoother the loss
function is, and thus the easier it is not to be stuck in a local minimum.
The division of a dataset into batches instead of updating everything
at once is not a problem.

8.3.2 Stochastic Gradient Descent

From a mathematical point of view the slope's estimate at the current position is called a gradient. A vanilla stochastic gradient descent
(SGD) can be written on three lines of code:

```
for epoch in range(num_epochs):
    gradient = compute_gradient(loss, data, W)
    W += -lr * gradient
```

You can see that this approach has weaknesses. For example, it is
easy to end up with a local minimum. Various strategies can be applied to try to overcome these limitations. A very popular technique is
to add momentum, a term representing movement history; this can be
likened to physical momentum and the effect is that it can escape from
local minima due to inertia.

The mathematics domain providing these tools is called optimization. Luckily, these so-called optimizers are already implemented
in deep-learning frameworks. However, it is still good to have some
idea about how they work because they are the reason NNs learn and
usually they have some parameters that might be tuned. Optimizers
are task dependent—there is no silver bullet. A good practice is to

start with the Adam [21] optimizer that combines the advantages of AdaGrad [22] and RMSProp [23], and usually achieves good results quickly.

8.3.2.1 Learning Rate

Learning rate ("lr" in the code above) is probably the single most important hyperparameter of a neural network. This parameter is a trade-off between not converging at all and training time. It affects how aggressively we update weights with a gradient. It is usually the first parameter to tune. Typically, this parameter is chosen from the 0.1–0.00001 range.

Usually, during training there is a learning rate scheduler. A common way to apply this is to decrease the learning rate slightly with each epoch. Recently, more advanced techniques have surfaced, like the triangular method described in [24], where a learning rate is changed according to a cyclic triangular function. In [25] a cyclic function is driven by cosine prescription (Fig. 22).

Training is usually done in a so-called minibatch in which, say, 32 (usually powers of 2) examples are feedforwarded, all gradients are computed and averaged, and then updated. Practice shows that a minibatch has several advantages. First, it reduces variance in the parameter update and can lead to more stable convergence. Second, this allows the computation to take advantage of highly optimized matrix operations that should be used for a well-vectorized computation of the cost and gradient.

To summarize, learning is possible not only because of backpropagation and the ability to compute gradient descent, but also because all operations inside NNs are differentiable (matrix multiplication is only multiplication and summation—both of which are differentiable; activation functions and loss functions can be chosen such that they are differentiable as well). The result of training is an architecture description plus a set of all weights.

One interesting and insightful view is to think about NNs as a tool to find an appropriate feature space transformation—a transformation where data points of the same class are close to each other and distinguishable from other classes.

8.3.3 Neural Networks vs. Deep Neural Networks

What does a deep neural network mean? We already know that NNs are DAGs of layers. There are input and output layers, but in between there might be hidden layers. In the historical origin of NNs they were shallow with only one or two hidden layers because such an NN was easier to train and compute. For reasons we described

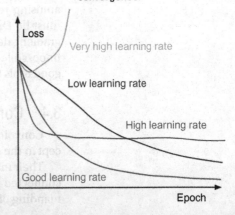

Fig. 22 Effect of various learning rates on loss function convergence.

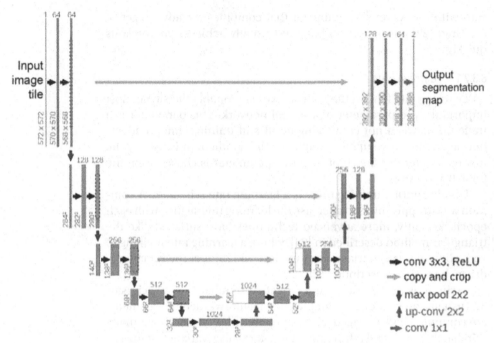

Fig. 23 A scheme of so-called U-Net [26]—an architecture widely used in biomedical image segmentation. It is more complex architecture compared to straightforward CNNs like VGG.

previously it is possible to design and train NNs with multiple hidden layers. There is no strict boundary between deep and shallow networks, but if you are using three or more layers, relax!—you are still doing deep learning (Fig. 23).

NN architectures are layered structures that can be stacked together (think LEGO blocks). This process can be creative and doesn't have to follow a simple string idea. For more complex architecture see, for example, U-Net [26] or YOLOv3 [27], which by the way makes for amusing reading. As mentioned before the only rules are that NNs must be DAGs, all building blocks must be differentiable (otherwise gradient descent optimization techniques can't be used—while not impossible to train it becomes much more difficult and we don't have good tools for this task), and the NN graph shouldn't have cycles.

8.4 Convolutional Neural Networks

Convolutional neural networks are another important tool/concept in the domain of NNs.

Their main advantage is that they use fewer parameters than a fully connected layer. Thus they are more robust to overfitting and less demanding of memory space.

As discussed in the previous sections the NN learns, filters, performs feature extraction, and makes predictions. CNNs are great for tasks where there is spatial information in the input data (e.g., images, audio, the order of words in a sentence).

8.4.1 What is a Convolution?

Convolution is defined as continuous data and as discrete data (the only difference is integration vs. summation). For computer science tasks the discrete, finite form is handier:

$$(f*g)[n] = \sum_{m=-M}^{M} f[n-m]g[m],$$

where M is half the size of the kernel. We can view convolution as computing a dot product with a kernel prescription for each possible position of the kernel in the input space. It is a kind of scanning window approach to searching for positions with the best response to a given kernel.

An intuitive way to imagine convolution is to regard it as a response to a given filter. The higher the data similarity to the kernel, the higher the response (see Figs. 24 and 25).

For example, let's say we are looking for a step-down edge in 1D data. We define our convolution filter as a simple $[-1, 1]$ (see Fig. 25 for a convolution response to the signal). Some code to generate Fig. 25 is:

```
1  import numpy as np # scientific computing library
2  import matplotlib.pyplot as plt # visualization library
3
4  x = [ 9,  9, 10,  9,  9,  5,  5,  9,  9,  7]
5  k = [-1, 1] # convolution kernel
6
7  plt.plot(x, drawstyle='steps-mid') # show input data
8  plt.scatter(range(len(x)), x) # draw single dots of data
9  c = np.convolve(x, k, 'valid') # do convolution
10 print(c)
11 # show convolution results shifted by 0.5 for more intuitive visualization
12 plt.plot(np.array(range(len(c))) + 0.5, c, marker='o')
13 plt.grid()
14 plt.legend(['data', 'convolution'])
15 plt.show()
```

It is straightforward to scale up the convolution into higher dimensions.

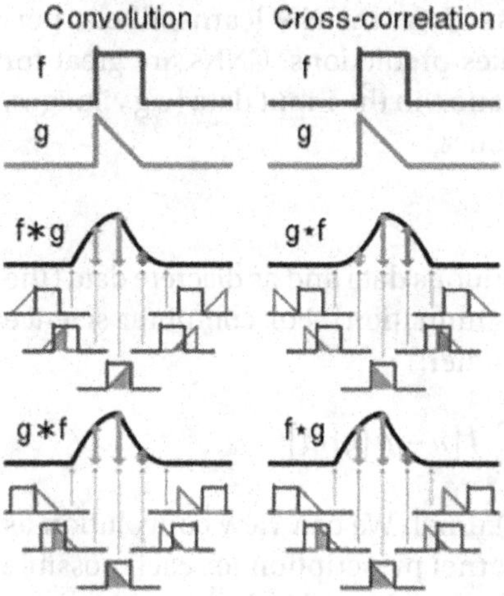

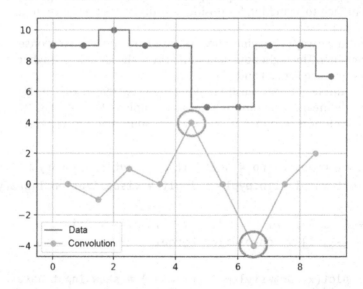

Fig. 24 A difference between convolution vs. cross-correlation. Correlation is a measure of similarity between two signals, and convolution is a measure of the effect of one signal on the other.

Fig. 25 In *blue* (dark gray in print version), we can see the visualized data vector *x*, in *orange* we can see a corresponding convolution with a kernel [−1, 1]. Notice the *green circle* (light gray in print version) around the maximum value of convolution response. One can easily see that in this place the downstepping minimum value is exactly in place of exactly the opposite trend—the upstepping edge.

8.4.2 A Convolution Layer

So now we know how convolution works. A convolution layer in neural networks does the same thing—a kernel slides over the input and results are written into the output matrix (feature map). The dimensionality of a kernel is given by the kernel size and by a number of channels in the input layer. For example, if we have an image of size 5×5×3 (where 3 stands for RGB channels) and a convolution layer with two 3×3 kernels, the weights in this layer are represented by 3×3×3 values (3×3 kernel times 3 input channels). The number 27 is derived because the input layer has 3 channels and there is a unique 3×3 matrix for each channel. In our example the weight matrix *W* will be $3\times3\times3\times2$ (kernel_h $\times$ kernel_w $\times$ num_input_channels $\times$ num_kernels). In comparison, a fully connected layer with 9 outputs will have $5\times5\times3\times9$ weights.

Note that by increasing an image size to $128\times128\times3$ the number of parameters in the convolution layer stays the same ($3\times3\times3\times2=54$ parameters), but the number of fully connected layers, when output should be $3\times3\times2$, is $128\times128\times3\times9 = \sim0.5$ M parameters. This might have a huge impact on memory requirements and provide a lot of space for overfitting. We can see that the concept of a convolution layer is crucial to solving problems with high-input volumes. Computer vision fulfills that (Fig. 26).

Two additional parameters are usually associated with convolutional layer computation—stride and padding. Stride defines the step the convolution kernel is moved (standard convolution has step 1 in

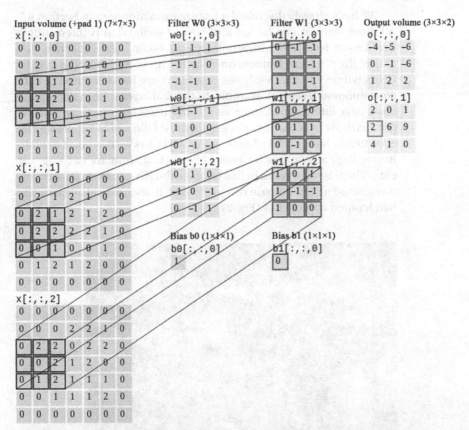

Fig. 26 A visualization of how convolutional layer works. On the left side, there is an input having three channels. Our convolution layer has two filters *(red ones—dark gray in print version)*. Each filter has its own values for each channel. The results are accumulated into two feature maps *(green ones—gray in print version)*.

both height and weight). It is possible to have, say, step 3 for convolution kernel 7×7, which means that the first convolution will be computed on coordinates (4, 4), the second on (4, 7), and so on. This will also reduce the output size by a factor of 3. The padding parameter defines the way in which border computations are managed. The first option is to do nothing such that stride = 1 and convolution kernel 7×7 produces outputs with height and weight smaller by 6 pixels (due to the nature of convolution). The second option is to require output to be the same size. In this case we can either fill the border with zeros (commonly used), or to prevent a potentially big ramp in the signal the border values will have the same values as the nearest pixel. The convolution layer might also be associated with the bias parameter for each kernel allowing output values to be moved by a constant.

8.4.3 Feature Extraction in Neural Networks

We have already discussed a feature extraction step in Section 4.7. All these features can be fed into NNs as well. What is interesting is that we can feed NNs with whole images, audio files, text series, etc., and let the NN learn features on its own. This is especially the case for CNNs where we can clearly see that filters are being learned in each convolutional layer. For example, in the first layer we can distinguish basic edge filters, blobs, corners, etc. In higher layers we can observe geometrical concepts (e.g., circles, squares) followed by others such as eyes, heads, hands, and wheels. In the last layers we can observe filters having high activations for entire objects (e.g., human, car, dog, cat, etc.). There is huge value in this because the NN can learn its own features based on the domain of a given task. It also usually outperforms handcrafted approaches (Fig. 27):

Fig. 27 Input image for VGG-16 [16] feature map visualization shown in Fig. 28.

```
1   import numpy as np
2
3   def convolution(X, kernels, biases, stride=1, padding='same'):
4       H, W, C = X.shape
5       kh, kw, in_channes1, num_kernels = kernels.shape
6
7       h_padding, w_padding = 0, 0
8       # padding same stands for padding input with zeros so
9       # the output has same heightxweight  as input when stride=1.
10      if padding == 'same':
11          h_padding, w_padding = kh/int(2), kw/int(2)
12
13      H_steps = (H - kh + 2*h_padding)/int(stride) + 1
14      W_steps = (W - kw + 2*w_padding)/int(stride) + 1
15      output = np.zeros((H_steps, W_steps, num_kernels))
16
17      for k in range(num_kernels):
18          output[:, :, k] += biases[k]
19
20          for ch in range(C):
21              channel = np.pad(X[:, :, ch], (h_padding, w_padding),
                                        mode='constant')
22              print(channel.shape)
23              for h in range(H_steps):
24                  for w in range(W_steps):
25                      hh, ww = h*stride, w*stride
26                      output[h, w, k] += np.sum(channel[hh:hh+kh, ww:ww+kw] *
                                                    kernels[:, :, ch, k])
27
28      return output
```

This code example of 2D convolution is a naive implementation. In a real deep-learning framework there will be a lot of different functions implementing convolution to better harness HW-specific operations and caches and using a faster algorithm based on parameters such as input size, kernel size, and number of inputs. One optimization is to modify the input matrix in such a way that convolution can be computed as a single matrix product. Another common technique is computing convolution as a multiplication in the Fourier domain, after appropriate padding (to prevent circular

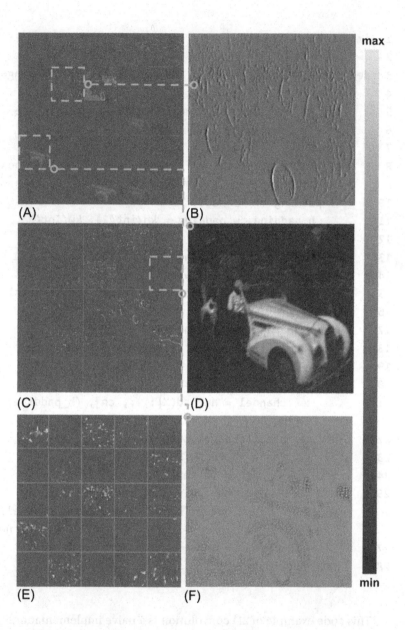

Fig. 28 Visualization of VGG-16 [16] feature maps when Fig. 27 is feed-forwarded. On the left side, first 25 feature maps after ReLU activation function are visualized. On the right side, the importance of each pixel is visualized for given feature map. For importance computation—so-called deconvolution is used, it was introduced by [28]. Reader can find interesting information there. (A) 25 out of 64 feature maps from 1st convolution layer. Feature map resolution is the same as input—224×224. (B) We can observe that this kernel is doing kind of vertical edge detection. (C) 25/128 feature maps from 3rd conv layer, resolution 128×128 after first max-pooling. (D) We can observe (given input image) that this kernel responds to the occurrence of *green color*. (E) 25/512 feature maps from 8th conv layer. Resolution is 32×32 after 3rd max-pooling. (F) Deeper the layer the more complex concepts are learned. This one might describe something like local pattern homogeneity/uniformity. See that it corresponds to areas like the white part of wheels, ground, white car body, etc.

convolution). Usually, a developer doesn't have to implement any of these functions because they are usually done automatically within the deep-learning framework.

8.4.4 Convolutional Neural Network Classifier Example

Let's now look at how to use PyTorch [29] (a deep-learning framework—we discuss other frameworks in Section 8.6) to train our own image classification. Our goal will be to classify images from the CIFAR-10 data set (as we've already described it in Section 4.6). In this example we won't do the real heavy lifting. The data set is relatively small as is the neural net. A regular notebook or desktop PC without GPU should make it possible to complete the training within 10–20 min. It is also possible to use GPU clouds. Many companies provide some trial credit (e.g., Google Cloud Platform). While speaking about Google at the time of writing they also provide a service called Colab [30], where some GPU resources are available for free:

```python
1  import torch
2  import torch.nn as nn
3  import torch.optim as opt
4  import torch.nn.functional as F
5  import torch.backends.cudnn as cudnn
6
7  import torchvision
8  import torchvision.transforms as transforms
9  import torchvision.models as models
10
11 class Net(nn.Module):
12     def __init__(self):
13         super(Net, self).__init__()
14         self.conv1 = nn.Conv2d(3, 8, 3)
15         self.pool = nn.MaxPool2d(2, 2)
16         self.conv2 = nn.Conv2d(8, 16, 3)
17         # 16 filters with output size 6x6
18         self.fc1 = nn.Linear(16 * 6 * 6, 100)
19         self.fc2 = nn.Linear(100, 84)
20         self.fc3 = nn.Linear(84, 10)
21
22     def forward(self, x):
23         x = self.pool(F.relu(self.conv1(x)))
24         x = self.pool(F.relu(self.conv2(x)))
25         x = x.view(-1, 16 * 6 * 6)
26         x = F.relu(self.fc1(x))
27         x = F.relu(self.fc2(x))
28         x = self.fc3(x)
29         return x
```

First, we need to import all the necessary modules (lines 1-9). The two main packages we are using are torch (PyTorch) and torchvision—utilities that make deep learning for computer vision even easier. Then we set out our neural net definition. Our net class is inherited from nn.Module and must employ the forward method. In constructor __init__ all building blocks are defined. Let us first look at the forward method that defines what our architecture will look like (see also Fig. 29). We have two blocks that are basically the same (lines 23-24)—a convolution layer followed by an activation function (relu) and then by pooling. On line 25 data are just transformed (flattened) from 4D (batch, height, width, channels) to 2D (batch, vector). This is necessary because on lines 26-27 a fully connected layer is used that expects

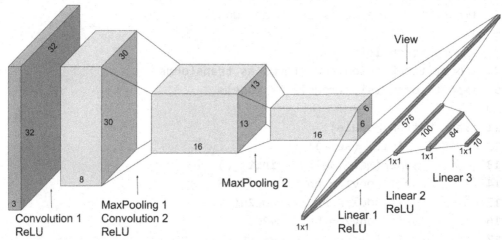

Fig. 29 Our example CNN architecture. The *green block* (gray in print version) represents the 32×32 pixels input image with three channels—red, green and blue. *Yellow blocks* (gray in print version) represent feature maps, results of convolution operations. *Red blocks* (dark gray in print version) represent vectorized feature maps. The *blue block* (dark gray in print version) is an output—10 numbers, each in a range [0, 1] representing probability of an image being one of 10 classes we have in CIFAR-10 dataset. The *arrows* between blocks represent operations. MaxPooling halves the spatial space. Convolution with kernel 3×3 without padding reduce spatial by 2 (e.g., from 32×32 to 30×30). ReLU has no effect on data shape but affects values. View just flattens feature maps into a vector of length 16×6×6=576. Notice that first convolution layer will have 3×3×3×8+8 (bias) parameters compared to first linear (fully connected) layer having 576×100 which is 257× more parameters! This is a huge opportunity for NN to overfit. To prevent that, modern CNN architectures prefers to be fully convolutional without fully connected layers (e.g., YOLO v3 [27] for object detection).

input from only the 1D vector. Fully connected layers are followed by the relu activation function.

As we now know what our network should look like we can better understand the constructor. On line 14 the first convolution layer is defined. It expects 3 channels on its input. It has 8 filters (kernels) and the kernel size is 3×3. On line 15 a 2×2 max pooling layer is defined (it will reduce the size of feature channels by two as it divides each channel into 2×2 blocks and returns the maximum value from these four). On line 16 a second 3×3 convolution layer is defined. It expects 8 channels on the input and returns 16 channels. The kernel size is 3×3. On lines 18–20 fully connected layers are defined. They are defined by the number of input neurons and the number of output neurons. In this case we need to compute how many neurons there will be on the input. This depends on all previous layers and the input size. We know for sure that there will be 16 channels. Their resolution is affected by input size (32×32). The first 3×3 convolution without padding is applied, which means the filter size will be 30×30. Then max pooling is applied resulting in a size of 15×15. On top of that, a second 3×3 convolution without padding is applied, which means the filter size is 13×13. Max pooling is once again applied, thus making the filter size 6×6. Thus the input of the first fully connected layer will be $16 \times 6 \times 6$ (see line 18). The number of output neurons (in our case 100) is a design choice. The last fully connected layer (line 20) must have 10 output neurons (as we have 10 classes).

On line 108 the loss function is chosen. In this case we are using cross-entropy loss $J = -\dfrac{1}{N}\left(\sum_{i=1}^{N} y_i \cdot \log\left(\hat{y}_i\right)\right)$, where N is number of

classes (in our case 10), y_i is the predicted probability of class i, and $\hat{y}_i$ is 1 when the ith class is the correct one. A perfect model would have zero loss. But this won't happen in practice. If you have zero loss you are either terribly overfitted or something is broken.

On line 109 an optimizer is defined. We are using a stochastic gradient descent (SGD) of learning rate of 0.1 and momentum of 0.9. A learning rate of 0.1 is pretty high (we are quite aggressive on training speed). If training diverges and loss starts to dramatically increase it might be time to consider learning rate reduction. As mentioned before, learning rate is an important hyperparameter.

The rest of the code is self-explanatory. Lines 79, 80 might be worthy of mention—where a data augmentation is added by doing random crop and horizontal flip (think why vertical flip wouldn't be such a good idea for our case).

```
30 def print_progress(b, b_sum, loss, acc):
31     print('{:d}/{:d} Loss: {:.3f} | Acc: {:.3%}'.format(
32         b, b_sum, loss, acc), end='\r')
33
34 def train(epoch, trainloader, optimizer, net, criterion):
35     print('\nEpoch: %d' % epoch)
36     net.train()
37     train_loss, correct, total = 0, 0, 0
38     for batch_idx, (inputs, targets) in enumerate(trainloader):
39         inputs, targets = inputs.to(device), targets.to(device)
40         optimizer.zero_grad()
41         outputs = net(inputs)
42         loss = criterion(outputs, targets)
43         loss.backward()
44         optimizer.step()
45
46         train_loss += loss.item()
47         _, predicted = outputs.max(1)
48         total += targets.size(0)
49         correct += predicted.eq(targets).sum().item()
50
51         print_progress(batch_idx, len(trainloader),
52             train_loss/(batch_idx+1), correct/total)
53
54 def test(epoch, testloader, optimizer, net, criterion):
55     print("\nValidation")
56     net.eval()
57     test_loss, correct, total = 0, 0, 0
58     with torch.no_grad():
59         for batch_idx, (inputs, targets) in enumerate(testloader):
60             inputs, targets = inputs.to(device), targets.to(device)
61             outputs = net(inputs)
62             loss = criterion(outputs, targets)
63
64             test_loss += loss.item()
65             _, predicted = outputs.max(1)
66             total += targets.size(0)
67             correct += predicted.eq(targets).sum().item()
68
69             print_progress(batch_idx, len(testloader),
70                 test_loss/(batch_idx+1), correct/total)
```

```
71  if __name__ == "__main__":
72      lr = 0.1
73
74      device = 'cuda' if torch.cuda.is_available() else 'cpu'
75
76      # Data
77      print('==> Preparing data..')
78      t_train = transforms.Compose([
79          transforms.RandomCrop(32, padding=4), # augmentation1
80          transforms.RandomHorizontalFlip(), # augmentation2
81          transforms.ToTensor(),
82          transforms.Normalize((0.5, 0.5, 0.5), (0.5, 0.5, 0.5)),
83      ])
84
85      t_test = transforms.Compose([
86          transforms.ToTensor(),
87          transforms.Normalize((0.5, 0.5, 0.5), (0.5, 0.5, 0.5)),
88      ])
89
90      trainset = torchvision.datasets.CIFAR10(
91          root='./data', train=True, download=True, transform=t_train)
92      trainloader = torch.utils.data.DataLoader(
93          trainset, batch_size=128, shuffle=True, num_workers=2)
94
95      testset = torchvision.datasets.CIFAR10(
96          root='./data', train=False, download=True, transform=t_test)
97      testloader = torch.utils.data.DataLoader(
98          testset, batch_size=100, shuffle=False, num_workers=2)
99
100     # Model
101     print('==> Building model..')
102     net = Net()
103     net = net.to(device)
104     if device == 'cuda':
105         net = torch.nn.DataParallel(net)
106         cudnn.benchmark = True
107
108     criterion = nn.CrossEntropyLoss()
109     optimizer = opt.SGD(net.parameters(), lr=lr, momentum=0.9)
110
111     for epoch in range(0, 20):
112         train(epoch, trainloader, optimizer, net, criterion)
113         test(epoch, testloader, optimizer, net, criterion)
```

Even with this simple NN we were able to achieve a validation accuracy slightly above 60% (60.69%) after 20 epochs. To start building intuition we suggest the reader does the following exercises:

- Try training without augmentation. What do you expect will happen?
- Try to experiment with architecture (add more convolution layers) and add padding.
- Try to experiment with optimizer parameters or try different optimizers (e.g., Adam).

8.4.5 Transfer Learning

It didn't take long for the deep-learning community to discover that when we are short of training data it is possible to start from a pretrained model used to solve a similar problem. For example, if you want to classify dog breeds, you can start with a network trained to classify breeds of cats. It is common to start from pretrained general classifiers (e.g., ImageNet) and modify only the last layer or layers of the network (changing the number of classes). This concept works surprisingly well because the lower layers of the network have learned to detect basic entities such as edges, corners, and blobs, while later layers detect ears, eyes, wheels, etc. Further into the network the next layers detect even higher concepts such as head and leg. Subsequent layers lead to detection of complete concepts such as a human, dog, and car. Ultimately, the network can "see" the relationships between concepts. So when we need to distinguish between dog breeds we can still use trained kernels for eye detection, ear detection, etc. We only need to tune the last layers (or the last couple of layers depending on the tasks, intuition, and a trial-and-error approach).

8.5 Recurrent Neural Networks

The recurrent neural network (RNN), discovered by John Hopfield in 1982, is used for operations on sequences (e.g., text, voice). It enables NNs to learn patterns over time (e.g., detecting actions in video sequences, speech detection in audio, etc.). RNNs use a connection from their output to their input to allow the NN to gain a concept of temporal memory. This can be imagined as copying the whole NN and adding the same architecture into the new structure, and then sharing weights (see Fig. 30). The RNN can send some signals/states based on what was already processed on the input.

Building on RNNs there is an improved concept called long short-term memory (LSTM). Such networks allow speech applications to be improved dramatically (they are also widely used for text processing like machine translation). LSTMs were also successfully used in combination with CNNs for image captioning [31]. It's also common

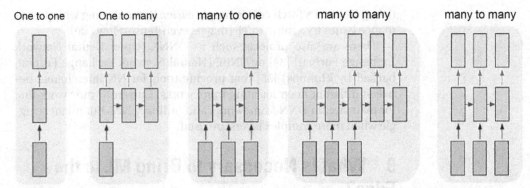

One to one   One to many   many to one   many to many   many to many

Fig. 30 An example of possible RNN architectures. *Green block* (light gray in print version) stands for a NN architecture and when multiple *green blocks* occur, it means that they share the same weights.

to combine CNNs with LSTMs for video streams for tracking or events detection (Fig. 31).

8.6 Deep-Learning Frameworks

Which framework to choose? That's the question. There are multiple possibilities and new ones are added each year. If asked to name the most common we would come up with TensorFlow [32], Caffe [33], and PyTorch [29].

TensorFlow (TF) is backed by Google. It is definitely the most used. It has a huge community and a fork called TensorFlow Lite focusing on mobile devices. Caffe seems to be in a decline. It was one of the first, widely used frameworks for CNNs.

PyTorch is a great tool for experiments. Compared with TensorFlow it is much easier to use to debug NN (as it boasts dynamic computational graph creation) and is commonly used in research. CNTDK is backed by Microsoft. The author's personal suggestion is to design, tune, and

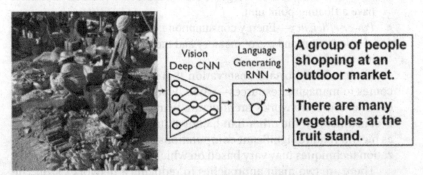

Vision Deep CNN   Language Generating RNN

A group of people shopping at an outdoor market.

There are many vegetables at the fruit stand.

Fig. 31 CNN augmented with LSTM is producing text captions for images published in Ref. [31].

debug NNs in PyTorch as it is simply easier. When dealing with performance issues try a move to TF (maybe even TensorFlow Lite).

There are also projects such as ONNX (Open Neural Network Exchange Format) [34] or NNEF (Neural Network Exchange Format, backed by Khronos) [35] that provide tools for NN interchange between different deep-learning frameworks. Currently, they work fine for mainstream DNN frameworks and ordinary NNs but might struggle when more complex layers are used.

9   What Is Necessary to Bring ML to the Edge?

Back in the day ML was heavily academic driven, where the primary goal was to make it work and push the boundaries of what was possible, no matter the cost. Engineers today have a completely different approach, especially in embedded application development where the industry is working to make its products smarter—but there are challenges:

- *Limited memory footprint*—For example, in the previous section we introduced a brief history of ImageNet. Even more advanced models than AlexNet (which has around 60 M parameters, which if represented as float32 have a size of 240 MB) don't have a negligible number of parameters nor a negligible memory footprint.
- *Limited computational resources*—During the model-training phase large computational clusters powered by cutting edge NVIDIA GPUs are usually used (e.g., the relatively affordable NVIDIA GTX 1080 Ti has 11.3 TFLOPs, which means it can process 11.3×10^{12} float operations per second). For example, MobileNet_v1_1.0_224 [36] needs 569 MACs/inference (MAC stands for multiply-accumulate operations). Therefore, there is clearly room for performance improvement if, for example, we switch from float to integer computations. Furthermore, switching from float to integer is essential because some devices don't have a floating-point unit.
- *Power efficiency*—Energy consumption is firmly linked to the number of calculations and use of memory and there is usually great interest in its reduction.

There's an important observation that needs to be made when it comes to managing resources—when using fully connected (FC) layers in a neural network's architecture usually a lot of memory space is consumed; on the other hand, convolutional layers are usually not as big but require significant computations. This suggests that optimization techniques may vary based on which layer is being optimized.

There are two main approaches to reducing a network's size. The first and most commonly applied relates to reducing representative precision. One option is called quantization (e.g., going from float32 representation of weights down to int8 or lower). Low-rank factorization is another option; this is where matrices/tensors are approximated with smaller ones that are used for computations before they are reconstructed back into the original format. If necessary, it is possible to further compress the model's size by applying encoding (e.g., Huffman encoding). The second approach is focused on making changes in architectures either by designing much smaller and/or computationally efficient NNs from the beginning or pruning NNs after or during a training phase. The use of these techniques is not exclusive, but they come into their own when used in combination to shrink the size and speed up inference. In addition, most techniques have a selectable degree of compression and thus can sacrifice some accuracy to meet given size/speed limits. The following sections discuss these techniques in more detail.

9.1 Quantization

The most important technique in the NN optimization domain is quantization [37]. Quantization does not change the architecture, instead it decreases the precision of weights and/or the activation functions. Most commonly seen is a decrease from float32 to int8 fixed-point precision—this alone reduces the memory footprint 4×. Note that for int8 multiplication you still need int32 registers.

The quantization of weights only reduces the size and might have no effect on inference time, although there's a possible speedup because size reduction leads to the model fitting better into faster memory/caches, and depending on the HW it might also be computed faster doing computations in INTs instead of FLOATs. To improve performance the quantization of activations is also necessary. Usually, weights and activations are represented using 8 bit, but if there is a bias term (like in a convolutional layer) it is usually represented using 32 bit.

9.2 Pruning

Pruning neural networks is not a new concept. Papers such as Lecun et al. "Optimal Brain Damage" [38] date back to 1990. Pruning assumes, as many results show, that neural networks are overparametrized and thus there is redundancy. Moreover, some neurons do not contribute significantly. There might also be "dead neurons" with outputs that are always zero resulting from too high a learning rate in combination with activation functions prone to this behavior. If we find a way to rank the neurons based on how much they contribute, we can then decide to remove the less valuable ones to save space and potentially speed it up.

Ranking can be done according to either the L1/L2 norm of their weights, the neurons mean activations on some reasonable validation data set, the number of times a neuron wasn't zero on a validation dataset, and other methods. After the pruning process accuracy drop is expected. The following step is commonly used to fine-tune the network to give the NN an opportunity to recover.

Speedup though is not guaranteed. Pruning usually results in irregular network connections that not only demand extra representation efforts, but also do not fit well with parallel computation. This is usually worthwhile only when NN size is the issue or it results in an NN with high sparsity, where overhead from sparse matrix computation is negligible. Another possibility is structured pruning [39] (e.g., removing whole convolution kernels, etc.).

9.3 Postprocessing vs. Dynamic Optimization

Back in the day, researchers tried the most straightforward approach—take a trained network, prune it, quantize weights, and see what happens. Even if this was done carefully, it was very frequently followed by a huge accuracy drop. Today, what is usually done is either *fine-tuning* or training where the pruning is scheduled, say, to start at the 50th epoch, and quantization is also added in the end. These techniques usually prolong the training phase but provide smaller and more computationally/ energy-efficient NNs without a dramatic drop in accuracy.

Another important observation for optimization, which might seem counterintuitive, is that starting with an optimized NN (quantized, pruned, and low-rank approximation applied) and training it from scratch doesn't usually work. It either ends up with poor accuracy (compared with an overparametrized NN) or it doesn't converge at all. These results are more observational since there are is no proper theory supporting these processes yet.

9.4 Low-Rank Factorization

The key idea behind low-rank factorization is to replace matrix multiplications with more matrix multiplications. Sounds counterintuitive, right? The reason this works is that these new matrices are smaller. The number of operations during matrix multiplication AB, $A \epsilon R^{M \times N}$, $B \epsilon R^{N \times O}$ is MNO multiplications and $M(N-1)O$ additions. For simplicity let's take it as just MNO multiply-accumulation operations (MACs).

A well-known matrix decomposition method is singular value decomposition (SVD). $A = U\Sigma V^*$, $U \epsilon R^{M \times M}$, $\Sigma \epsilon R^{M \times N}$, $V \epsilon R^{N \times N}$. Σ is a diagonal matrix, and diagonal entries are called singular values. By construction these singular values can be placed in descending order. The compression technique is based on the fact that we can take only the first k singular values, thus reducing matrix sizes into $U \epsilon R^{M \times k}$, $\Sigma \epsilon R^{k \times k}$, $V \epsilon R^{k \times N}$. The magic happens when we do matrix multiplication: $U \Sigma V B$. We start with $V B$, which is kNO MACs, and end up with matrix $D \epsilon R^{k \times O}$. We can have $U \Sigma$ precomputed as matrix $C \epsilon R^{M \times k}$, then the CD multiplication will cost MkO MACs. The whole $U \Sigma V B$ with $U \Sigma$ precomputed cost is $kNO + MkO = kO(N + M)$.

Consider the last fully connected layer of VGG [16]. It has an input size of 4096 and an output size of 1000. What happens here is $Wx + b$, $W \epsilon R^{1000 \times 4096}$, $x \epsilon R^{4096 \times 1}$, $b \epsilon R^{1000 \times 1}$, where W is the weight matrix, x is input, and b is the bias term. When W is decomposed a reasonable k is chosen. For $k = 800$ there is almost zero gain because the original multiplication will cost 4.096 M MACs, and with decomposition it will be 4.076 M. For $k = 400$ we need only 2.038 M MACs—a 2× savings. The memory requirements will also be smaller. From NM (4.096 M) weights going down to $kN + kM$ (for $k = 400$; it is the same as 2.038 M MACs). The reason memory is the same as MACs is that we are multiplying by a vector, thus $O = 1$.

This approach is good to follow as it has a fine-tuning phase compensating for a possible drop in accuracy. In the end one FC layer will be replaced by two smaller layers. The first has an input size of 4096, an output size of k, and no bias term. The second layer has an input size of k, an output size of 1000, and a bias term b.

Other techniques take this idea even further. There are more optimal approaches than SVD—a good example is Fastfood kernel decomposition [40, 41]. It is applied only to matrix multiplications. When we speak about convolution we are dealing with tensor multiplication (tensor is a more general concept than matrices and vectors). 4D tensor multiplication usually takes place in convolution. Similar to SVD decomposition are Tucker decomposition [42] and CP decomposition [43]. However, they take place in the higher dimensional space used for NN compression.

9.5 Architecture Design

As is known, fully connected layers consume a lot of memory, but convolutions stress the processing resources. Another way to save resources is to design the architecture with optimality in mind. In the field of CNNs there are networks, such as FD-MobileNet [44] and ShuffleNet [45], doing exactly this. One recent trend is not to use fully connected layers as they are prone to overfitting and have many parameters resulting in a big memory footprint.

There are several ideas related to convolution layers. Probably one of the first was published in [46]. This involved an NN architecture called SqueezeNet where 1×1 convolutional kernels are applied before more complex 3×3 convolution layers to reduce the number of input channels. The authors showed comparable accuracy with AlexNet, but with fewer parameters. At first sight 1×1 convolutional kernels might seem like a nonsense. What they do though is to combine feature maps in the linear way. Usually, an activation is applied on top of them adding more nonlinearity into the system.

Another idea is the depthwise-separable filter, which was introduced as part of the MobileNets architecture [36]. The trick is in dividing the standard convolution layer into two steps. Instead of doing N-times convolution on all input features (where N is the number of filters) a convolution is done only once followed then by N 1×1 convolutions to end up with N separate feature maps. For example, a 3×3 depthwise-separable convolution (as used in [36]) uses between eight and nine times less computation than standard convolutions with only a small reduction in accuracy.

For further reading we suggest a paper about SqueezeNext that focuses on hardware-aware NN design [47].

10 Edge Learning/Training

We have discussed ML on the edge only in terms of inference. But what about edge training as well? There might be some strong motivation such as privacy or connectivity (either connectivity is not present or has bandwidth limitations).

How do you update your NN on the edge? What if, for example, the number of classes in the network needs to change (e.g., a face recognition system where a new face is added). You could perform the training again from scratch, although this is probably not worth the effort nor even possible. Some approaches allow you to do a small portion of training—so-called fine-tuning—but this usually comes with issues like overfitting to new examples.

The most commonly used approach is to use an NN as a feature extraction tool. We can do this by removing the last layer (usually softmax in the case of classification) and end up with some alternate output layer. With this application in mind we can intentionally design an architecture such that we have, for example, 128 neurons in the penultimate layer thus having a descriptor of 128 numbers. We can take these numbers as a feature descriptor and feed it into some lightweight ML algorithms such as SVM and k-NN.

Let's look intuitively at how this approach works. Consider that the last layer was fully connected with softmax activation. This means one dot product plus a nonlinearity function have been used to properly classify the output of the penultimate layer. Therefore, when we train a classifier on top of the penultimate layer we have the same expressive power, but the training of, say, SVM is much faster and easier to do than retraining the whole NN.

Another frequently used technique is to freeze most of the layers and just fine-tune the remaining layers (e.g., last layer) on a small data sample. However, it is important to remember that we must restrict ourselves to inference when it comes to NN deployment, but when we need to carry out training we need to support efficient computation of backpropagation, which complicates things a lot. Currently, most edge inference engines don't support training.

Other more advanced approaches are built on top of concepts like k-shot learning [48] [49]. This is a hot research area in which new classes are learned based only on k examples, where k is a small number (sometimes even 1).

Bringing ML to the edge is not an easy task, but the potential impact on improving or creating so far unimaginable products is enormous. An increase in the number of research projects on these topics would greatly help to do this as would a number of open-source projects and raising embedded community interest. As stated at the beginning of this chapter this field is changing each day. We should expect thrilling times in the future when embedded learning meets machine learning.

References

[1] NumPy Homepage, http://www.numpy.org/, 2018. [Online] [Cited 27.7.18].
[2] A. Hodges, Alan Turing Scrapbook—Turing Test, http://www.turing.org.uk/scrapbook/test.html, 2018. [Online] [Cited: 31.08.14].
[3] R. Rifkin, A. Klautau, In defense of one-vs-all classification, J. Mach. Learn. Res. 5 (2004) 101–141.
[4] V.N. Vapnik, A.Y. Chervonenkis, On uniform convergence of the frequencies of events to their probabilities, Theor. Prob. Appl. 16 (1971) 264–280.
[5] S. Arlot, A. Celisse, A survey of cross-validation procedures for model selection, Stat. Surv. 4 (2010) 40–79.
[6] A. Krizhevsky, I. Sutskever, G. Hinton, ImageNet Classification With Deep Convolutional Neural Networks NIPS, Neural Information Processing Systems, Lake Tahoe, Nevada, 2012.
[7] C.M. Bishop, Pattern Recognition and Machine Learning (Information Science and Statistics). https://dl.acm.org/citation.cfm?id=1162264, 2006. [Online] [Cited 9.10.18].
[8] T. Hofmann, B. Schölkopf, A.J. Smola, Kernel methods in machine learning, Ann. Stat. 36 (2008) 1171–1220.
[9] G. Varoquaux, et al., Scikit-learn: Machine Learning Without Learning the Machinery. https://dl.acm.org/citation.cfm?id=2786995, 2015. [Online] [Cited: 9.10.18].
[10] Bentley, Jon L n.d. Multidimensional binary search trees used for associative searching. Commun. ACM, Vol. 18, pp. 509–517.

[11] S. Magnussen, E. Tomppo, The k-nearest neighbor technique with local linear regression, Scand. J. For. Res. 29 (2014) 120–131.

[12] L. Breiman, Random forests, Mach. Learn. 45 (2001) 5–32.

[13] Y. Freund, R.E. Schapire, A short introduction to boosting, J. Jpn. Soc. Artif. Intell. 14 (5) (1999) 771–780.

[14] T. Chen, C. Guestrin, XGBoost: A Scalable Tree Boosting System arXiv: Learning, 2016, pp. 785–794.

[15] L. Fei-Fei, J. Deng, K. Li, ImageNet: constructing a large-scale image database, J. Vis. 9 (2010) 1037.

[16] K. Simonyan, A. Zisserman, Very deep convolutional networks for large-scale image recognition, arXiv, Comput. Vis. Pattern Recognit. (2015).

[17] C. Szegedy, et al., Going deeper with convolutions, arXiv, Comput. Vis. Pattern Recognit. (2015) 1–9.

[18] K. He, et al., Deep residual learning for image recognition, arXiv, Comput. Vis. Pattern Recognit. (2016) 770–778.

[19] F. Luan, et al., Deep photo style transfer, arXiv, Comput. Vis. Pattern Recognit. (2017) 6997–7005.

[20] K. He, G. Gkioxari, P. Dollár, R. Girshick, Mask R-CNN, in: 2017 IEEE International Conference on Computer Vision (ICCV), 2018, pp. 2980–2988.

[21] D.P. Kingma, J. Ba, Adam: a method for stochastic optimization, arXiv, Learning (2015).

[22] J.C. Duchi, E. Hazan, Y. Singer, Adaptive subgradient methods for online learning, J. Mach. Learn. Res. 12 (2011).

[23] T. Tieleman, G. Hinton, Lecture 6.5—RMSProp, COURSERA: Neural Networks for Machine Learning, 2012.

[24] L.N. Smith, Cyclical learning rates for training neural networks, arXiv, Comput. Vis. Pattern Recognit. (2017) 464–472.

[25] I. Loshchilov, F. Hutter, SGDR: stochastic gradient descent with warm restarts, arXiv, Learning (2017).

[26] O. Ronneberger, P. Fischer, T. Brox, U-Net: convolutional networks for biomedical image segmentation, arXiv, Comput. Vis. Pattern Recognit. (2015) 234–241.

[27] J. Redmon, A. Farhadi, YOLOv3: an incremental improvement, arXiv, Comput. Vis. Pattern Recognit. (2018).

[28] M.D. Zeiler, R. Fergus, Visualizing and understanding convolutional networks, arXiv, Comput. Vis. Pattern Recognit. (2014) 818–833.

[29] A. Paszke, S. Gross, S. Chintala, G. Chanan, E. Yang, Z. DeVito, Z. Lin, A. Desmaison, L. Antiga, A.S. Lerer, Automatic Differentiation in PyTorchl, NIPS-W, 2017.

[30] Google Colab, [Online]. https://colab.research.google.com/.

[31] O. Vinyals, et al., Show and tell: a neural image caption generator, arXiv, Comput. Vis. Pattern Recognit. (2015) 3156–3164.

[32] TensorFlow, n.d. [Online] https://www.tensorflow.org.

[33] Y. Jia, et al., Caffe: convolutional architecture for fast feature embedding, arXiv: Comput. Vis. Pattern Recognit. (2014) 675–678.

[34] Open Neural Network Exchange Format, [Online]. https://onnx.ai/.

[35] Neural Network Exchange Format, [Online]. https://www.khronos.org/nnef.

[36] A.G. Howard, et al., MobileNets: efficient convolutional neural networks for mobile vision applications, arXiv, Comput. Vis. Pattern Recognit. (2017).

[37] B. Jacob, et al., Quantization and training of neural networks for efficient integer-arithmetic-only inference, arXiv, Learning (2018) 2704–2713.

[38] Y. LeCun, J.S. Denker, S.A. Solla, Optimal brain damage, Adv. Neural Inf. Process. Syst. (1990) 598–605.

[39] S. Anwar, K. Hwang, W. Sung, Structured pruning of deep convolutional neural networks, ACM J. Emerg. Technol. Comput. Syst. 13 (2017) 32.

[40] Q.V. Le, T. Sarlos, A.J. Smola, Fastfood: approximating kernel expansions in loglinear time, arXiv: Learning (2013).

[41] Z. Yang, et al., Deep fried convnets, arXiv, Learning (2015) 1476–1483.
[42] Y.-D. Kim, et al., Compression of deep convolutional neural networks for fast and low power mobile applications, arXiv, Comput. Vis. Pattern Recognit. (2016).
[43] V. Lebedev, et al., Speeding-up convolutional neural networks using fine-tuned CP-decomposition, arXiv, Comput. Vis. Pattern Recognit. (2015).
[44] Z. Qin, et al., FD-MobileNet: improved MobileNet with a fast Downsampling strategy, arXiv, Comput. Vis. Pattern Recognit. (2018).
[45] X. Zhang, et al., ShuffleNet: an extremely efficient convolutional neural network for mobile devices, arXiv, Comput. Vis. Pattern Recognit. (2018).
[46] M. Motamedi, D. Fong, S. Ghiasi, Fast and energy-efficient CNN inference on IoT devices, arXiv, Distrib. Parallel Clust. Comput. (2016).
[47] A. Gholami, et al., SqueezeNext: hardware-aware neural network design, Neural Evolut. Comput. (2018) 1638–1647.
[48] A. Santoro, et al., One-shot learning with memory-augmented neural networks, arXiv, Learning (2016).
[49] A. Graves, G. Wayne, I. Danihelka, Neural turing machines, Neural Evolut. Comput. (2014).

Further Reading

[50] C. Cortes, V. Vapnik, Support-vector networks, Mach. Learn. 20 (1995) 273–297.
[51] P. Isola, et al., Image-to-image translation with conditional adversarial networks, arXiv: Comput. Vis. Pattern Recognit. (2017) 5967–5976.
[52] G.E. Zitzewitz, Survey of Neural Networks in Autonomous Driving, 2017.
[53] D. Mishkin, J. Matas, All you need is a good init, arXiv, Learning (2016).
[54] T.C. Henderson, N. Boonsirisumpun, Issues related to parameter estimation in model accuracy assessment, Procedia Comput. Sci. 18 (2013) 1969–1978.
[55] V. Sze, et al., Efficient processing of deep neural networks: a tutorial and survey, arXiv, Comput. Vis. Pattern Recognit. 105 (2017) 2295–2329.
[56] Google Colab, https://colab.research.google.com/.
[57] A. Krizhevsky, Learning Multiple Layers of Features from Tiny Images, [Online] 2009 https://www.cs.toronto.edu/~kriz/learning-features-2009-TR.pdf, 2009.

APPENDIX

PERFORMANCE ANALYSIS USING NXP's i.MX RT1050 CROSSOVER PROCESSOR AND THE ZEPHYR™ REAL-TIME OPERATING SYSTEM

A benchmark study to understand performance advantages as compared to Linux BSP on i.MX 6UL Processors

Florin Leotescu, Marius Cristian Vlad, and Michael C. Brogioli

A.1 Introduction

Software and hardware performance analysis is integral to the evaluation and design of embedded systems. Such analysis helps to understand the limitations of a system, identify performance bottlenecks, and determine how well the system is performing in comparison with other devices. Performance analysis can be done using custom software benchmarking applications that execute specific algorithms, which will deliver performance statistics about the system under test, design, and development. Examples of such benchmarks are the SPEC CPU benchmarks, designed to provide performance measurements that can be used to compare compute-intensive workloads on different computer systems.[1] EEMBC is another group of benchmarks, predominantly targeted at embedded computing.[2] EEMBC benchmark suites are developed by working groups of members who share an interest in developing clearly defined standards for measuring the performance and energy efficiency of embedded processor implementations, from IoT edge nodes to next-generation advanced driver-assistance systems.

[1] https://www.spec.org/benchmarks.html.
[2] https://www.eembc.org.

603

In addition to the use of standardized benchmarks, like those mentioned earlier, system developers often also elect to implement micro-benchmarks that focus on a very small, or critical, feature of the system. While not intended to characterize broader system-level performance, microbenchmarks can be a very useful tool when focusing on specific system components. For example, microbenchmarks can be used to analyze the time required to create threads of execution within a given system. While this does not characterize the performance of the entire system, nor the system under load of a given target application, it can be used to provide fine-grained insights into specific aspects of the system.

It should be noted, however, that the use of benchmarking and micro-benchmarking can only go so far. Many embedded solutions vendors do not open up the underlying hardware design of their solution, nor very often provide access for system users to their system-level software or source code. As such, benchmarking and microbenchmarking are limited in terms of analyzing and comparing features between systems.

This section provides a real-world example of the use of micro-benchmarks to perform an analysis of differing hardware and software solutions that are critical to embedded systems design. Specifically, a performance analysis is presented comparing the Zephyr™ OS running on the NXP i.MX RT1050 crossover processor, based on the Arm® Cortex®-M7 core, and the Linux BSP running on the NXP i.MX 6UL applications processor, based on the Arm Cortex-A7 core. This analysis is performed via the use of custom microbenchmarks for various system components, including but not limited to thread creation, use of mutexes, and memory allocation, all of which are fundamental contexts to modern high-performance embedded systems design.

Noting the differences between Zephyr™ OS (a tiny open-source RTOS for IoT) and Linux (an open-source monolithic Unix-like computer operating system kernel), it is important to recognize that this comparison is not fully an "apples to apples" comparison.[3] Rather, this study is intended to provide embedded designers with a set of exemplary microbench-marks to compare hardware and software solutions when executing the same tasks. It is left to the reader or system developer to extrapolate how these system-level tasks relate to their overall target application. To evaluate the performance difference between the two solutions, certain synthetic microbenchmarks were developed specifically to evaluate the time between dynamic memory allocation and deallocation, mutex lock and unlock, thread creation, thread joining, and context switching.

In summary, the results of this performance analysis showed that the Zephyr™ OS (running on an i.MX RT1050 crossover processor) improved overall system responsiveness and ultimately reduced costs of the IoT and embedded systems development. The aforementioned

---

[3] https://en.wikipedia.org/wiki/Linux_kernel.

tasks executed much faster on the Zephyr™ OS + i.MX RT1050 solution, compared with the Linux + i.MX 6UL solution. The following sections explain the methodology used to derive the results of the comparison.

A.2 Configuration Information

The first configuration analyzed in the study is the NXP i.MX RT1050. The i.MX RT1050 is a crossover processor that combines the high-performance and high level of integration of an application processor with the ease of use and real-time functionality of a microcontroller. The i.MX RT1050 runs on the Arm® Cortex®-M7 core at 600 MHz.[4]

This device is fully supported by NXP's MCUXpresso Software and Tools, a comprehensive and cohesive set of free software development tools for Kinetis, LPC, and i.MX RT microcontrollers. MCUXpresso SDK also includes project files for Keil MDK and IAR Embedded Workbench for Arm.[5]

Configuration #1 also includes the Zephyr™ operating system. Zephyr™ is a small real-time and scalable operating system for connected, resource-constrained devices supporting multiple architectures and released under the Apache License 2.0.[6,7]

A.2.1 Summary of Configuration #1: i.MX RT1050 Configuration—Hardware and Software

---

[4] https://www.nxp.com/products/processors-and-microcontrollers/arm-based-processors-and-mcus/i.mx-applications-processors/i.mx-rt-series/i.mx-rt1050-crossover-processor-with-arm-cortex-m7-core:i.MX-RT1050.

[5] https://www.nxp.com/support/developer-resources/software-development-tools/mcuxpresso-software-and-tools/mcuxpresso-integrated-development-environment-ide:MCUXpresso-IDE.

[6] https://en.wikipedia.org/wiki/Zephyr_(operating_system).

[7] https://www.zephyrproject.org.

- Development board: MIMXRT1050-EVK
  - Processor: MIMXRT1052DVL6A Arm® Cortex®-M7 core
  - Number of cores: 1
  - Core frequency: 600 MHz
  - Board schematic: SCH-29538 REV A1
- OS name: Zephyr OS 1.11.99
  - OS type: real-time
  - Zephyr OS web page

The second configuration analyzed in the study is a combination of the NXP i.MX 6UL hardware with the Linux operating system. The i.MX6 UltraLight is a high-performance and efficient processor family featuring an ARM A7 core operating at speeds up to 696 MHz at the time of writing. The i.MX6 UltraLite applications processor includes an integrated power management module that reduces the complexity of the external power supply and simplifies power sequencing. Each processor in this family provides various memory interfaces, including 16-bit LPDDR2, DDR3, DDR3L, raw and managed NAND flash, NOR flash, eMMC, Quad SPI, and a wide range of other interfaces for connecting peripherals, such as WLAN, Bluetooth™, GPS, displays, and camera sensors. The software running on the i.MX6 UL is Linux Board Support Package release Linux BSP - kernel 4.9.88-imx_4.9.88_2.0.0_ga. As discussed in greater detail later, unlike the Zephyr™ OS of the first configuration, this is not a real-time variant of Linux.

A.2.2 Summary of Configuration # 2: i.MX 6UL Configuration

- Development board: MCIMX6G2CVM05AB
  - Processor: i.MX6UL: i.MX 6UltraLite Processor, based on Arm Cortex-A7 core
  - Number of cores: 1
  - Core frequency: 528 MHz
  - Board schematic: SCH-29163 REV A2
- OS name: Linux BSP - kernel 4.9.88-imx_4.9.88_2.0.0_ga
  - OS type: nonreal-time

A.3   Scope of Analysis

As mentioned in the introduction, this work is intended to evaluate and compare the performance of the i.MX RT1050 EVK with the Zephyr™ OS and the i.MX 6UL EVK with the Linux BSP. The goal being to determine any potential performance gaps between the MIMXRT1050-EVK board, equipped with an embedded ARM SoC, and a similar board equipped with an application processor. Due to the fact that the closest CPU speed configuration to the i.MX RT1050 EVK was found with the i.MX 6UL EVK, we selected the i.MX 6UL development board for best comparison.

In addition, the Zephyr™ OS was selected over other real-time operating systems because it is free and very comprehensive, developed as a collaborative project, and supported by an active open-source community. While both Zephyr™ and Linux OSs can exhibit real-time characteristics, the Zephyr™ OS was originally designed to fully abide with traditional RTOS principles, whereas Linux has traditionally served larger workloads for the desktop and server spaces. Furthermore, at the time of writing, Linux requires additional patches to abide to traditional RTOS principles.

To obtain comparable results, despite the known operating system differences, focus was placed on using the same application peripheral interface (API) for the custom microbenchmarks used in this study. The microbenchmarks were developed in C language and made use of the Pthreads API library (POSIX API library). In the case of the Zephyr™ OS, the available API version was POSIX PSE52, which according to Zephyr™ community documentation, implements only partial support for the full POSIX specifications.

The microbenchmarks perform memory allocation and deallocation, mutex lock and unlock, thread creation, thread joining, context switching, and record the time spent on each of these actions.

To determine the time spent performing the tasks, we used the POSIX *clock_gettime ()*for the Linux + i.MX 6UL EVK solution. For the Zephyr™ OS, running on i.MX RT1050 EVK, we used the *TIMING_INFO_ PRE_READ ()*function instead of *clock_gettime()*. Due the nature of the OS scheduler on Zephyr™, which is beyond the scope of this study, the *clock_gettime()*function generates inconsistent timing values, and because of the fact that Zephyr™ source code also uses the TIMING_INFO_ PRE_READ() function, the decision was made to continue with it.

A.3.1   Microbenchmark #1: Dynamic Memory (Heap) Allocation and Deallocation Benchmark

In C programming language, dynamic memory allocation refers to performing manual memory management via a group of functions in the C standard library. The C programming language manages memory statically, automatically, or dynamically.

Static variables are allocated in main memory, along with the executable code, and persist for the lifetime of the program. The automatically managed variables or local variables are allocated on the stack and they come and go as functions are called and as functions exit. The size of the memory allocation for the static and local variables is defined at the compile-time, except for variable-length arrays. If the required size is not known until runtime (e.g., if data of arbitrary size is being read from the user or from a disk file), then using fixed-size data objects is inadequate. In this situation, dynamic memoryallocation solves the problem—memory is more explicitly managed, typically by allocating it in large regions of free spacecalled heap (Fig. 1).

In other words, heap is a memory region of the computer which is managed manually by the programmer (in the case of C language). In other programming languages, for example, Java, memory is managed automatically.

To manage heap memory location in C under Linux, the *malloc ()* and *free ()*functions are used (there are also *new ()*and *delete ()*functions on C++). The *malloc*function is used for allocating a space into this memory, and *free*is used to deallocate it. In the case of the Zephyr™ OS these functions are named k_malloc() and k_free(), which do the same thing as malloc() and free(). For this analysis, the microbenchmark was developed around these functions and was named

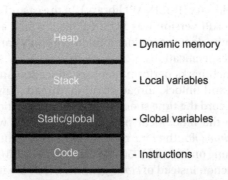

Fig. 1 Application memory organization.

heap_bench. The purpose was to measure the time for allocating and deallocating heap memory. Behind the scenes, the benchmark allocates 4 bytes (sizeof(int)) for 1000 iterations of the heap allocation loop, and then deallocates the same allocated memory via the second "heap deallocation" loop. For the Linux BSP, each loop of allocation and deallocation time was measured using a *time_get_time (),* which is a wrapper function on top of clock_get_time (). For the Zephyr™ OS, the *TIMING_INFO_PRE_READ ()* function was used.

```
...
//Linux BSP - heap allocation //
for (i = 0; i < ITERATIONS; i++) {
                time_get_time(&start);
                pointer[i]= malloc(sizeof(int));
                *pointer[i] = 0xdeadbeef;
                time_get_time(&stop);
                diff = time_get_diff(&stop, &start);
                total += diff;
}
printf("Only call time function: %.1f ns\n", (total) / (double) ITERATIONS);
...
//Linux BSP - heap deallocation //
for (i = 0; i < ITERATIONS; i++) {
                time_get_time(&start);
                free(pointer[i]);
                time_get_time(&stop);
                diff = time_get_diff(&stop, &start);
                total += diff;
}
printf("Average heap free time: %.1f ns\n", (total) / (double)ITERATIONS);
...
...
// Zephyr OS - heap allocation //
for (i = 0; i < ITERATIONS; i++) {
                TIMING_INFO_PRE_READ();
                heap_malloc_start_time = TIMING_INFO_OS_GET_TIME();
                pointer[i]= k_malloc(sizeof(int));
                *pointer[i] = 0xdeadbeef;
                TIMING_INFO_PRE_READ();
                heap_malloc_end_time = TIMING_INFO_OS_GET_TIME();
...
...
//Zephyr OS - heap deallocation//
for (i = 0; i < ITERATIONS; i++) {

                TIMING_INFO_PRE_READ();
                heap_free_start_time = TIMING_INFO_OS_GET_TIME();
                k_free(pointer[i]);
                TIMING_INFO_PRE_READ();
                heap_free_end_time = TIMING_INFO_OS_GET_TIME();
...
```

At the end of the heap allocation and heap deallocation loops, the final average allocation and deallocation times were calculated for a given loop, as can be seen in the source code above.

A.3.2   Microbenchmark #2: Thread Creation and Joining Benchmark

In computer science, a thread of execution is a small sequence of programmed instructions that can be managed independently by a scheduler, the scheduler being part of the operating system in this context. The implementation of threads and processes differ between operating systems, but in most cases a thread is a component of a process. Multiple threads can exist within one process, executing concurrently and sharing resources, like memory, across threads, while different processes do not share these resources. In particular, the threads of a process share its executable code and the values of its variables at any given time.[8] Fig. 2 depicts two processes, each one having one or multiple threads.

In this comparison, the process is the running benchmark, named *thread_bench*, which spawns multiple threads using the *pthread_create* ()POSIX function. It creates 2000 threads and measures the time of creation for all 2000 threads. At the end of thread creation, the time recorded is divided by the number of threads created, giving the average time to create a thread.

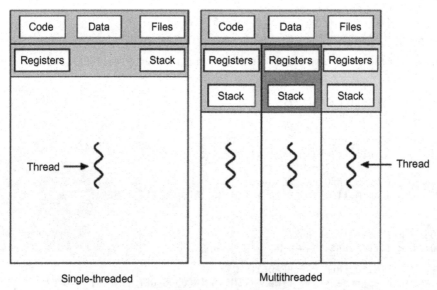

Fig. 2 Single threaded process model vs. multithreaded process model.

[8] https://en.wikipedia.org/wiki/Thread_(computing).

```
...
for (i = 0; i < ITERATIONS; i++) {
            if (pthread_attr_init(&attr[i]) != 0) {
                fprintf(stderr, "pthread_attr_init!\n");
                exit(EXIT_FAILURE);
            }
            if (posix_memalign(&stacks[i], sysconf(_SC_PAGESIZE), MAX_STACK_SIZE) != 0) {
                fprintf(stderr, "Failed to allocate aligned memory\n");
                exit(EXIT_FAILURE);
            }
            if (pthread_attr_setstack(&attr[i], stacks[i], MAX_STACK_SIZE) != 0) {
                fprintf(stderr, "Failed pthread_attr_setstack!\n");
                exit(EXIT_FAILURE);
            }
            time_get_time(&start);
            if (pthread_create(&threads[i], &attr[i], test_function, NULL) != 0) {
                fprintf(stderr, "Failed to create thread!\n");
                exit(EXIT_FAILURE);
            }
            time_get_time(&stop);
#ifdef DEBUG
            fprintf(stdout, "Created thread_id %d\n", i);
#endif
            diff = time_get_diff(&stop, &start);
            total += diff;
}
fprintf(stdout, "Average pthread_create time: %.1f ns\n", (total/(double)ITERATIONS));
...
```

This benchmark also measures join time using the *pthread_join ()* function, which synchronizes the parent thread by pausing its execution, until the child thread terminates.

```
...
for (i = 0; i < ITERATIONS; i++) {
            if (threads[i]) {

                    time_get_time(&start);
                    if (pthread_join(threads[i], NULL) < 0) {
                        fprintf(stdout, "Failed to join thread\n");
                        exit(EXIT_FAILURE);
                    }
                    time_get_time(&stop);
                    diff = time_get_diff(&stop, &start);
                    total += diff;
#ifdef DEBUG
                    fprintf(stdout, "thread %d, joined\n", i);
#endif
            }
    }
    fprintf(stdout, "Average pthread_join time: %.1f ns\n",(total/(double)ITERATIONS));
...
```

A.3.3 Microbenchmark #3: Mutex Lock and Unlock Benchmark

In computer science, mutual exclusion is a concurrency control method dedicated to prevent race conditions between two, or multiple, threads. A race condition is a behavior of a system where two independent workflows are modifying in a shared resource, which is used to generate the output of the system. Making an analogy to the real world, we can consider two mechanics (two threads) who are jointly assembling a car engine. They assemble the engine in parallel, however, some of the subcomponents must be assembled in some specific order to ensure that the engine will work properly. Each mechanic has their own part of the engine to assemble. To be sure that components are mounted in the correct order, each mechanic should have exclusive ownership to the relevant portion of the engine during the critical sections of assembly. This exclusive ownership could be associated with the mutex lock, where mutex is our car engine. Freeing the engine could be associated with mutex unlock.

The mutex lock and unlock benchmark, named *mutex_bench*, measures the time of these two actions 1000 times. At the end, it calculates the average time for locking and unlocking a mutex variable. To execute mutex lock and unlock, we used the *pthread_mutex_lock ()* and *pthread_mutex_unlock ()*functions of the Pthread API library. Below are some code samples of the benchmark which measures lock and unlock timings.

```
...
//Linux BSP//
for (i = 0; i < nr_iterations; i++) {
        time_get_time(&start);
        pthread_mutex_lock(&lock);
        time_get_time(&stop);
        delta = time_get_diff(&stop, &start);
        total_lock += delta;
        time_get_time(&start);
        pthread_mutex_unlock(&lock);
        time_get_time(&stop);
        delta = time_get_diff(&stop, &start);
        total_unlock += delta;
}
    fprintf(stdout, "Average time for locking a mutex: %.8f ns\n",
                (double) total_lock/ (double) nr_iterations);
    fprintf(stdout, "Average time for unlocking a mutex: %.8f ns\n",
                (double) total_unlock/ (double) nr_iterations);
...

...
//Zephyr OS//
for (i = 0; i < nr_iterations; i++) {
            TIMING_INFO_PRE_READ();
```

```
mutex_lock_start_time = TIMING_INFO_OS_GET_TIME();
pthread_mutex_lock(&lock);
TIMING_INFO_PRE_READ();
mutex_lock_end_time = TIMING_INFO_OS_GET_TIME();
    total_lock += (mutex_lock_end_time -mutex_lock_start_time);
TIMING_INFO_PRE_READ();
mutex_unlock_start_time = TIMING_INFO_OS_GET_TIME();
pthread_mutex_unlock(&lock);
TIMING_INFO_PRE_READ();
mutex_unlock_end_time = TIMING_INFO_OS_GET_TIME();
    total_unlock += (mutex_unlock_end_time -
                            mutex_unlock_start_time);
...
```

A.3.4   Microbenchmark #4: Context Switching Benchmark

In computing, a context switch is the process of storing the state of a process of a thread, so that it can be restored and then resume execution from the same point later. This allows multiple processes to share a single CPU and is an essential feature of a multitasking operating system.

The precise meaning of the phrase "context switch" varies significantly in usage. In a multitasking context, it refers to the process of storing the system state for one task, so that task can be paused and another task resumed. A context switch can also occur as the result of an interrupt, such as when a task needs to access disk storage, freeing up CPU time for other tasks. Some operating systems also require a context switch to move between user-mode and kernel-mode tasks. The process of context switching can have a negative impact on system performance, although the size of this effect depends on the nature of the switch being performed (Fig. 3).[9]

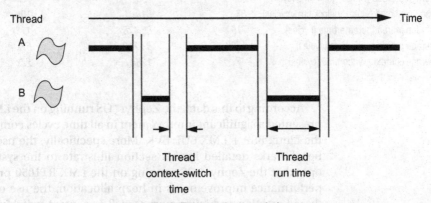

Fig. 3 Thread contextswitch.

[9] https://en.wikipedia.org/wiki/Context_switch.

This benchmark measures the context switch time by creating two threads, which are continuously context switched 500,000 times. During context switch time, the benchmark records elapsed time which is then divided by the number of context switches to generate the average time for a context switch.

A.4 Analysis Results

Figs. 4 and 5 contain the scores reported by the aforementioned benchmarks. Three iterations were performed for each benchmark. As can be seen, the results using the Zephyr™ OS are deterministic. Each time you execute the benchmark on the Zephyr™ OS with i.MX RT1050 EVK, the results will be the same (Fig. 4).

With the Linux BSP running on the i.MX 6UL, the results varied from run to run with a deviation from average values of up to 9% (Fig. 5).

The table below highlights the average time calculated from these benchmark iterations. The average time here is calculated in cycles (lower is better).

OS	Zephyr™ OS 1.11.99	Linux BSP 4.9.88-imx_4.9.88_2.0.0_ga	Difference (as a multiple)
Board name	i.MX RT1050 EVK	i.MX 6UL EVK	–
CPU cores	1	1	–
Core frequency (MHz)	600	528	–
Average heap malloc time (cycles)	1001	11,499	11x
Average heap free time (cycles)	1126	4870	4x
Average pthread_mutex_lock time (cycles)	53	799	15x
Average pthread_mutex_unlock time (cycles)	83	818	10x
Average pthread_create time (cycles)	719	85,478	118x
Average pthread_join time (cycles)	1702	89,219	52x
Average context switch time (cycles)	47	1284	27x

According to this data, the Zephyr™ OS running on the i.MX RT1050 presented a significant improvement in all time cycles compared with the Linux BSP + i.MX 6UL EVK. More specifically, the use of micro-benchmarks detailed in this section illustrate to the system developer that the Zephyr™ OS running on the i.MX RT1050 provides key performance improvements in heap allocation, the use of mutexes, thread creation and join times, as well as context switching. As most embedded solutions developers will appreciate, these are often considered key building blocks in the overall design and implementation of solutions and system-level applications.

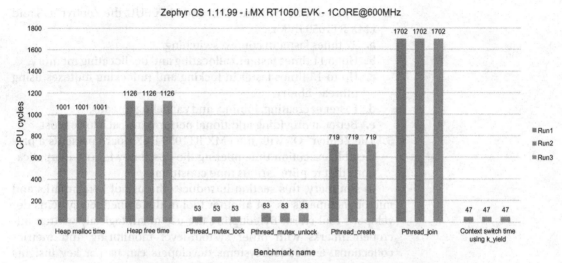

Fig. 4 Benchmark results on the i.MX RT1050 EVK with the Zephyr™ OS.

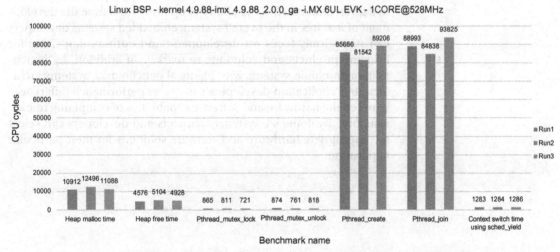

Fig. 5 Benchmark results onthe i.MX 6UL EVK with the Linux BSP.

A.5   Summary and Conclusions

Key contributions of this section are detailed below:

1. A performance analysis was completed by running custom micro-benchmarks on two different hardware and software solutions.
2. Benchmarks were developed around a common API to ensure comparable results.
3. Different functions were used for collecting elapsed time: *clock_get_time()*on Linux and *TIMING_INFO_PRE_READ* on the Zephyr™ OS.

4. Compared with the Linux BSP and i.MX 6UL, the Zephyr™ OS and i.MX RT1050 EVK is:
   a. 27 times faster in context switching.
   b. Up to 11 times faster in allocating and deallocating memory.
   c. Up to 15 times faster in locking and unlocking mutexes using pthread library.
   d. Faster at creating, joining, and canceling threads.
   e. Better at providing additional performance at a lower cost.
5. The Zephyr™ OS with the i.MX RT1050 EVK board presents a predictable execution time offering the possibility for use in applications that require various time constrains.

In summary, this section introduces the use of benchmarks and microbenchmarks as yet another tool in the embedded systems developer's tool chest. By coupling the use of strategically written microbenchmarks with other system-level monitoring and metrics collections, embedded systems developers can garner key insights into the performance of various hardware and software components within a given solution, as well as across competing solutions within in a given market. With the ability to optimize, and tune the development of features in the overall system, embedded systems developers can strategically focus on development and optimization times for bringing products and solutions to market. In addition, by benchmarking multiple systems with identical benchmarks, systems architects and application developers can assess performance differences between the hardware and software capabilities of competing market solutions. By doing so, systems architects and developers can select the appropriate hardware and software solutions for their particular application.

INDEX

Note: Page numbers followed by *f* indicate figures and *t* indicate tables.

Printed in the United States
by Baker & Taylor Publisher Services